# Strigolactones

**Scrivener Publishing**
100 Cummings Center, Suite 541J
Beverly, MA 01915-6106

*Publishers at Scrivener*
Martin Scrivener (martin@scrivenerpublishing.com)
Phillip Carmical (pcarmical@scrivenerpublishing.com)

# Strigolactones

## Emerging Plant Hormones

Edited by

**Tariq Aftab**

*Department of Botany, Aligarh Muslim University, India*

and

**Kaiser Iqbal Wani**

*Department of Botany, Aligarh Muslim University, India*

Scrivener Publishing

WILEY

This edition first published 2025 by John Wiley & Sons, Inc., 111 River Street, Hoboken, NJ 07030, USA
and Scrivener Publishing LLC, 100 Cummings Center, Suite 541J, Beverly, MA 01915, USA

For more information about Scrivener publications please visit www.scrivenerpublishing.com.



**Wiley Global Headquarters**
111 River Street, Hoboken, NJ 07030, USA

For details of our global editorial offices, customer services, and more information about Wiley products visit us at www.wiley.com.



***Library of Congress Cataloging-in-Publication Data***

ISBN 978-1-394-30279-6

Front cover image courtesy of Adobe Firefly
Cover design by Russell Richardson

Set in size of 11pt and Minion Pro by Manila Typesetting Company, Makati, Philippines

Printed in the USA

10 9 8 7 6 5 4 3 2 1

# Contents

# Preface

Strigolactones, a fascinating and rapidly evolving class of plant hormones, have garnered significant attention in plant biology over the past decade. Initially discovered for stimulating the germination of parasitic plants, strigolactones are now recognized as key regulators of numerous plant processes, including growth, development, and response to environmental stresses. Their multifaceted nature and wide-ranging impact on plant physiology make strigolactones a critical study area for researchers aiming to enhance crop yield, resilience, and overall agricultural productivity.

This edited volume provides a comprehensive overview of the current state of knowledge on strigolactones, exploring their biosynthesis, signaling mechanisms, and practical applications in agriculture. The book collects contributions from leading experts in the field, offering a diverse and in-depth perspective on the various roles that strigolactones play in plant biology.

The chapters in this volume cover a broad spectrum of topics, from the molecular and genetic basis of strigolactone biosynthesis to their interactions with other phytohormones and environmental factors. We delve into the regulatory functions of strigolactones in plant architecture, including shoot branching, root development, and leaf senescence, as well as their involvement in stress responses such as drought, salinity, and pathogen attack. Furthermore, the book highlights recent advancements in our understanding of strigolactones, signaling pathways and the potential for genetic engineering to manipulate these hormones for crop improvement.

This book is a valuable resource for researchers, students, and professionals in plant sciences, agronomy, and related fields. By consolidating the latest research and offering practical insights, it should inspire new investigations and applications that harness the power of strigolactones to address some of the most pressing challenges in agriculture today.

The editors hope that this book will catalyze further research and innovation in the field of plant hormone biology, ultimately contributing to the development of more resilient and productive agricultural systems.

We are grateful to the contributing authors for their dedication and expertise, which made possible this comprehensive exploration. We extend our thanks to the reviewers who have provided invaluable feedback and guidance throughout the preparation of this volume, and to Martin Scrivener and Scrivener Publishing for their support and publication.

**Tariq Aftab**
**Kaiser Iqbal Wani**
*Aligarh Muslim University, India*
October 2024

# 1

# Strigolactones: Journey from Rhizospheric Chemoattractants to Plant Growth Regulators

**Kaiser Iqbal Wani, M. Naeem and Tariq Aftab***

*Department of Botany, Faculty of Life Sciences, Aligarh Muslim University, Aligarh, India*

## *Abstract*

Strigolactones are a broad class of plant hormones that have progressed from being identified solely as chemoattractants in the rhizosphere to becoming important regulators of plant growth and development. Originally identified by their ability to promote the germination of seeds from parasitic plants such as *Striga*, *Orobanche*, and *Phelipanche*, strigolactones are now widely recognized as a novel class of plant hormones that play an important role in plant growth regulation. They play an important role in developing symbiotic associations with microbes such as rhizobia and arbuscular mycorrhizal fungi (AMF), which facilitate nutrient absorption, especially under nutrient-scarce conditions. Their main influence on root development, shoot branching, and overall plant architecture has been demonstrated by numerous studies conducted in the last two decades. This novel group of hormones aids plants in integrating environmental cues for the regulation of many development processes by influencing the transport and signaling of other hormones, especially auxins. Their involvement in a wide range of growth and developmental processes in plants, under both normal and stressed conditions, highlights their potential for agricultural applications, including enhancing crop resilience and productivity in challenging conditions.

***Keywords***: Strigolactone, *Striga*, *Orobanche*, rhizospheric chemoattractant, agriculture

**Corresponding author*: tarik.alig@gmail.com

Tariq Aftab and Kaiser Iqbal Wani (eds.) Strigolactones: Emerging Plant Hormones, (1–12) © 2025 Scrivener Publishing LLC

## 1.1 Introduction

In biology, hormones are a class of signaling molecules secreted by specific glands in organisms and transported to target organs via the circulatory system to regulate physiology and behavior. Traditionally associated with glands in animals, these molecules impart an action that is beyond simple chemical secretions either locally diffused to nearby cells or acting on the cells themselves that secrete them. Once these hormones reach their target cells, they bind with specific receptor proteins, and signal transduction pathways induce a cascade of cellular changes. Plants also have their version of such essential molecules; even though they lack dedicated hormone-secreting glands like animals, they are produced by the plant itself in trace amounts, the compounds being referred to as plant hormones. Plants are similarly dependent on the use of a varied array of natural factors that include, among others, water, oxygen, light, and a range of minerals and nutrients. The contribution of these exogenous factors to the growth of plants is considered essential. Alongside these exogenous factors, however, are intrinsic factors that help in their development. There exist factors intrinsic to the plants that are more specifically called plant hormones or phytohormones. They exercise their regulatory powers at very low concentrations either locally, in the cells adjacent to their origin, or upon distant tissues or organs. Darwin and Francis (1880) proposed the term plant hormones from the phenomenon of phototropism, which is the bending of a plant in the direction of light. Thiamann gave the term "phytohormones" to an organic substance of natural occurrence in plants.

Plant hormones play a crucial role in controlling different aspects of plant development and growth and reactions to environmental stimuli during different stages of the life cycle. The production of these signaling molecules occurs in specific tissues of plants and are then transported to other parts to exert their functions. The complex interactions between different hormones provide a framework necessary to coordinate plant physiological processes for optimal development in a changing environment. Of the many plant hormones, five are commonly regarded as "the big five": auxins, gibberellins, cytokinins, ethylene, and abscisic acid (ABA). These hormones regulate a wide range of functions, frequently influencing several events at the same time. Auxins promote growth [1], cytokinins regulate cell division, which aids in the formation of new plant organs [2], gibberellins contribute to stem elongation [3], ethylene orchestrates processes such as ripening [4], and ABA regulates moisture levels by mediating stomatal behavior [5]. The intricate interplay of hormones constitutes a complex mechanism, many of which are still unidentified.

During the last 15 years or so, several classes of new chemical compounds that have hormone-like activity have been independently discovered. The name for this new class of phytohormones is strigolactones. In this intricate domain relating to plant–soil interactions, strigolactones are complex compounds with far-reaching consequences for plants and the environment surrounding them, particularly in rhizospheres. Initially identified as rhizospheric chemoattractants [6], these small chemical molecules came out to be versatile signaling molecules governing a wide array of activities integrally related to plant growth, development, and environmental adaptation [7]. This chapter briefly reviews the history of these multi-faceted novel phytohormones, from their detection in root exudates to key plant growth regulators.

## 1.2 Brief History of Strigolactones

This is the sixth decade since the first pivotal publication elucidating the isolation and characterization of strigol as an effective seed germination stimulant for *Striga lutea* [8]. Since then, a number of key discoveries have been made concerning strigolactones, as indicated in Figure 1.1 in the form of a timeline. The work by Cook *et al.* was seminal not only in aiding the discovery of a group of compounds known as strigolactones but also in establishing their activity in stimulating the germination of seeds belonging to parasitic plants such as *Striga*. More recently, their identification as key players in mediating the plant–arbuscular mycorrhizal fungi symbiosis [9] and their activity as shoot-derived endogenous plant hormones have driven new research opportunities [10, 11]. This structurally related group of compounds collectively called strigolactones occupies

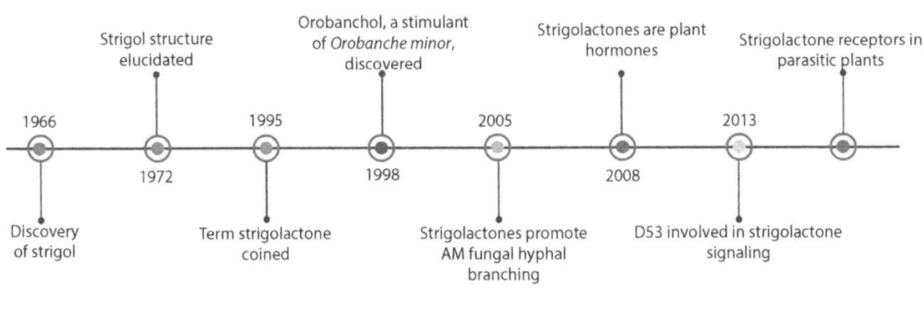


**Figure 1.1** Milestone discoveries in strigolactones.

a prominent position in the scientific world today for their multifaceted involvement in rhizosphere signaling processes and their potential in parasitic weed management. This newfound understanding of their regulatory functions as plant hormones unraveled many intricacies of plant physiological responses.

Among various applications, strigolactones are most commonly used to control parasitic weeds. Parasitic weeds from the *Striga*, *Phelipanche*, and *Orobanche* genera pose significant agricultural threats in many regions around the world, necessitating immediate action [12]. The field has made significant progress, culminating in the organization of the inaugural international strigolactone congress in Wageningen in March 2015.

The first strigolactone, (+)-strigol, was discovered in 1966 and extracted from cotton root exudates as already discussed at the beginning. However, it was not until nearly two decades later that its structure was explained [13–15]. The ability of strigol to induce seed germination of the *Striga* and *Orobanche* species is significant. Initially, strigol was the only naturally occurring strigolactone known. After 1990, novel strigolactones were isolated from various root exudates, including sorgolactone from sorghum [16], orobanchol from red clover [17], and solanacol from tobacco [18].

Strigolactones are found in trace amounts in root exudates, making it challenging to determine their structures and assign stereochemistry [19]. All strigolactones have a fundamental structure consisting of an annulated system of three rings known as the ABC scaffold, which is connected to a butenolide ring via an enol ether bridge. Two families of strigolactones have been identified: one with (+)-strigol-like stereochemistry at the BC junction and the other with (−)-orobanchol-like BC stereochemistry. Across all natural strigolactones, the D-ring configuration at the C-2′ position is consistently R.

As the interest in strigolactones is growing day by day, new information about their biological properties is emerging. It has been discovered that strigolactones function as branching factors for arbuscular mycorrhizal (AM) fungi [9, 20] and as inhibitors of bud outgrowth and shoot branching, with the latter involving an interplay with auxins [10, 11]. Furthermore, they have a significant impact on plant architecture (both above- and below-ground), playing a key role in its regulation. These signaling molecules are now recognized as a new class of plant hormones with a promising future.

## 1.3 What is the Origin of the Name Strigolactone?

Strigolactones were first reported in root exudates because of their ability to induce the germination of seeds from the parasitic plant *Striga*. *S. hermonthica*, also known as purple or giant witchweed, is a notable species in this plant family. The name "Striga" originated from the observation of these witchweeds by subsistence farmers in Africa. These plants would germinate unexpectedly, destroying the crops. The term "witchweed" was derived from this idea because farmers thought that these plants germinated mysteriously similar to how a witch would appear all of a sudden without an introduction. The scientific term *Striga* originates from ancient Rome, where it was referred to as some sort of mythological witch due to her vicious nature, specifically her malice toward others, especially that of children. *Striga* species are part of the broomrape family (Orobanchaceae), and the majority of them grow as parasites on other plants.

The lactone component of strigolactones indicates its chemical structure. Lactones are cyclic esters formed by the condensation of a carboxylic acid and an alcohol group within a single molecule. Strigolactones, in particular, contain two lactone rings. The chemical variations to the core structure and differences in stereochemical configurations cause variation within the strigolactone family [21]—for example, strigol and orobanchol are frequently cited examples of A and B rings undergoing oxidation, resulting in distinct stereochemical differences between the B and C rings.

## 1.4 Diverse Strigolactone Functions: From Single-Celled Alga to Terrestrial Plants

Strigolactones have been reported in simple single-celled algae and primitive terrestrial plants like liverworts and mosses [22]. Their early function probably included intercellular signaling and control of growth and differentiation in primitive plants—for example, strigolactones are produced in liverworts, mosses, and the alga *Chara coralline*; they induce the outgrowth of rhizoids there. In filamentous mosses, like *Physcomitrella patens*, they regulate the branching of protonema and the growth of filaments in neighboring colonies [23]. As a result, strigolactones help neighboring organisms coordinate their growth and competition—a principle that also applies to higher plants, regulating the growth of roots and shoots. As terrestrial colonization progressed millions of years ago, fungal symbioses emerged.

Certain liverworts formed symbiotic relationships with mycorrhizal fungi, though the role of strigolactones in this interaction is unknown and requires further investigation. With the evolution of vascular plants came complex patterns of shoot branching and the possibility of long-distance strigolactone transport. The use of strigolactones by witchweeds, which they extract from host plants, is the most recent evolutionary adaptation in strigolactone history.

## 1.5 Strigolactones in Ferns and Mosses: Rhizosphere Signals or Phytohormones?

Genetic and physiological studies have reported that strigolactones act as plant hormones in angiosperms [24, 25]. Moreover, in most plant species, strigolactones are secreted by roots into the rhizosphere, hence acting as a signal that attracts AM fungi, making nutrients most available for uptake [26, 27]. This phenomenon is observed in a variety of plant groups, including bryophytes, indicating a long-standing evolutionary relationship [28]. This raises the question of whether the original function of strigolactones was primarily hormonal regulation or as a rhizospheric signal.

Non-seed plants lacking essential components of strigolactone signaling, such as D14 and SMXL7-type proteins, may need an alternative mechanism for perception and response [29]—for example, *Marchantia polymorpha* has lost the ability to produce strigolactones due to the absence of CCD8 and MAX1 enzymes [30]. If strigolactones primarily served as developmental or hormonal signals in this species, the loss of synthesis ability would likely lead to evolutionary selection pressure against it. Consequently, *M. polymorpha* has lost its capability to attract AM fungi, unlike other *Marchantia* species that still retain this ability.

What about the role of strigolactones in mosses? As already known, *P. patens* produces carlactone, which suggests the presence of strigolactone-like compounds [22, 31]. Furthermore, normal development in *P. patens* is dependent on carlactone biosynthesis. But what if *P. patens* is the exception? Here strigolactones appear to regulate plant growth in a way similar to quorum sensing, which could be viewed as a limited developmental response to what is essentially a signal in the rhizosphere. Given that *P. patens* has also lost the capability for AM symbiotic association, unlike its immediate relatives, the use of strigolactones as a developmental signal in this species may be an independent evolutionary adaptation unique to it or its immediate relatives.

## 1.6 Agricultural Loss and Root Parasitic Plants

Root parasitic plants of Orobanchaceae family are one of the most important agricultural problems in Southern Europe, Africa, and Asia, where there are a number of obligate parasites of this family that require the availability of a host for development and reproduction. They include Broomrapes, genera *Orobanche*, *Phelipanche*, and witchweeds, genus *Striga* [32–34]. Notably, witchweeds are a real danger to African food security [35]. Indeed this weed infests close to 50 million hectares of agricultural land across the continent, especially to the detriment of 300 million farmers who stand to lose over $7 billion USD worth of economies annually [36]. Parasitic plants can infect a wide range of plant species, but staple crops such as cereals and legumes are especially vulnerable. They germinate due to strigolactones exuded by the roots of host crop plants (Figure 1.2) and cause severe agricultural loss. The seed production of a

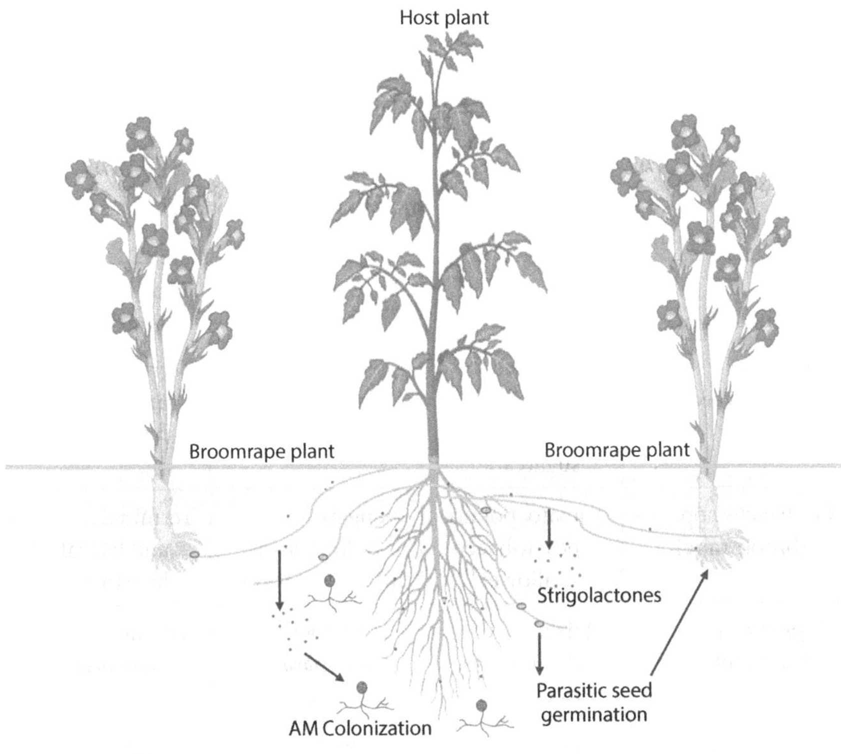


**Figure 1.2** Germination of parasitic seeds belonging to Broomrapes and enhanced AM colonization by strigolactones exuded from host plant roots in the rhizosphere.

single *Striga* plant may vary from 10,000 to over 200,000 seeds, and these seeds could remain dormant in the soil for up to 20 years [37]. The control of their propagation is a daunting task because the weight of these seeds is very small, about 7 μg. They are easily dispersed by wind, water, animals, and the human population as well as agricultural equipment [38, 39]. Furthermore, cases of infection of crops previously considered to be non-host species by *Striga* have been reported [40].

For poor farmers without access to efficient control techniques, the threat posed by parasitic plants is particularly serious, while pesticides, ethylene, methyl bromide, and stringent quarantine regulations were necessary in the United States to successfully eradicate infestations of *Striga asiatica* [32, 41]. These solutions are frequently unavailable to or prohibitively expensive for African farmers. As an alternative, they turn to less successful techniques like hand weeding, intercropping, and more fertilizer [42]. Farmers are occasionally compelled to give up their land or crops because there are no sustainable alternatives [35]. Therefore, there is a pressing need for novel, approachable strategies to eradicate these parasitic weeds.

Some of the known parasitic plants belonging to the Orobanchaceae family, along with their host plants, are given below in Table 1.1.

**Table 1.1** Some known parasitic plants belonging to the Orobanchaceae family and their host plants.

| **Parasitic plant** | **Host plant** | **Type of parasitism** | **Type of haustoria** |
|---|---|---|---|
| *Striga* spp. (witchweeds) | Maize, rice, sorghum, millet, sugarcane | Obligate hemiparasite Infects the roots | Terminal and lateral haustoria |
| *Orobanche* spp. (broomrape) | Tomato, potato, pea, tobacco, sunflower | Obligate holoparasite Infects the roots | Terminal and lateral haustoria |
| *Triphysaria versicolor* | Maize, *Arabidopsis*, rice, tomato | Facultative hemiparasite Infects the roots | Lateral haustoria |
| *Phtheirospermum* | Arabidopsis, Medicago | Facultative hemiparasite Infects the roots | Lateral haustoria |

## 1.7 Strigolactones and Agriculture: Why Do They Hold the Key?

The Green Revolution introduced us to shorter, high-yield cereal plants that thrive in intensive farming with fertilizers and pesticides. These shorter varieties require less energy for stem growth and allocate more toward producing seeds by reducing their sensitivity to gibberellins or their production [43, 44]. In a future with finite global phosphate reserves and most nitrogen fertilizers requiring vast amounts of fossil fuel, it will become more and more important to breed crop varieties for their ability to scavenge nutrients more efficiently. It could also become necessary to foster symbiotic associations between plants and soil microbes to maximize plant growth. Variation in plant strigolactones opens up opportunities for breeding crop varieties with better nutrient efficiency and/or symbiotic properties [45, 46]. The structure of roots and shoots is controlled by the exuded strigolactones, offering an opportunity for plant breeding improvement. Concurrent efforts can be made to minimize witchweed parasitism, particularly in Africa, where it causes maximum losses to subsistence farmers. This could be achieved through the use of either plant breeding techniques or strigolactone analogs that trigger self-destruction—the so-called suicidal germination—of witchweed seeds before crop planting. In that respect, strigolactone research gives significant promise to finding solutions in connection with the important issues of crop improvement and agricultural management.

## References


1. Tivendale, N.D. and Millar, A.H., How is auxin linked with cellular energy pathways to promote growth? *New Phytol.*, *233*, 6, 2397–2404, 2022.
2. Yang, W., Cortijo, S., Korsbo, N., Roszak, P., Schiessl, K., Gurzadyan, A., Meyerowitz, E., Molecular mechanism of cytokinin-activated cell division in Arabidopsis. *Science*, *371*, 6536, 1350–1355, 2021.
3. Zhang, H., Wang, W., Huang, J., Wang, Y., Hu, L., Yuan, Y., Wu, B., Role of gibberellin and its three GID1 receptors in *Jasminum sambac* stem elongation and flowering. *Planta*, *255*, 1–16, 2022.
4. Hu, S., Liu, L., Li, S., Shao, Z., Meng, F., Liu, H., Wang, Q., Regulation of fruit ripening by the brassinosteroid biosynthetic gene SlCYP90B3 via an ethylene-dependent pathway in tomato. *Hort. Res.*, *7*, 163, 2020.

5. Hasan, M.M., Gong, L., Nie, Z.F., Li, F.P., Ahammed, G.J., Fang, X.W., ABA-induced stomatal movements in vascular plants during dehydration and rehydration. *Environ. Exp. Bot.*, *186*, 104436, 2021.
6. Lanfranco, L., Fiorilli, V., Gutjahr, C., Partner communication and role of nutrients in the arbuscular mycorrhizal symbiosis. *New Phytol.*, *220*, 4, 1031–1046, 2018.
7. Aliche, E.B., Screpanti, C., De Mesmaeker, A., Munnik, T., Bouwmeester, H.J., Science and application of strigolactones. *New Phytol.*, *227*, 4, 1001–1011, 2020.
8. Cook, C.E., Whichard, L.P., Turner, B., Wall, M.E., Egley, G.H., Germination of witchweed (*Striga lutea* Lour.): isolation and properties of a potent stimulant. *Science*, *154*, 3753, 1189–1190, 1966.
9. Akiyama, K., Matsuzaki, K.I., Hayashi, H., Plant sesquiterpenes induce hyphal branching in arbuscular mycorrhizal fungi. *Nature*, *435*, 7043, 824–827, 2005.
10. Gomez-Roldan, V., Fermas, S., Brewer, P.B., Puech-Pagès, V., Dun, E.A., Pillot, J.P., Rochange, S.F., Strigolactone inhibition of shoot branching. *Nature*, *455*, 7210, 189–194, 2008.
11. Umehara, M., Hanada, A., Yoshida, S., Akiyama, K., Arite, T., Takeda-Kamiya, N., Yamaguchi, S., Inhibition of shoot branching by new terpenoid plant hormones. *Nature*, *455*, 7210, 195–200, 2008.
12. Rubiales, D., Broomrape threat to agriculture. *Outlooks Pest Manag.*, *31*, 3, 141–145, 2020.
13. Cook, C.E., Whichard, L.P., Wall, M., Egley, G.H., Coggon, P., Luhan, P.A., McPhail, A.T., Germination stimulants. II. Structure of strigol, a potent seed germination stimulant for witchweed (*Striga lutea*). *JACS*, *94*, 17, 6198–6199, 1972.
14. Coggon, P., Luhan, P.A., McPhail, A.T., Crystal and molecular structure of the germination stimulant strigol by X-ray analysis. *J. Chem. Soc. Perkin Trans.*, *2*, 4, 465–469, 1973.
15. Brooks, D.W., Bevinakatti, H.S., Powell, D.R., The absolute structure of (+)-strigol. *J. Org. Chem.*, *50*, 20, 3779–3781, 1985.
16. Hauck, C., Müller, S., Schildknecht, H., A germination stimulant for parasitic flowering plants from *Sorghum bicolor*, a genuine host plant. *J. Plant Physiol.*, *139*, 4, 474–478, 1992.
17. Yokota, T., Sakai, H., Okuno, K., Yoneyama, K., Takeuchi, Y., Alectrol and orobanchol, germination stimulants for *Orobanche minor*, from its host red clover. *Phytochem.*, *49*, 7, 1967–1973, 1998.
18. Xie, X., Kusumoto, D., Takeuchi, Y., Yoneyama, K., Yamada, Y., Yoneyama, K., 2′-Epi-orobanchol and solanacol, two unique strigolactones, germination stimulants for root parasitic weeds, produced by tobacco. *J. Agric. Food. Chem.*, *55*, 20, 8067–8072, 2007.
19. Zwanenburg, B. and Pospíšil, T., Structure and activity of strigolactones: new plant hormones with a rich future. *Mol. Plant*, *6*, 1, 38–62, 2013.

20. Parniske, M., Arbuscular mycorrhiza: the mother of plant root endosymbioses. *Nat. Rev. Microbiol.*, *6*, 10, 763–775, 2008.
21. Nomura, T., Seto, Y., Kyozuka, J., Unveiling the complexity of strigolactones: exploring structural diversity, biosynthesis pathways, and signaling mechanisms. *J. Exp. Bot.*, *75*, 4, 1134–1147, 2024.
22. Delaux, P.M., Xie, X., Timme, R.E., Puech-Pages, V., Dunand, C., Lecompte, E., Séjalon-Delmas, N., Origin of strigolactones in the green lineage. *New Phytol.*, *195*, 4, 857–871, 2012.
23. Proust, H., Hoffmann, B., Xie, X., Yoneyama, K., Schaefer, D.G., Yoneyama, K., Rameau, C., Strigolactones regulate protonema branching and act as a quorum sensing-like signal in the moss *Physcomitrella patens. Development*, *138*, 8, 1531–1539, 2011.
24. Bouwmeester, H.J., Fonne-Pfister, R., Screpanti, C., De Mesmaeker, A., Strigolactones: plant hormones with promising features. *Angew. Chem. Int. Ed.*, *58*, 37, 12778–12786, 2019.
25. Wani, K.I., Zehra, A., Choudhary, S., Naeem, M., Khan, M.M.A., Castroverde, C.D.M., Aftab, T., Mechanistic insights into strigolactone biosynthesis, signaling, and regulation during plant growth and development. *J. Plant Growth Regul.*, 40, 1–17, 2020.
26. Umehara, M., Strigolactone, a key regulator of nutrient allocation in plants. *Plant Biotechnol.*, *28*, 5, 429–437, 2011.
27. Mitra, D., BE, G.S., Khoshru, B., De Los Santos Villalobos, S., Belz, C., Chaudhary, P., Mohapatra, P.K.D., Impacts of arbuscular mycorrhizal fungi on rice growth, development, and stress management with a particular emphasis on strigolactone effects on root development. *Commun. Soil Sci. Plant Anal.*, *52*, 14, 1591–1621, 2021.
28. Ruyter-Spira, C. and Bouwmeester, H., Strigolactones affect development in primitive plants. The missing link between plants and arbuscular mycorrhizal fungi? *New Phytol.*, *195*, 4, 730–733, 2012.
29. Bythell-Douglas, R., Rothfels, C.J., Stevenson, D.W., Graham, S.W., Wong, G.K.S., Nelson, D.C., Bennett, T., The complex origins of strigolactone signalling in land plants, bioRxiv, 102715, 2017.
30. Walker, C. and Bennett, T., Reassessing the evolution of strigolactone synthesis and signalling, bioRxiv, 228320, 2017.
31. Lopez-Obando, M., Guillory, A., Boyer, F.D., Cornu, D., Hoffmann, B., Le Bris, P., Bonhomme, S., The *Physcomitrium* (*Physcomitrella*) *patens* PpKAI2L receptors for strigolactones and related compounds function via MAX2-dependent and-independent pathways. *Plant Cell*, 33, 11, 3487–3512, 2021.
32. Parker, C., Parasitic weeds: a world challenge. *Weed Sci.*, *60*, 2, 269–276, 2012.
33. Parker, C., The parasitic weeds of the Orobanchaceae, in: *Parasitic Orobanchaceae: Parasitic mechanisms and control strategies*, pp. 313–344, Springer Berlin Heidelberg, Berlin, Heidelberg, 2013.

34. Spallek, T., Mutuku, M., Shirasu, K., The genus *Striga*: a witch profile. *Mol. Plant Pathol.*, *14*, 9, 861–869, 2013.
35. Parker, C., Observations on the current status of *Orobanche* and *Striga* problems worldwide. *Pest Manag. Sci.*, *65*, 453–459, 2009.
36. Ejeta, G., The Striga scourge in Africa: a growing pandemic, in: *Integrating new technologies for Striga control: towards ending the witch-hunt*, pp. 3–16, 2007.
37. Hearne, S.J., Control—the *Striga conundrum. Pest Manag. Sci.*, *65*, 5, 603–614, 2009.
38. Dafaallah, A.B., Biology and physiology of witchweed (*Striga* spp.): A review. *Int. J. Acad. Multidiscip. Res.*, *3*, 10, 42–51, 2019.
39. Berner, D.K., Cardwell, K.F., Faturoti, B., Ikie, F.O., Williams, O.A., Relative roles of wind, crop seeds, and cattle in dispersal of *Striga* spp. *Plant Dis.*, *78*, 4, 402–406, 1994.
40. Abbasher, A.A., Zaroug, M.S., Zahran, E.B., Sauerborn, J., Germination, attachment and development of *Striga hermonthica* (Del.) Benth induced by host and non-host crops. *J. Agric. Sci.*, *20*, 3, 2019.
41. Epée Missé, P.T., Comparative Analysis on the Management of the Parasitic Weed Striga in the USA, Australia and Kenya, (Novembre 27, 2018), 2018, https://dx.doi.org/10.2139/ssrn.3231805.
42. Teka, H.B., Advance research on *Striga* control: A review. *Afr. J. Plant Sci.*, *8*, 11, 492–506, 2014.
43. Gao, S. and Chu, C., Gibberellin metabolism and signaling: targets for improving agronomic performance of crops. *Plant Cell Physiol.*, *61*, 11, 1902–1911, 2020.
44. Castro-Camba, R., Sánchez, C., Vidal, N., Vielba, J.M., Plant development and crop yield: The role of gibberellins. *Plants*, *11*, 19, 2650, 2022.
45. Kelly, J.H., Tucker, M.R., Brewer, P.B., The strigolactone pathway is a target for modifying crop shoot architecture and yield. *Biology*, *12*, 1, 95, 2023.
46. Marro, N., Lidoy, J., Chico, M.Á., Rial, C., García, J., Varela, R.M., López-Ráez, J.A., Strigolactones: New players in the nitrogen–phosphorus signalling interplay. *Plant Cell Environ.*, *45*, 2, 512–527, 2022.

# 2

# Nature, Structural Diversity, Biosynthetic Pathway, and Strigolactone Transport in Plants

**Ishrat Mehmood[1]*, Kaiser Iqbal Wani[2] and Tariq Aftab[2]**

[1]*Department of Biophysics, Panjab University, Chandigarh, India*
[2]*Department of Botany, Aligarh Muslim University, Aligarh, India*

***Abstract***
Strigolactones (SLs) are a class of phytohormones with diverse roles in plants which range from regulation of plant growth and development and their interactions with the environment. Their original roles, associated with seed germination of root parasitic plants of broomrapes, were extended later to include the regulation of shoot branching inhibitions, root development, and plant–microbe interactions. The SLs hold great importance in the agricultural sector, particularly in combatting root parasitic weeds via suicidal germination and other means. Recent research has revealed their involvement in inter-plant signaling, adding a layer to their ecological significance. This chapter is about the chemical nature, structural diversity, biosynthetic pathway, and naming protocol of SLs. Canonical SLs comprise a tricyclic lactone ring system coupled to a fourth ring by an enol-ether bridge. However, the non-canonical SL variants exhibit structural deviations, influencing their biological activities. The naming protocol of different SLs ensures clarity and precision, facilitating communication in scientific conversations. The biosynthetic pathway involves enzymatic steps, including carotenoid cleavage, cyclization, and modification, mediated by enzymes like CCD7, CCD8, MAX1, and D27. Investigations for understanding their nature, diversity, and biosynthesis offer valuable insights and hold the key for agricultural and ecological applications, emphasizing their potential to address global challenges.

***Keywords*:** Strigolactones, biosynthesis, nature, transport, canonical, non-canonical, *Striga*

**Corresponding author*: ishratmehmood943@gmail.com

Tariq Aftab and Kaiser Iqbal Wani (eds.) Strigolactones: Emerging Plant Hormones, (13–34) © 2025 Scrivener Publishing LLC

## 2.1 Introduction

Strigolactones (SLs) represent a class of phytohormones that have emerged to take central roles in the complex web of signaling mechanisms controlling various aspects of plant growth, development, and environmental interactions. Derived from carotenoids, they are secreted into the rhizosphere, where they interact with root parasitic plants and symbiotic fungi. These small molecules were primarily identified for their role in stimulating the germination of root parasitic plants but have extended to include many physiological processes not only within host plants but also extending into non-host plants [1]. Their key role which gives them immense significance is their role in the regulation of plant architecture [2]. They participate in shoot branching and root development, senescence, and overall plant morphology. These signaling compounds negatively regulate axillary bud outgrowth, thus inhibiting the formation of lateral branches for the adaptability of the plant to environmental changes [3].

Strigolactones are linked to plant–microbe interactions, particularly in the context of symbiosis and parasitism. In the case of symbiotic associations, these compounds act as signaling molecules in the rhizosphere that attract arbuscular mycorrhizal fungi (AMF) to establish a mutualistic association with plant roots. They act as rhizosphere signaling molecules, attracting AMF hyphae toward the roots and stimulating the branching of fungal hyphae in the soil [4]. On the other side, they facilitate parasitic relationship between certain plants and root parasitic weeds. They arbitrarily facilitate the invasion of host plant roots by inducing parasitic relationships between certain plants and root parasite weeds, leading to nutrient and water deficits.

The significance of SLs extends even further. Recent studies [5] suggest that these compounds mediate immediate inter-plant signaling, acting as allelochemicals that convey information about the local environment. This form of chemical communication allows plants to respond collectively to threats, such as herbivore attacks or impending stress conditions, by adjusting their growth and defence strategies. Also, the historical context of SL discovery is interesting to know.

The story unfolds against a backdrop of scientific curiosity, agricultural challenges, and the pursuit of knowledge that has reshaped our understanding of plant biology. The journey of SLs began in the mid-20th century when researchers were grappling with the enigmatic phenomenon of unwanted seed germination of root parasitic plants [6]. Scientists sought to unravel the mystery of how these parasitic plant species, such as *Striga* and *Orobanche*, detect the presence of host plants and initiate the germination process. The role play between parasitic plants and their hosts posed significant challenges for agriculture, leading researchers to seek answers. In the early 1960s, scientists made a pivotal breakthrough when Italian researchers led by Pietro C. Rubiales identified a compound in root exudates of host plants that stimulated the germination of seeds of parasitic plants which then parasitize the roots of host plants. This compound, later named strigol, marked the first glimpse into the world of SLs. The term "strigolactone" was coined years later when additional related compounds were discovered, broadening the scope of these signaling molecules [7, 8].

The discovery of SLs was not merely an academic pursuit; it held profound implications for agriculture. Root parasitic weeds, witchweeds, or *Striga* species, were known to cause devastating losses in crop yields, particularly in Africa, where they posed a serious threat to staple crops like sorghum and millet [9]. Understanding the signaling mechanisms behind their germination provided a potential avenue to develop strategies to control these parasitic plants and mitigate agricultural losses. As research on SLs progressed, scientists uncovered their role not only in parasitic plant interactions but also in various other aspects of plant biology.

The 21st century has witnessed an explosion of research on SLs, revealing their involvement in shoot branching, root development, plant–microbe interactions, responses to abiotic stress, and even inter-plant communication. In 2008, the identification of the first SL receptor, DWARF14 (D14), by a collaborative effort led by researchers from the United States and Japan, marked a significant milestone [10, 11]. This discovery provided insights into the molecular mechanisms through which SLs exert their effects on plant growth and development. What began as a quest to understand the mechanisms behind parasitic plant germination evolved into a broader exploration of plant signaling, architecture, and ecological interactions.

## 2.2 Chemical Nature of Strigolactones

Strigolactones (SLs) constitute a class of sesquiterpene lactones initially identified for their role in promoting seed germination of parasitic weeds, including *Striga* and *Orobanche* species. These compounds feature a tricyclic lactone structure (ABC tricyclic ring) linked to a fourth ring (the D ring or α, β-unsaturated furan ring) through an enol-ether bridge. SLs are synthesized from carotenoids, which are themselves made from terpene or isoprene building blocks. Structurally, SLs consist of four rings (A, B, C, and D), with the D-ring being the key characteristic feature. This related group of compounds is defined as substances with a butenolide (D-ring) structure [12, 13].

The fundamental structure of SLs consists of three interconnected rings, designated as rings A, B, and C. A six-membered ring containing an enol, ether, or lactone moiety is known as Ring A, while Ring B is a five-membered ring typically bearing a ketone functional group. Ring C, the third ring, is also five-membered, connected to ring B by an oxygen atom. Their tricyclic lactone ring structure serves as the molecular basis for a variety of biological activities. A varied tail that extends from ring A is used to identify various SL compounds. Because of this tail's high sensitivity to modifications, several SL variations with unique biological roles have been identified. The tail consists of a polyenyl side chain, characterized by a series of double bonds. The length and arrangement of these double bonds vary among different SLs, contributing to the specificity of their interactions with receptors and downstream signaling cascades [14].

- Ring A
  Ring A is a six-membered ring and is characterized by the presence of either lactone or an enol ether moiety. Whereas the lactone configuration has a cyclic ester group, the enol ether configuration has a double bond between carbon atoms, forming an oxygen–carbon double bond. The choice between these two configurations influences the chemical properties of ring A, which, in turn, contributes to SLs' overall adaptability [12].
- Ring B
  Ring B is a five-membered ring that has a ketone functional group that is connected to ring A. The biological activity of the SLs is dependent on the ketone group, which also influences how SLs interact with receptors and downstream signaling pathways. The presence of the ketone group in ring B sets SLs apart from other classes of plant hormones and imparts specific chemical properties to this ring, contributing to the overall structure and functionality of the tricyclic lactone ring system [15].
- Ring C
  Ring C is another five-membered ring connected to ring B by an oxygen atom that completes the tricyclic lactone ring system. One important structural feature that contributes to the stability and conformational flexibility of strigolactones is the linkage between rings B and C. As a connecting element, the oxygen atom influences the molecule's overall

three-dimensional structure and facilitates its linkage with other molecules in biological systems [16].

In addition to being SLs' structural hallmark, the tricyclic lactone ring system also influences their functional characteristics. The distinct configuration of atoms in this ring system gives SLs certain chemical and physical characteristics that influence their solubility, stability, and reactivity. In addition, the tricyclic lactone ring system plays a vital role in the detection and binding of SLs to their receptors, which starts downstream signaling cascades that control various physiological processes in plants [17].

## 2.3 Naming Protocol for Strigolactones

Stereoisomers have several stereogenic (chiral) centers; strigol, for instance, contains three of these centers, and there are eight possible stereoisomers. The International Union of Pure and Applied Chemistry (IUPAC) recommends the so-called CIP system, named after the proposers Cahn, Ingold, and Prelog, for the notation of the configuration at a stereogenic center [18]. The sensation of chirality is indicated by the stereochemical descriptors R and S in the CIP system. Strict abstract requirements must be adhered to assign the appropriate description to a particular chiral center. It is frequently challenging to imagine a structure that corresponds with such R and S descriptors since the R, S nomenclature is based on abstract principles. These days, the ChemDraw application makes it easy to produce these descriptions. The majority of scientists want an "at a glance" classification of the perfect stereochemistry. To do this, a notation known as (+)-3aR,5S,8bS,2′R)-strigol is utilized, which describes the stereochemical interaction with a parent reference structure. The prefixes ent, which stands for enantiomer, and epi, which is an acronym for epimer, denote the opposite configuration at a particular atom, indicating the association with (+)-3aR,5S,8bS,2′R)-strigol. Given that (+)-strigol is the first completely studied and verified structure, using it as the reference chemical makes sense. Due to the three stereogenic (chiral) centers in strigol, eight possible stereoisomers exist. Eight stereoisomers of 5-deoxy-4-hydroxy-strigol, also known as orobanchol, are known to exist. These may be named unequivocally using the R and S descriptors or by prefixing the word orobanchol with the relevant information to indicate the relative stereochemistry concerning (+)-strigol [19, 20].

For the ent/epi approach, it is required to select a parent molecule or reference substance. The naming of SLs was straightforward before the structural rectification of orobanchol: the stereochemistry of all other SLs was tied to (+)-strigol, which was the logical parent molecule. However, there were two choices available which either adhere to the naming convention that used (+)-strigol as the parent compound or utilize the newly discovered structure for natural orobanchol as the parent compound for the orobanchol family [21, 22]. Because both approaches are being used, things might get complicated. This approach has few advantages and might be difficult for people who are not familiar with stereochemistry. Using both (+)-strigol and (−)-orobanchol as standards, Scaffidi and colleagues proposed an alternate name and notation in the structural correlation of GR24 stereoisomers. As a result, several stereoisomers were given two names: for example, ent-2′-epi-5-deoxystrigol is also known as 4-deoxyorobanchol [23]. This is a complex way for people who are not as knowledgeable with stereochemical difficulties, and it offers little additional benefit.

## 2.4 Natural Diversity of Strigolactones

Naturally occurring SLs in plants are carotenoid-derived hormones or their precursors that control a vast array of growth and physio-biochemical processes in plants [24, 25]. Based on the diversity in chemical structures, they can be classified into two major groups:

a. Strigol and its related compounds containing ABC-ring system connected to the methyl butenolide D-ring through an enol-ether bridge are called *canonical SLs* (Figures 2.1 and 2.2).
b. Certain germination stimulant compounds which lack the ABC ring system but possess the D-ring have been characterized, and such compounds are called *non-canonical SLs* (Figure 2.3).

Through purification bioassays, approximately 25 canonical SLs have been identified and isolated from root exudates thus far [24, 26]. Except for 5-deoxystrigol, which was first described as the hyphal branching stimulant for AM fungi [4], the majority were identified as germination stimulants of root parasitic weeds. Carlactone (CL) [27] is the most basic non-canonical SL and acts as a common precursor for the biosynthesis of SL. The root exudates of a number of plants have been shown to contain carlactonoic acid (CLA), an oxidized metabolite of CL [28, 29]. Non-canonical SLs participate in chemical communications in the rhizosphere along with canonical SLs.

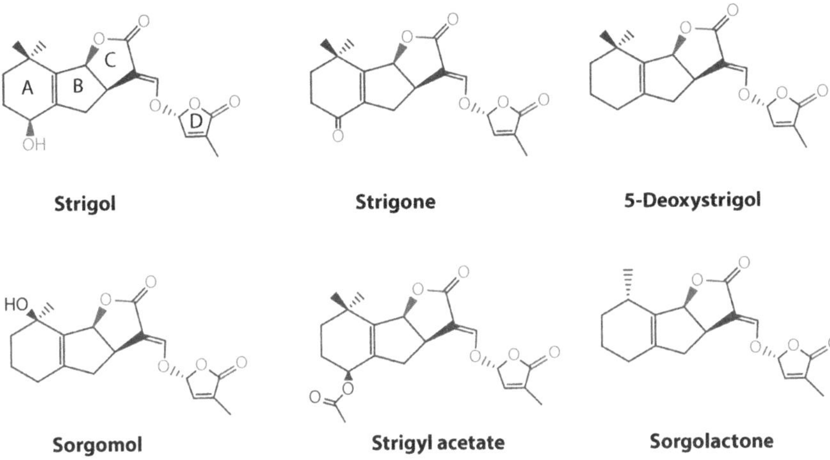


**Figure 2.1** Chemical structures of some representative strigol-type strigolactones.

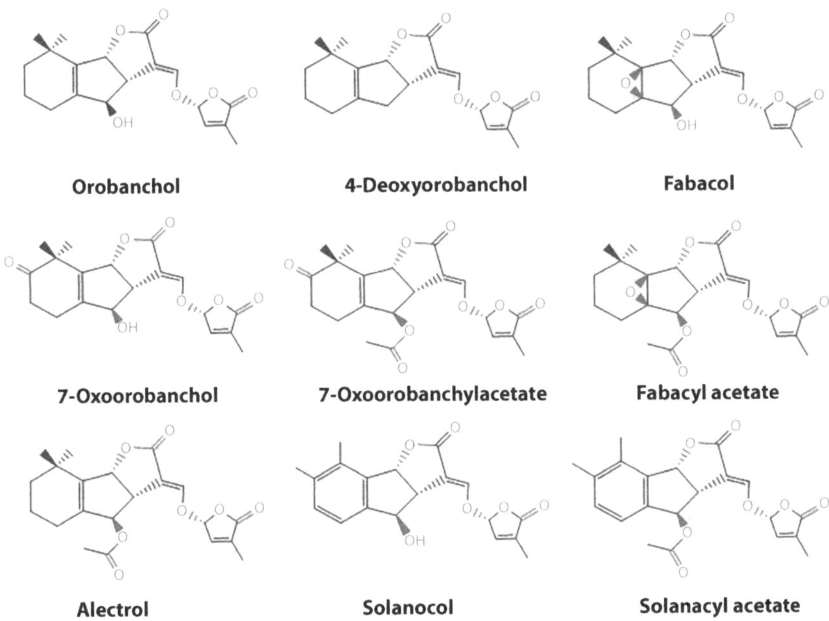


**Figure 2.2** Chemical structures of some representative orobanchol-type strigolactones. Note: While drawing the molecular structures, stereochemistry is shown by plain lines connecting carbon atoms in the plane of the page. The wedges project out of the plane of the page, with dashes projected into the plane of the page.

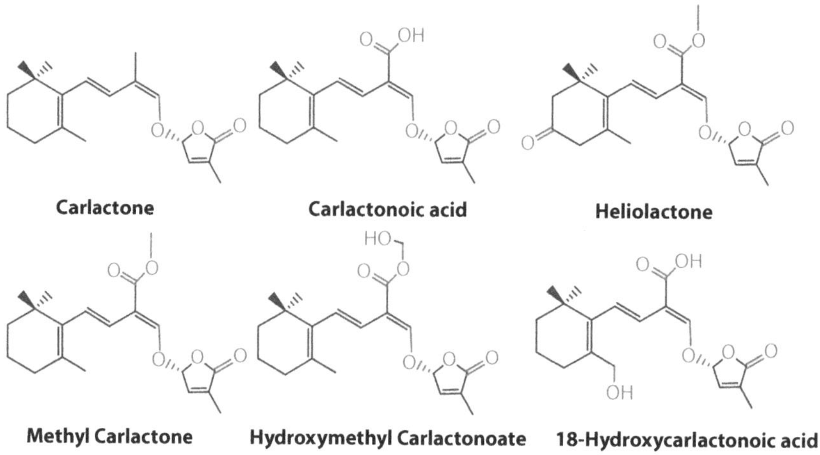


**Figure 2.3** Chemical structures of some representative non-canonical strigolactones.

### 2.4.1 Canonical Strigolactones

Canonical strigolactones (SLs) are classified into two categories based on the stereochemistry of the C-ring: strigol-type (illustrated in Figure 2.1) and orobanchol-type (illustrated in Figure 2.2) SLs, with β- and α-oriented C-ring stereochemistry, respectively [25]. Typically, plants predominantly produce either strigol-type or orobanchol-type SLs; however, certain species, such as tobacco, synthesize both types [30]. Gymnosperms like *Gingko biloba*, *Pinus thunbergii*, *Cryptomeria japonica*, and the lycophyte *Selaginella moellendorfi* have been reported to produce orobanchol-type SLs [26, 29]. While many angiosperms are also orobanchol-type SL producers, some, including cotton and strawberry, exclusively produce strigol-type SLs. Within the Poaceae family, sorghum is a strigol-type SL producer, whereas rice produces orobanchol-type SLs [11]. It has recently been demonstrated that some cultivars of sorghum resistant to *Striga* produce orobanchol, a powerful branching stimulant for AM fungus but a feeble germination stimulant for *Striga* [31], although these cultivars are resilient to *Striga* infestation but still act as suitable hosts for AM fungi. The chemical structures of some of the common canonical strigolactones are shown in Figure 2.1 (strigol type) and Figure 2.2 (orobanchol type).

### 2.4.2 Non-Canonical Strigolactones

Despite reports that strigol is exuded by maize, a primary host of *Striga*, Jamil and others [32] used LC–MS/MS analysis to detect non-canonical SLs, including strigol, in the maize root exudates. Two new germination stimulants, SL1 and SL2, were reported instead. Likewise, no known canonical SL that potently stimulates the germination of broomrapes has been found in the root exudates of *Avena strigosa* [33]. A great deal of work has been done to identify and elucidate the germination stimulants generated by black oats and maize. Two non-canonical SLs, zealactone [34] and avenaol [33], have been extracted from their respective root exudates.

The stereochemical structure of avenaol, which has been reported in *A. strigosa* root exudates, has been elucidated by total synthesis. It has shown strong activity on *Phelipanche ramosa*. It is produced from precursor CL via hydroxylation, elimination, and cyclization (Figure 2.4). The non-canonical SLs, although lacking the A, B, or C rings, nevertheless have the enol-ether-D ring moiety, which is obligatory for SLs to function biologically. Several non-canonical SLs have been identified as being produced by maize; among them, SL2 is zealactone [35], while SL1 is a combination of at least two isomers. Canonical SLs are not known to be produced by sunflower plants. Heliolactone, a non-canonical SL, has been extracted from sunflower root exudates, another *Orobanche* host. However, non-canonical SLs rapidly break down during purification and storage and are thus less stable than their canonical counterparts.

The first non-canonical SL to be reported is carlactone, an SL biosynthetic precursor. The amount of CL in plant tissues appears to be fairly high compared with canonical SLs and is unaffected by phosphate limitation, which indicates that the response to phosphate deficiency is regulated in steps after CL formation [36]. The exudates from the roots of several plants have been shown to contain CLA, an oxidized metabolite of CL [37]. Based on these findings, CLA likely plays a role in rhizosphere chemical communications with neighboring plants or microbes.

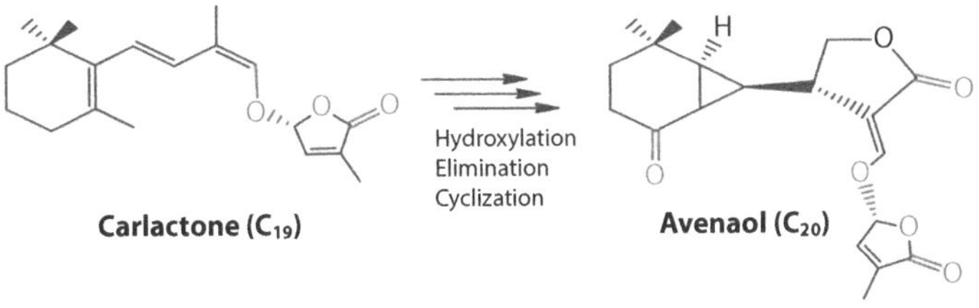


**Figure 2.4** Conversion of carlactone into avenaol.

The simplest canonical SL, 5DS, and a non-canonical SL, lotuslactone, are both exuded by *Lotus japonicus*, a model legume plant [38]. Except for acetoxyl derivatives, which have two extra carbons, such as orobanchyl acetate, and sorgolactone, a C18 SL, most canonical SLs are C19 compounds. Non-canonical SLs, on the other hand, are C20 compounds because they have an extra carbon atom (excluding CL and CLA). Methyl carlactonoate (MeCLA), its isomers, and their hydroxyl derivatives appear to be the source of these non-canonical SLs as this extra carbon most likely originates from the ester methyl group [28]. Plants of the Solanaceae family, including potato, tomato, aubergine, and tobacco, release many isomers of presumptive didehydro-orobanchol in addition to the canonical and non-canonical SLs [16, 28, 39]. The structures of these isomers are yet unknown and need to be elucidated.

## 2.5 Biosynthetic Pathway of SLs

Strigolactones are biosynthesized via the carotenoid pathway, as demonstrated by studies using carotenoid biosynthetic inhibitors and carotenoid-deficient mutants. These observations were made before SLs were identified as a novel class of plant hormones influencing shoot branching. Afterward, a group of recessive mutants with increased shoot branching allowed researchers to gain a much better understanding of SL biosynthesis. These mutants, which include pea *ramosus* (*rms*), rice *dwarf/high-tillering dwarf* (*d/htd*), petunia *decreased apical dominance* (*dad*), and *Arabidopsis more axillary growth (max)*, were intended to find a novel shoot branching regulator substitute for conventional hormones like auxin and cytokinin. The *Carotenoid Cleavage Dioxygenase 7* (*CCD7*) and *CCD8* mutants were identified as SL-deficient. These included pea *rms1* and *rms5*, rice *d17* and *d10*, and *Arabidopsis max3* and *max4* mutants [10, 11]. Then, it was discovered that rice *D27* and *Arabidopsis AtD27*, two new iron-binding proteins, were involved in the production of SL. In 2012, Alder *et al.* [40] used recombinant *D27*, *CCD7*, and *CCD8* proteins from rice, peas, and *Arabidopsis* to conduct an excellent *in vitro* biochemical investigation. They demonstrated how these three enzymes sequentially convert all-*trans*-β-carotene via 9-*cis*-β-carotene and 9-cis-β-apo-10′-carotenal to CL [27] (Figure 2.5). A thorough evaluation of the catalytic properties of CCD7, CCD8, and D27 enzymes was recently conducted [40]. Eventually, an LC–MS/MS study confirmed that CL was the endogenous molecule in rice and *Arabidopsis* [36]. Furthermore, feeding studies using stable isotope-labeled CL showed that CL is transformed into

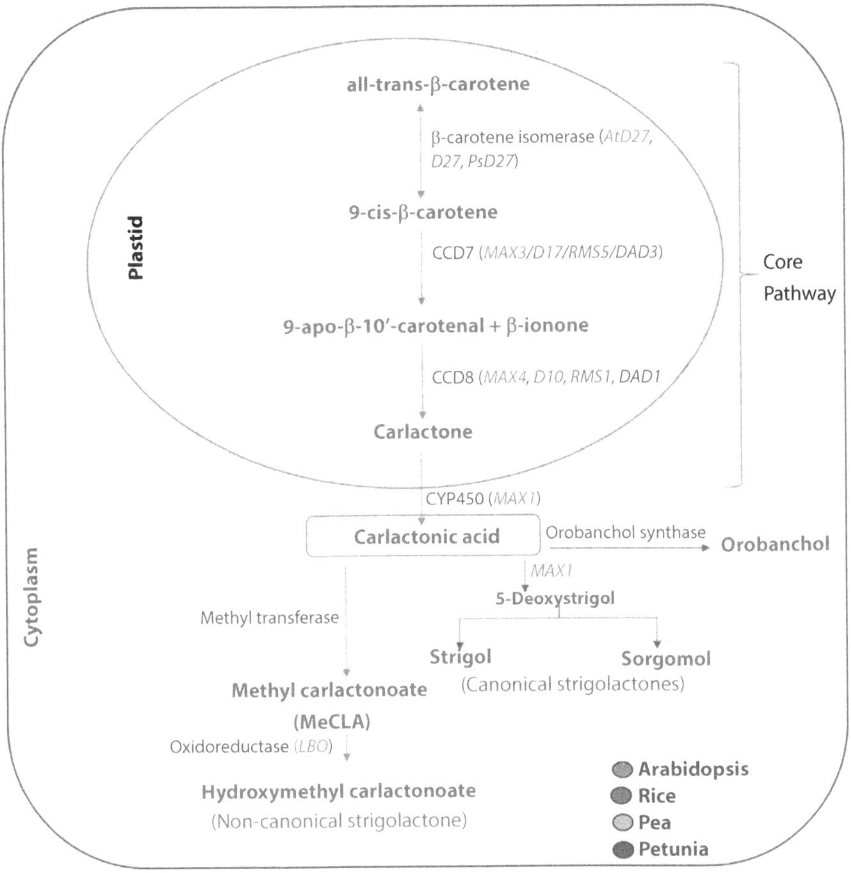


**Figure 2.5** Biosynthetic pathway of SLs.

orobanchol and 4-deoxyorobanchol (4DO), two endogenous classical SLs in rice, indicating that CL represents an intermediary of SL production *in vivo*. Significantly, it was shown that the endogenous CL's absolute configuration at the C-11 location in the D-ring—which equates to C-2′ in natural SLs—was (11*R*) [41]. The stereochemistry of SLs is therefore thought to depend critically on the stereospecificity of the CCD8-catalyzed reaction that yields (11*R*)-CL.

In vascular plants, cytochrome P450 oxygenases belonging to the CYP711A subfamily are responsible for converting CL into both canonical and non-canonical SL forms. *Arabidopsis* was the first plant in which CYP711A was reported to be involved in SL biosynthesis. The CYP711A1-defective *max1* mutants of *Arabidopsis* have been reported to display the hyper-branching phenotype, as demonstrated by *max3* and *max4* [42]. According to grafting studies, MAX1 acts after MAX3 and MAX4. Furthermore, there is a significant rise in the endogenous CL level in the *max1* mutants, indicating that CL functions as a substrate for MAX1 enzyme [36, 43]. As demonstrated in *Arabidopsis*, the MAX1 recombinant protein produced in yeast microsomes catalyzes three step oxidations of the methyl group at the C-19 position to a carboxylic acid by converting CL into CLA via 19-hydroxy-CL [41] (Figure 2.5). The substrate specificity of MAX1 may also be required for the effective synthesis of the (2′R)-configured SLs since MAX1 favored (11*R*)-CL over (11*S*)-CL. In contrast, it has been shown that, in rice, CYP711A2/Os900 changes CL into 4DO, and CYP711A3/Os1400 then catalyzes the further hydroxylation of 4DO to produce ORO. CYP711A subfamily in vascular plants probably catalyzes the common reaction of converting CL to CLA. Additionally, they demonstrated that whereas CYP711A17v1/v3 of *Selaginella moellendorffii* and CYP711A18 of *Zea mays* could catalyze the A2- and A3-type reactions, respectively, the conversion of CL to 4DO via CLA (CYP711A2-type) and 4DO to ORO (CYP711A3-type) was not ubiquitous [28]. According to these findings, CYP711As are crucial for the production of both canonical and non-canonical SLs.

A summary of different proteins and their respective genes involved in the biosynthesis and signaling mechanism of SLs is given in Table 2.1.

**Table 2.1** Summary of different proteins and genes involved in the biosynthesis and signaling pathway of SLs.

| | | **Gene name** | | | | |
|---|---|---|---|---|---|---|
| **Role** | **Protein name** | ***Arabidopsis*** | **Rice** | **Pea** | **Petunia** | **References** |
| Biosynthesis | All-trans-β-carotene isomerase | *AtD27* | *D27* | | | [27, 44, 45] |
| | CCD7 | *MAX3* | *D17* | *RMS5* | *DAD3* | [27, 43, 46, 47] |
| | CCD8 | *MAX4* | *D10* | *RMS1* | *DAD1* | [27, 48, 49–51] |
| | CYP711 | *MAX1* | *Carlactone Oxidase* | | *PhMAX1* | [43, 46, 52] |
| Signaling | α/β-Hydrolase | *AtD14* | *D14* | | *DAD2* | [53–55] |
| | F-box protein | *MAX2* | *D3* | *RMS4* | *PhMAX2A PhMAX2B* | [46, 54, 56–58] |
| | Class I Clp ATPase protein | | *D53* | | | [59, 60] |

## 2.6 Strigolactone Transport

### 2.6.1 Movement of SL from Root to Shoot

Endogenous hormone transport systems have developed in plants to regulate development and respond to environmental changes [60]. Exuded from roots into the soil, SLs serve as symbiotic signals for AM fungi, particularly in low-phosphorus or low-nitrogen environments. In addition, SLs serve as long-range signals from roots to shoots. In petunia, pea, and *Arabidopsis*, grafting experiments using *dad*, *rms*, and *max* mutants have indicated that locally produced SLs in shoots are sufficient to inhibit lateral bud elongation, as evidenced by the normal phenotype displayed by WT scions which were grafted onto SL-deficient mutant rootstocks [41]. The hyper-branching phenotype of SL-deficient mutant scions was, however, demonstrated to be rescued by WT rootstocks, indicating that SLs or their precursor(s) generated in roots can migrate shoots in an acropetal manner. It is proposed that root-derived SLs in the shoot act as signals that integrate soil nutrient availability and shoot architecture since SLs are significantly formed in roots under nitrogen- and phosphate-limited conditions [61–63].

Potential possibilities exist for the mobile form of SLs. Since CL is the only known intermediate that is both upstream and downstream of MAX4, CL might function as a mobile signal. In 2005, it has been demonstrated how MAX1 rootstocks could restore the shoot branching phenotype of *max4* scions [43]. Furthermore, in the *max1* mutant, CL accumulated at a very high level [37]. Additionally, it was shown that the shoot branching phenotype of *Atd27* scions could be restored by *max1* rootstocks but not *max4* ones, indicating that 9-*cis*-β-carotene and 9-*cis*-β-apo-10′-carotenal are not transportable signals. Because plastids are the organelles that carry out the production of SLs from all-*trans*-β-carotene to CL, these physiological results make sense [44, 45] (Figure 2.6). Furthermore, it was shown that *LATERAL BRANCHING OXIDOREDUCTASE* (*lbo*) rootstocks were able to replicate the branching phenotype of *max1* scions, indicating that CLA, methyl carlactonoate (MeCLA), and/or their unidentified metabolites may function as long-range mobile signals. The rosette branches of the *lbo* mutant were similarly drastically decreased when grafting *lbo* scions onto WT rootstocks. This suggests that 1′-OH-MeCLA, which is synthesized from MeCLA by LBO, is one of the potential long-distance signals from the root to the shoot [64, 65]. A hint to take into account regarding

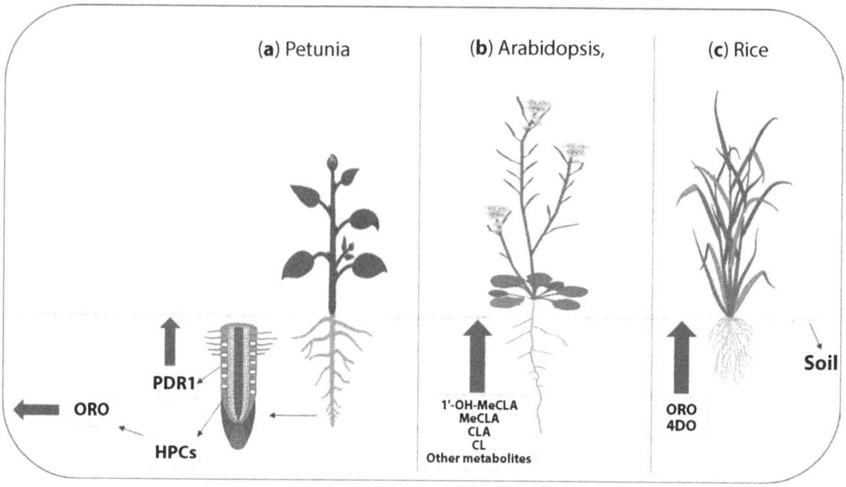


**Figure 2.6** Transport of SLs. (a) In petunia, it was proposed that asymmetrically localized PDR1 is involved in both SL export from hypodermal cells (HPCs) in roots and SL transport from root tips toward the shoot via the hypodermis. The substrates of PDR1 are unknown, although orobanchol (ORO) is one of the options. (b) Mobile signals from roots to shoots may be conveyed via several SL biosynthetic intermediates downstream of MAX4/CCD8, according to grafting investigations conducted with the *Arabidopsis max*, *atd27*, and *lbo* mutants. (c) In rice, roots were the source of transport for root-fed stable isotope labels 4DO and ORO, which may have occurred via cell-to-cell mobility.

the migration of SLs is provided by the LC–MS/MS analysis of the xylem sap in addition to physiological tests. Several SLs, including ORO, were found in the xylem sap of tomatoes and *Arabidopsis*. On the other hand, Xie *et al.* observed that xylem saps taken from a variety of plant species, including tomato and *Arabidopsis*, did not include any recognized SLs, such as ORO and SL biosynthetic intermediates like CL, CLA, and MeCLA [66]. Additionally, it was verified that stable isotope-labeled ORO and 4-deoxyorobanchol (4DO) administered to rice roots were found in shoots but not in the xylem sap within 20 h, indicating that rice uses directed cell-to-cell movement rather than xylem-mediated transport of SLs [41] (Figure 2.6). It is possible, however, that unknown chemicals connected to SL are propagating through the xylem from the roots to the shoots. When in combination, grafting and LC–MS/MS investigations imply that many molecular species of SLs and their biosynthetic intermediates, rather than just one, might represent mobile signals. Determining which chemicals are the real endogenous signals that go from roots to shoots will be crucial.

### 2.6.2 Transport of SL via ABCG/PDR Pathway

In 2012, reverse genetic analysis displayed the involvement of G-type ABC transporter PLEIOTROPIC DRUG RESISTANCE1 (PDR1) in the transfer of SLs in petunia. At the hypodermal passage cells (HPCs) in roots, where the hyphae of AM fungus enter during mycorrhization, there was a strong expression of *PDR1*. In response to phosphate shortage and AM fungal invasion, *PDR1* expression in roots was likewise elevated, suggesting that *PDR1* is involved in the release of SLs from HPCs into the rhizosphere. The *pdr1* mutant showed reduced interaction with AM fungus and dramatically lowered ORO levels in root exudates [67]. A subsequent subcellular localization study showed that *PDR1* is located at the outer lateral membrane of HPCs above the root tip and the apical membrane of hypodermal cells at the root tip. Exogenously administered isotope-labeled GR24 moved through the *pdr* mutant much more slowly than it did in the WT. *PDR1* is thought to have two roles: exporting SL toward the soil from HPCs above the root tip and transporting SL shootward from root tips where SL is biosynthesized because CAROTENOID CLEAVAGE DIOXYGENASE-8 (*DAD1/CCD8*) is expressed in the cortex cells of the root tip [68] (Figure 2.6). More recently, it was shown that *MtABCG59/MtPDR23*, a homolog of *PDR1*, is expressed in *Medicago truncatula*, a plant without HPCs, in both the cortical cells and the root tip. Under phosphate-limiting circumstances, *MtABCG59* transcript levels rise, in line with those of petunia *PDR1*. Interestingly, it was demonstrated that the *MtABCG59*-defective mutants decreased the colonization rate of AM fungus in comparison to WT, as was seen in the *pdr1* mutant [69]. These results imply that, despite the anatomical variations in roots, PDR-type transporters are involved in the export of SL from roots during mycorrhization in Solanaceae and Fabaceae species.

Apart from the SL transport root tissue, *PDR1* and its homologs could also be involved in the aerial portions (Figure 2.6). It was observed that the *pdr1* mutant exhibited a hyper-branching phenotype [67]. Furthermore, a near homolog of *PDR1* in tobacco (*Nicotiana tabacum*), *NtPDR6*, exhibits more secondary stems with lateral branches on its knockdown line than it does on the wild type. It has been demonstrated that *PDR1* is expressed in the vasculature and nodes close to leaf axils but not in dormant buds [41]. *NtPDR6* expression was also seen in nodes located around the base of the stem [70]. Their expression patterns and mutant phenotypes suggest that the PDR-type transporters prevent shoot branching in Solanaceae species by acting as SL transporters. It is yet unknown, however, which compounds *PDR1* substrates are. Since the *pdr1* mutant resulted in lower

ORO levels in root exudates, orobanchol is one of the more viable options [67, 71]. On the other hand, *PDR1* could be a multi-substrate transporter for SLs and the intermediates involved in their biosynthesis. Determining the substrate selectivity of *PDR1* and its orthologs is essential to advance our knowledge of the SL transport system. Moreover, it is interesting to investigate if the transport of SLs by PDR-type transporters is universal in other plants, given that these transporters are retained beyond Solanaceae and Fabaceae species [72, 73].

## 2.7 Conclusion

In conclusion, the study of SLs has revealed the domain of plant chemical signaling and their chemical nature which is further characterized by a butanolide ring and various substituents and their structural diversity through canonical and non-canonical forms. This structural variability explains their potential for diverse biological activities and interactions within ecosystems. The naming protocol established for SLs provides a standardized framework for identification and classification, facilitating communication and understanding across scientific communities. Additionally, new studies on PDR1 have demonstrated that the control of plant growth and plant–fungal symbiosis is largely dependent on the transport of SLs, in addition to its production and signal perception. However, the need for SL transporters is widespread among plants. Additionally, numerous questions remain regarding the mobility of strigolactones (SLs). It is particularly unclear which endogenous molecules are mobile within plants, similar to the uncertainties surrounding their biosynthesis. Evidence for ABCG/PDR-mediated SL transport exists only within the Solanaceae and Fabaceae families. Both local and long-distance transport of SLs, as well as their exudation from roots, are crucial for their hormonal functions and interactions with AM fungi. SL production is finely regulated by nutritional conditions and other plant hormones. Investigating the molecular mechanisms governing SL transport across various plant species will enhance our understanding of how plants effectively utilize SLs as signaling molecules. Understanding the nature, structural diversity, and biosynthetic pathways of SLs is pivotal not only for fundamental research but also for practical applications. As research in this field continues to evolve, further exploration into the multifaceted roles and mechanisms of SLs promises to deepen our understanding of plant biology and offer novel solutions to global challenges in agriculture and environmental management.

## References


1. Soliman, S., Wang, Y., Han, Z., Pervaiz, T., El-Kereamy, A., Strigolactone in plants and their interaction with the ecological microbiome in response to abiotic stress. *Plants, 11*, 24, 3499, 2022.
2. Wu, F., Gao, Y., Yang, W., Sui, N., Zhu, J., Biological functions of strigolactones and their crosstalk with other phytohormones. *Front. Plant Sci., 13*, 821563, 2022.
3. Khuvung, K., Silva Gutierrez, F.A., Reinhardt, D., How strigolactone shapes shoot architecture. *Front. Plant Sci., 13*, 889045, 2022.
4. Akiyama, K., Matsuzaki, K.I., Hayashi, H., Plant sesquiterpenes induce hyphal branching in arbuscular mycorrhizal fungi. *Nature, 435*, 7043, 824–827, 2005.
5. Kong, C.H., Xuan, T.D., Khanh, T.D., Tran, H.D., Trung, N.T., Allelochemicals and signaling chemicals in plants. *Mol., 24*, 15, 2737, 2019.
6. Wani, K.I., Zehra, A., Choudhary, S., Naeem, M., Khan, M.M.A., Castroverde, C.D.M., Aftab, T., Mechanistic insights into strigolactone biosynthesis, signaling, and regulation during plant growth and development. *J. Plant Growth Regul.*, 39, 1, 1–17, 2020.
7. Xiong, G., Wang, Y., Li, J., Action of strigolactones in plants, in: *The Enzymes*, vol. 35, pp. 57–84, Academic Press, San Diego, CA, 2014.
8. Wakabayashi, T., Nakayama, M., Kitano, Y., Homma, M., Miura, K., Takikawa, H., Sugimoto, Y., Discovery of strigol synthase from cotton (*Gossypium hirsutum*): the enzyme behind the first identified germination stimulant for *Striga*. *Plants People Planet*, 5, 4, 282–295, 2023.
9. David, O.G., Ayangbenro, A.S., Odhiambo, J.J., Babalola, O.O., *Striga hermonthica*: A highly destructive pathogen in maize production, *Environ. Chall., 8*, 100590, Elsevier, 2022.
10. Gomez-Roldan, V., Fermas, S., Brewer, P.B., Puech-Pagès, V., Dun, E.A., Pillot, J.P., Rochange, S.F., Strigolactone inhibition of shoot branching. *Nature, 455*, 7210, 189–194, 2008.
11. Umehara, M., Hanada, A., Yoshida, S., Akiyama, K., Arite, T., Takeda-Kamiya, N., Yamaguchi, S., Inhibition of shoot branching by new terpenoid plant hormones. *Nature, 455*, 7210, 195–200, 2008.
12. Cao, Y.Y., Chen, Y.C., Guo, S.H., Gan, X.Y., Tian, L., Huang, L.Q., Yuan, Y., Research progress in strigolactones and application prospect in medicinal plants. *China J. Chin. Mater. Med., 48*, 12, 3132–3139, 2023.
13. Kee, Y.J., Ogawa, S., Ichihashi, Y., Shirasu, K., Yoshida, S., Strigolactones in rhizosphere communication: multiple molecules with diverse functions. *PCP, 64*, 9, 955–966, 2023.
14. Li, X.D., Tan, Z.Z., Wu, D., Zhao, X., Dong, L.L., Zhan, M.L., Zhao, K.H., Structure Reveals the Impact of Surface Charge Distribution on the Phase Separation and Aggregation of Slr0280. *Biochem., 62*, 13, 2021–2028, 2023.

15. Guercio, A.M., Palayam, M., Shabek, N., Strigolactones: diversity, perception, and hydrolysis. *Phytochem. Rev.*, *22*, 2, 339–359, 2023.
16. Wakabayashi, T., Moriyama, D., Miyamoto, A., Okamura, H., Shiotani, N., Shimizu, N., Sugimoto, Y., Identification of novel canonical strigolactones produced by tomato. *Front. Plant Sci.*, *13*, 1064378, 2022.
17. Quinodoz, P., Lumbroso, A., Lachia, M., Screpanti, C., Rendine, S., Horoz, B., De Mesmaeker, A., Stereoselective Synthesis and Biological Profile of All Stereoisomers of Lactam Analogues of Strigolactones GR24 and GR18. *Helv. Chim. Acta*, *106*, 1, e202200145, 2023.
18. Baumann, P., About the Misleading Use of Stereodescriptors in Labeling the Stereoisomers of Milnacipran and Levomilnacipran. *J. Clin. Psychopharmacol.*, *39*, 6, 673–674, 2019.
19. Carlsson, A.C.C., Karlsson, S., Munday, R.H., Tatton, M.R., Approaches to Synthesis and Isolation of Enantiomerically Pure Biologically Active Atropisomers. *Acc. Chem. Res.*, *55*, 20, 2938–2948, 2022.
20. Spasov, A.A., Iezhitsa, I.N., Vassiliev, P.M., Ozerov, A.A., Agarwal, R., *Pharmacology of drug stereoisomers*, vol. 76, Springer Nature, San Diego, CA, 2022.
21. Ueno, K., Sugimoto, Y., Zwanenburg, B., The genuine structure of alectrol: end of a long controversy. *Phytochem. Rev.*, *14*, 835–847, 2015.
22. Zwanenburg, B. and Pospíšil, T., Structure and activity of strigolactones: new plant hormones with a rich future. *Mol. Plant*, *6*, 1, 38–62, 2013.
23. Scaffidi, A., Waters, M.T., Sun, Y.K., Skelton, B.W., Dixon, K.W., Ghisalberti, E.L., Smith, S.M., Strigolactone hormones and their stereoisomers signal through two related receptor proteins to induce different physiological responses in Arabidopsis. *Plant Physiol.*, *165*, 3, 1221–1232, 2014.
24. Xie, X., Yoneyama, K., Yoneyama, K., The strigolactone story. *Annu. Rev. Phytopathol.*, *48*, 93–117, 2010.
25. Al-Babili, S. and Bouwmeester, H.J., Strigolactones, a novel carotenoid-derived plant hormone. *Annu. Rev. Plant Biol.*, *66*, 161–186, 2015.
26. Xie, X., Structural diversity of strigolactones and their distribution in the plant kingdom. *J. Pestic. Sci.*, *41*, 4, 175–180, 2016.
27. Alder, A., Jamil, M., Marzorati, M., Bruno, M., Vermathen, M., Bigler, P., Al-Babili, S., The path from β-carotene to carlactone, a strigolactone-like plant hormone. *Sci.*, *335*, 6074, 1348–1351, 2012.
28. Yoneyama, K., Mori, N., Sato, T., Yoda, A., Xie, X., Okamoto, M., Nomura, T., Conversion of carlactone to carlactonoic acid is a conserved function of MAX 1 homologs in strigolactone biosynthesis. *New Phytol.*, *218*, 4, 1522–1533, 2018.
29. Yoneyama, K., Xie, X., Yoneyama, K., Kisugi, T., Nomura, T., Nakatani, Y., McErlean, C.S., Which are the major players, canonical or non-canonical strigolactones? *J. Exp. Bot.*, *69*, 9, 2231–2239, 2018.

30. Xie, X., Yoneyama, K., Kisugi, T., Uchida, K., Ito, S., Akiyama, K., Yoneyama, K., Confirming stereochemical structures of strigolactones produced by rice and tobacco. *Mol. Plant*, *6*, 1, 153–163, 2013.
31. Akiyama, K., Ogasawara, S., Ito, S., Hayashi, H., Structural requirements of strigolactones for hyphal branching in AM fungi. *PCP*, *51*, 7, 1104–1117, 2010.
32. Jamil, M., Charnikhova, T., Houshyani, B., van Ast, A., Bouwmeester, H.J., Genetic variation in strigolactone production and tillering in rice and its effect on *Striga hermonthica* infection. *Planta*, *235*, 473–484, 2012.
33. Kim, H.I., Kisugi, T., Khetkam, P., Xie, X., Yoneyama, K., Uchida, K., Yoneyama, K., Avenaol, a germination stimulant for root parasitic plants from *Avena strigosa*. *Phytochemistry*, *103*, 85–88, 2014.
34. Xie, X., Kisugi, T., Yoneyama, K., Nomura, T., Akiyama, K., Uchida, K., Yoneyama, K., Methyl zealactonoate, is a novel germination stimulant for root parasitic weeds produced by maize. *J. Pestic. Sci.*, *42*, 2, 58–61, 2017.
35. Charnikhova, T.V., Gaus, K., Lumbroso, A., Sanders, M., Vincken, J.P., De Mesmaeker, A., Bouwmeester, H.J., Zealactones. Novel natural strigolactones from maize. *Phytochemistry*, *137*, 123–131, 2017.
36. Seto, Y., Sado, A., Asami, K., Hanada, A., Umehara, M., Akiyama, K., Yamaguchi, S., Carlactone is an endogenous biosynthetic precursor for strigolactones. *PNAS*, *111*, 4, 1640–1645, 2014.
37. Abe, S., Sado, A., Tanaka, K., Kisugi, T., Asami, K., Ota, S., Nomura, T., Carlactone is converted to carlactonoic acid by MAX1 in Arabidopsis and its methyl ester can directly interact with AtD14 *in vitro*. *PNAS*, *111*, 50, 18084–18089, 2014.
38. Xie, X., Mori, N., Yoneyama, K., Nomura, T., Uchida, K., Yoneyama, K., Akiyama, K., Lotuslactone, a non-canonical strigolactone from *Lotus japonicus*. *Phytochemistry*, *157*, 200–205, 2019.
39. Wu, F. and Tanksley, S.D., Chromosomal evolution in the plant family Solanaceae. *BMC Genomics*, *11*, 1–11, 2010.
40. Jia, K.P., Baz, L., Al-Babili, S., From carotenoids to strigolactones. *J. Exp. Bot.*, *69*, 9, 2189–2204, 2018.
41. Mashiguchi, K., Seto, Y., Yamaguchi, S., Strigolactone biosynthesis, transport and perception. *Plant J.*, *105*, 2, 335–350, 2021.
42. Stirnberg, P., van De Sande, K., Leyser, H.O., MAX1 and MAX2 control shoot lateral branching in Arabidopsis. *Development*, 129, 8, 1131–1141, 2002.
43. Booker, J., Sieberer, T., Wright, W., Williamson, L., Willett, B., Stirnberg, P., Leyser, O., MAX1 encodes a cytochrome P450 family member that acts downstream of MAX3/4 to produce a carotenoid-derived branch-inhibiting hormone. *Dev. Cell*, *8*, 3, 443–449, 2005.
44. Lin, H., Wang, R., Qian, Q., Yan, M., Meng, X., Fu, Z., Wang, Y., DWARF27, an iron-containing protein required for the biosynthesis of strigolactones, regulates rice tiller bud outgrowth. *Plant Cell*, *21*, 5, 1512–1525, 2009.

45. Waters, M.T., Brewer, P.B., Bussell, J.D., Smith, S.M., Beveridge, C.A., The Arabidopsis ortholog of rice DWARF27 acts upstream of MAX1 in the control of plant development by strigolactones. *Plant Physiol.*, *159*, 3, 1073–1085, 2012.
46. Drummond, R.S., Martínez-Sánchez, N.M., Janssen, B.J., Templeton, K.R., Simons, J.L., Quinn, B.D., Snowden, K.C., *Petunia hybrida* CAROTENOID CLEAVAGE DIOXYGENASE7 is involved in the production of negative and positive branching signals in petunia. *Plant Physiol.*, *151*, 4, 1867–1877, 2009.
47. Zou, J., Zhang, S., Zhang, W., Li, G., Chen, Z., Zhai, W., Zhu, L., The rice HIGH-TILLERING DWARF1 encoding an ortholog of Arabidopsis MAX3 is required for negative regulation of the outgrowth of axillary buds. *Plant J.*, *48*, 5, 687–698, 2006.
48. Arite, T., Iwata, H., Ohshima, K., Maekawa, M., Nakajima, M., Kojima, M., Kyozuka, J., DWARF10, an RMS1/MAX4/DAD1 ortholog, controls lateral bud outgrowth in rice. *Plant J.*, *51*, 6, 1019–1029, 2007.
49. Simons, J.L., Napoli, C.A., Janssen, B.J., Plummer, K.M., Snowden, K.C., Analysis of the DECREASED APICAL DOMINANCE genes of petunia in the control of axillary branching. *Plant Physiol.*, *143*, 2, 697–706, 2007.
50. Snowden, K.C., Simkin, A.J., Janssen, B.J., Templeton, K.R., Loucas, H.M., Simons, J.L., Klee, H.J., The Decreased apical dominance1/Petunia hybrida CAROTENOID CLEAVAGE DIOXYGENASE8 gene affects branch production and plays a role in leaf senescence, root growth, and flower development. *Plant Cell*, *17*, 3, 746–759, 2005.
51. Sorefan, K., Booker, J., Haurogné, K., Goussot, M., Bainbridge, K., Foo, E., Leyser, O., MAX4 and RMS1 are orthologous dioxygenase-like genes that regulate shoot branching in Arabidopsis and pea. *Genes Dev.*, *17*, 12, 1469–1474, 2003.
52. Zhang, Y., Van Dijk, A.D., Scaffidi, A., Flematti, G.R., Hofmann, M., Charnikhova, T., Bouwmeester, H.J., Rice cytochrome P450 MAX1 homologs catalyze distinct steps in strigolactone biosynthesis. *Nat. Chem. Biol.*, *10*, 12, 1028–1033, 2014.
53. Arite, T., Umehara, M., Ishikawa, S., Hanada, A., Maekawa, M., Yamaguchi, S., Kyozuka, J., d14, a strigolactone-insensitive mutant of rice, shows an accelerated outgrowth of tillers. *PCP*, *50*, 8, 1416–1424, 2009.
54. Hamiaux, C., Drummond, R.S., Janssen, B.J., Ledger, S.E., Cooney, J.M., Newcomb, R.D., Snowden, K.C., DAD2 is an α/β hydrolase likely to be involved in the perception of the plant branching hormone, strigolactone. *Curr. Biol.*, *22*, 21, 2032–2036, 2012.
55. Nakamura, H., Xue, Y.L., Miyakawa, T., Hou, F., Qin, H.M., Fukui, K., Asami, T., Molecular mechanism of strigolactone perception by DWARF14. *Nat. Commun.*, *4*, 1, 2613, 2013.
56. Beveridge, C.A., Ross, J.J., Murfet, I.C., Branching in pea (action of genes Rms3 and Rms4). *Plant Physiol.*, *110*, 3, 859–865, 1996.

57. Ishikawa, S., Maekawa, M., Arite, T., Onishi, K., Takamure, I., Kyozuka, J., Suppression of tiller bud activity in tillering dwarf mutants of rice. *PCP*, *46*, 1, 79–86, 2005.
58. Stirnberg, P., Furner, I.J., Ottoline Leyser, H.M., MAX2 participates in an SCF complex which acts locally at the node to suppress shoot branching. *Plant J.*, *50*, 1, 80–94, 2007.
59. Jiang, L., Liu, X., Xiong, G., Liu, H., Chen, F., Wang, L., Li, J., DWARF 53 acts as a repressor of strigolactone signalling in rice. *Nature*, *504*, 7480, 401–405, 2013.
60. Vanstraelen, M. and Benková, E., Hormonal interactions in the regulation of plant development. *Ann. Rev. Cell Dev. Biol.*, *28*, 463–487, 2012.
61. Umehara, M., Hanada, A., Magome, H., Takeda-Kamiya, N., Yamaguchi, S., Contribution of strigolactones to the inhibition of tiller bud outgrowth under phosphate deficiency in rice. *PCP*, *51*, 7, 1118–1126, 2010.
62. Kohlen, W., Charnikhova, T., Liu, Q., Bours, R., Domagalska, M.A., Beguerie, S., Ruyter-Spira, C., Strigolactones are transported through the xylem and play a key role in shoot architectural response to phosphate deficiency in nonarbuscular mycorrhizal host Arabidopsis. *Plant Physiol.*, *155*, 2, 974–987, 2011.
63. de Jong, M., George, G., Ongaro, V., Williamson, L., Willetts, B., Ljung, K., Leyser, O., Auxin and strigolactone signalling are required for modulation of Arabidopsis shoot branching by nitrogen supply. *Plant Physiol.*, *166*, 1, 384–395, 2014.
64. Brewer, P.B., Yoneyama, K., Filardo, F., Meyers, E., Scaffidi, A., Frickey, T., Beveridge, C.A., LATERAL BRANCHING OXIDOREDUCTASE acts in the final stages of strigolactone biosynthesis in Arabidopsis. *PNAS*, *113*, 22, 6301–6306, 2016.
65. Yoneyama, K., Akiyama, K., Brewer, P.B., Mori, N., Kawano-Kawada, M., Haruta, S., Nomura, T., Hydroxyl carlactone derivatives are predominant strigolactones in Arabidopsis. *Plant Direct*, *4*, 5, e00219, 2020.
66. Xie, X., Wang, G., Yang, L., Cheng, T., Gao, J., Wu, Y., Xia, Q., Cloning and characterization of a novel Nicotiana Tabacum ABC transporter involved in shoot branching. *Physiol. Plant.*, *153*, 2, 299–306, 2015.
67. Kretzschmar, T., Kohlen, W., Sasse, J., Borghi, L., Schlegel, M., Bachelier, J.B., Martinoia, E., A petunia ABC protein controls strigolactone-dependent symbiotic signalling and branching. *Nature*, *483*, 7389, 341–344, 2012.
68. Sasse, J., Simon, S., Gübeli, C., Liu, G.W., Cheng, X., Friml, J., Borghi, L., Asymmetric localizations of the ABC transporter PaPDR1 trace paths of directional strigolactone transport. *Curr. Biol.*, *25*, 5, 647–655, 2015.
69. Banasiak, J., Borghi, L., Stec, N., Martinoia, E., Jasinski, M., The full-size ABCG transporter of *Medicago truncatula* is involved in strigolactone secretion, affecting arbuscular mycorrhiza. *Front. Plant Sci.*, *11*, 18, 2020.
70. Xie, X., Yoneyama, K., Kisugi, T., Nomura, T., Akiyama, K., Asami, T., Yoneyama, K., Strigolactones are transported from roots to shoots, although not through the xylem. *J. Pestic. Sci.*, *40*, 4, 214–216, 2015.

71. Wakabayashi, T., Hamana, M., Mori, A., Akiyama, R., Ueno, K., Osakabe, K., Sugimoto, Y., Direct conversion of carlactonoic acid to orobanchol by cytochrome P450 CYP722C in strigolactone biosynthesis. *Sci. Adv.*, *5*, 12, eaax9067, 2019.
72. Borghi, L., Kang, J., Ko, D., Lee, Y., Martinoia, E., The role of ABCG-type ABC transporters in phytohormone transport. *Biochem. Soc. Trans.*, *43*, 5, 924–930, 2015.
73. Nuruzzaman, M., Zhang, R., Cao, H.Z., Luo, Z.Y., Plant pleiotropic drug resistance transporters: transport mechanism, gene expression, and function. *J. Integr. Plant Biol.*, *56*, 8, 729–740, 2014.

3

# Unveiling the Strigolactone Signaling Pathway: From Receptors to Responses

**Rachel Helmich**

*Plastomics, Inc., Saint Louis, MO, USA*

***Abstract***

Strigolactones (SLs), the most recently discovered category of plant hormones, were initially identified for their function in parasitic and symbiotic relationships in the rhizosphere. They play an important function in governing plant growth and development as well as stress tolerance, especially osmotic stress and nutrient deprivation. Insights into the SL biosynthesis route have been uncovered in the last years as have numerous essential components of the SL signaling pathway. Strigolactones are sensed by an α/β hydrolase DWARF-14 (D14) receptor. The SL receptor is remarkable because it serves as both a receptor as well as an enzyme, capable of hydrolyzing its ligand SL. However, the signaling cascade from SL perception to different phenotypes remains to be completely characterized. The processes for SL perception and signaling are similar to those previously identified for other phytohormones in several areas, but they appear to be distinct in the realm of plants in other aspects. This chapter is a comprehensive exploration of the intricate world of SL signaling in plants. It delves into the molecular mechanisms that govern SL perception, transduction, and downstream responses in plants. In addition, the biological functions of SLs, their relationship to other phytohormones during plant development, and their resistance to constantly changing environments are all covered in this chapter. Finally, light is shed on the latest breakthroughs in the potential agricultural uses and biotechnological implications of SL research.

***Keywords*:** Strigolactone, signaling, perception, DWARF-14, abiotic stress, phytohormone

*Email*: rachelhelmich@outlook.com

Tariq Aftab and Kaiser Iqbal Wani (eds.) Strigolactones: Emerging Plant Hormones, (35–68) © 2025 Scrivener Publishing LLC

## 3.1 Introduction

Strigolactones (SLs) make up a class of phytohormones first described from root exudates in 1966 when Cook *et al.* extracted (+)-strigol from cotton (*Gossypium hirsutum*) roots; the isolated SL was capable of stimulating germination in the parasitic weed *Striga lutea*, though cotton is not a typical host plant of the species [1, 2]. As suggested by the word "Striga", SLs were named after the genus, a genus that most likely arose somewhere between Ethiopia's Simien Mountain range and Sudan's Nubian Hills. Unsurprisingly, this is the same region from which one of its common host species, *Sorghum bicolor* (L.), was originally domesticated [3]. Members of this parasitic genus are commonly referred to as "witchweeds", while the broader family they belong to is Orobanchaceae, which is more broadly known as the "broomrape" family. Though not all *Striga* spp. are pathogenic, some species are among the most catastrophic weeds to parasitize agricultural lands, wreaking devastating havoc to croplands spanning millions of hectares, of which the primary areas of impact are in African countries. Multiple billions of dollars of damage occur annually due to the intensity of their parasitism [4]. In 1956, *Striga* was discovered in the United States along the Carolinas, which was the breaking point that prompted a great need in the United States to elucidate the germination nature of the weed before croplands became impacted. By 1966, it was already understood that *Striga* seeds do not germinate until the host plants produce some sort of stimulating chemical, but Cook *et al.* were the first to actually isolate and define the said chemical [1]. They named the isolated stimulant "strigol" and determined that its molecular formula was the same as gibberellic acids (GA), though not identical to known gibberellins nor could GA trigger *Striga* germination while strigol could [1].

Two decades later, in 1986, root exudate was extracted from *S. bicolor* (L.) for characterization of the strigol compounds being emitted into the soil by a host species [2]. The analysis concluded that *Striga* is capable of registering the SL concentration present in the soil in order to determine its ability to viably connect to a host for germination along with support for further growth and development (seeds die <2 weeks post-germination if no host is present) [2]. SLs were known as "strigol-type compounds" until 1994 when they were officially termed "SLs" in a study that found low biosynthesis levels of SLs correlated with higher resistance to parasitism in different *S. bicolor* genotypes [5]. From first inception up until the mid-2000s, SLs were mostly known for their ability to stimulate parasitic seed germination simply because most plant species biosynthesize SLs in such low levels that they were essentially imperceptible in most plant species until better testing was developed.

The first major shift in the understanding of SL functionality came in 2005 when Akiyama *et al.* investigated the nature of "branching factors" that *Lotus japonicus* noticeably secretes into the rhizosphere in order to attract arbuscular mycorrhizal fungi (AMF) prior to AMF hyphal branching induction [6]. This branching factor matched the structure of the naturally occurring SL 5-deoxy-strigol when analyzed by spectroscopy and chemical synthesis [6]. AMFs enhance nutrient uptake from the rhizosphere, particularly phosphorus, which is a critical nutrient for general plant growth and development. AMF symbioses are ancient relationships believed to have begun >500 million years ago [7]. Building upon that, both the model species *Arabidopsis thaliana* and the bryophyte *Physcomitrium patens* secrete SLs, suggesting that the SL metabolic pathways likely predate the divergence between vascular plants and bryophytes [8].

The groundbreaking 2008 study of Gomez-Roldan *et al.* was the final nail in the coffin to the old notion that relegated SLs to simple roles as stimulatory molecules with functions only in germination and AMF associations [9]. SLs are biosynthesized by carotenoid cleavage dioxygenases (CCDs), and in their study, they showed that garden pea (*Pisum sativum*) *ccd8* mutants were SL-deficient and confirmed that the deficiency caused a reduction in mutant ability to form mycorrhizal associations that was reversible upon exposure to the synthetic SL GR24 [9]. Though valuable observations, the most significant outcome of the study was that the GR24 application caused the *ccd8* mutants in *A. thaliana* to display inhibited shoot branching, while mutants of the downstream response protein MORE AXILLARY GROWTH2 (MAX2) proved to be insensitive [9]. Because this combination of results revealed the expanded role of SLs to include an internal signaling mechanism, SLs could finally be denoted as a new class of phytohormones rather than just as vaguely described signaling molecules, 42 years after first being described [9]. Figure 3.1 denotes some of the roles that have been ascribed to SLs in the years since.

In the 15 years since SL phytohormone classification, we have greatly expanded our understanding of SL biosynthesis, signaling, repression, and physiological functions. SL biosynthesis is directed by a combination of CCD7, CCD8, cytochrome P450 (MAX1), and DWARF27 (D27, a carotenoid isomerase) [10]. The signaling pathway produces particular perplexity to its study due to the way in which the receptor acts quite differently than other phytohormone receptors behave by how it acts as a single-turnover enzyme to irreversibly bind to and hydrolyze the SL at a slow turnover rate [11, 12]. Thus, the purpose of this review is to bring

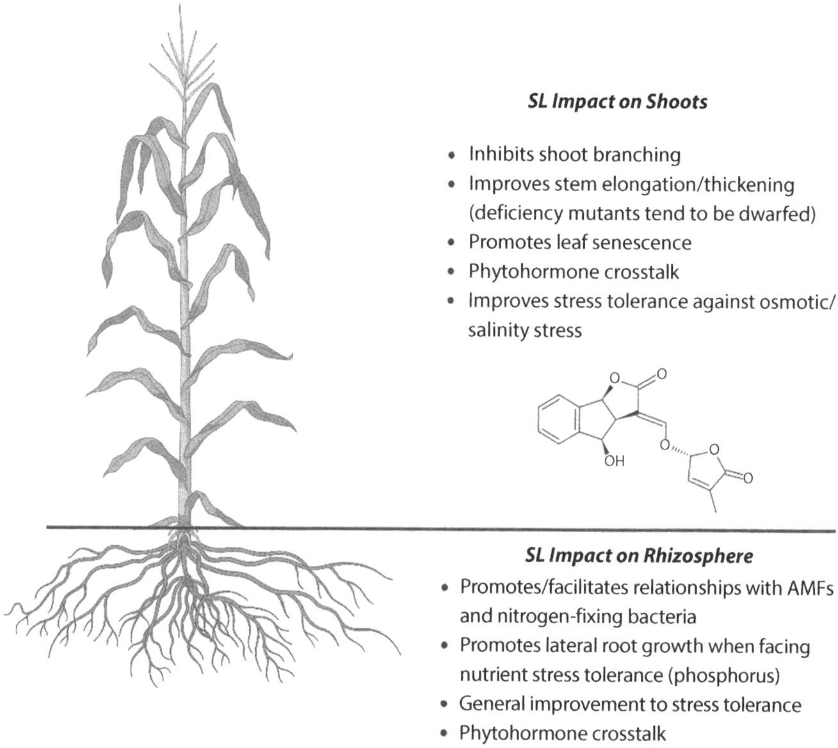


**Figure 3.1** Some impacts that SLs have on plant physiology (image produced using BioRender).

to the forefront the current understanding of SL perception, transduction, and transcriptional regulation while also noting the ways by which environmental factors impact SL signaling and further look into how this knowledge can be applied to the agricultural and biotechnological industries. Finally, this review will consider the challenges that the study of SL signaling presents along with considerations of perpetuating questions yet to be answered.

## 3.2 The Strigolactone Perception

Secondary metabolites called "phytohormones" are intrinsic regulators of plant growth and development that act individually or collaborate antagonistically, synergistically, or varying measures of both to maintain internal homeostasis against the highly variable external environment [13].

Including SLs, there are now 10 recognized classes of phytohormones: **auxins (IAA), cytokinins (CK), gibberellins (GA), ethylene (ETH), abscisic acid (ABA)**, brassinosteroids (BR), salicylic acid (SA), jasmonic acid (JA), karrikins (KAR)—the bolded classes are the "classical 5" phytohormones that were the first to be identified and so have the most extensive bodies of literature available [14]. Each phytohormone class can be defined by the regions within a plant where members are typically produced along with a general list of shared major functions (e.g., CKs can be generally stated to be transported via the vasculature, xylem and phloem, and they stimulate cellular division to regulate growth and development [15, 16]) [13]. In the early 20th century, studies were driven by questions being asked about phytohormone influence on plant physiology; those questions expanded to discern specifics around biosynthesis and how biosynthesis pathways differ between phytohormones, and then those questions further delved into questions about the nuance of mechanisms behind phytohormone metabolic processes and transportation [17]. Three decades ago, questions further shifted to focus on signaling pathways to elucidate the means by which phytohormones are perceived and their signals transduced/repressed [17]. Most phytohormone signaling pathways are regulated by the ubiquitin–proteasome system (UPS); however, IAAs and JAs use F-box proteins directly for reception, while GA and SLs must bind to α/β-hydrolases to then be able to interact with their F-box proteins [18]. SL signaling adds greater complexity compared to other signaling pathways due to the receptor being an active hydrolase [18]; therefore, the SL must be converted into a bioactive form by its receptor before it can be used by the plant—thus, the signaling pathway is a dynamic process [19].

MAX2, an F-box protein, was first noted to play a role in *A. thaliana* shoot branching by Stirnberg *et al.* nearly a quarter-century ago, though at the time the SL's role was known only as an unidentified mediator protein that used a Skp, Cullin, F-box containing complex (SCF) type E3 ligase for protein ubiquitination and degradation as the mechanism for controlling shoot branching, leaf, and hypocotyl growth [20]. Less than a decade later, SLs were identified as those proteins Stirnberg *et al.* had implicated as their shoot branching inhibitor [9], and additional mutant deficiency studies in rice (*Oryza sativa*) pointed to DWARF14 (D14) as the α/β-hydrolase enzyme receptor responsible for SL perception in *O. sativa* [19]. Building upon background from previous mutant studies, the D14 receptor confirmation was done by comparing multiple WT against multiple *O. sativa* mutants: *d10*, *d14*, and a double *d10-d14* mutant [19]. The concluding results found that *d10* and *d14* both exhibited dwarfism and increased tillering, while the double mutant showed no significant difference from the individual mutants; however, the *d14* mutant produced more of the naturally occurring SL 2′-*epi-5-deoxystrigol* (*epi*-5DS) than the WT and was insensitive to phenotypic changes when GR24, a synthetic SL, was applied, whereas *d10* was sensitive to GR24, and decreased tillering and increased height occurred following application [19]. D14 orthologs continue to be discovered in new species, but a significant difference has been noted in *Striga hermonthica* reception. *S. hermonthica* does have the D14-MAX2 orthologs (ShD14 and ShMAX2, respectively); however, *atd14*-mutants (the loss-of-function mutants) did not see a negative impact on germination ability, while the parasitic clade of HYPOSENSITIVE TO LIGHT (HTL) orthologs ShHTL4-9 clade is responsive to SLs and ShHTL7 is crucial to *S. hermonthica* germination. Despite using a different but closely related receptor, MAX2 is conserved in downstream signaling activity [20].

Understanding the mechanisms of two receptors D14 and HTL cannot be separated from work on the karrikins (KAR) signaling pathway due to the similar butanolide moiety structure SLs and karrikins share [20, 21]. It quickly became clear that the SL receptors D14 and the parasitic HTL clade needed to be distinguished as only participating in SL signaling and not in the KAR signaling pathway. Structural differences are key. To date, ~17 natural SLs are capable of germinating Orobanchaceae spp., and those SLs can be subdivided into two separate stereochemical classes: "strigol" and "orobanchol" [20]. Both classes share the same tricyclic lactone (ABC-) enol-ether butanolide (D-ring) ring but diverge in that the stereochemistry of the BC junction is flipped in orobanchol SLs [22]. Additionally, the R-configuration of the C-2′ position in the D-ring appears to result in greater *S. hermonthica* germination than an S-configuration does—the R-configuration is found in all natural SLs [20], but additional studies found that stereochemistry limits C and D rings from being at the active site, instead presenting the potential for direct nucleophilic attack of the D lactone ring [21, 23]. Aside from a few non-canonical SLs lacking the full ABC-ring, the ABC ring appears to be important to stimulate germination, as seen by the canonical SLs and comparison of different synthetic analogs [20]. However, to broadly state, SL activity does not necessitate the full presence of ABC rings [24] (see Figure 3.2 for the ABC-D ring scaffold and structures of known SLs and synthetic analogues).

The receptor DWARF-like (D14L), also known as KARRIKIN INSENSITIVE 2 (KAI2) or HTL, is an α/β-hydrolase KAR receptor found in *S. sativa* and *S. hermonthica* and has additionally been connected to *A. thaliana* germination [20, 21]. D14 and D14L share ~50% of their amino acid sequence identities, of which the Ser–Asp–His catalytic triad is conserved between them and their homologs, and both receptors must

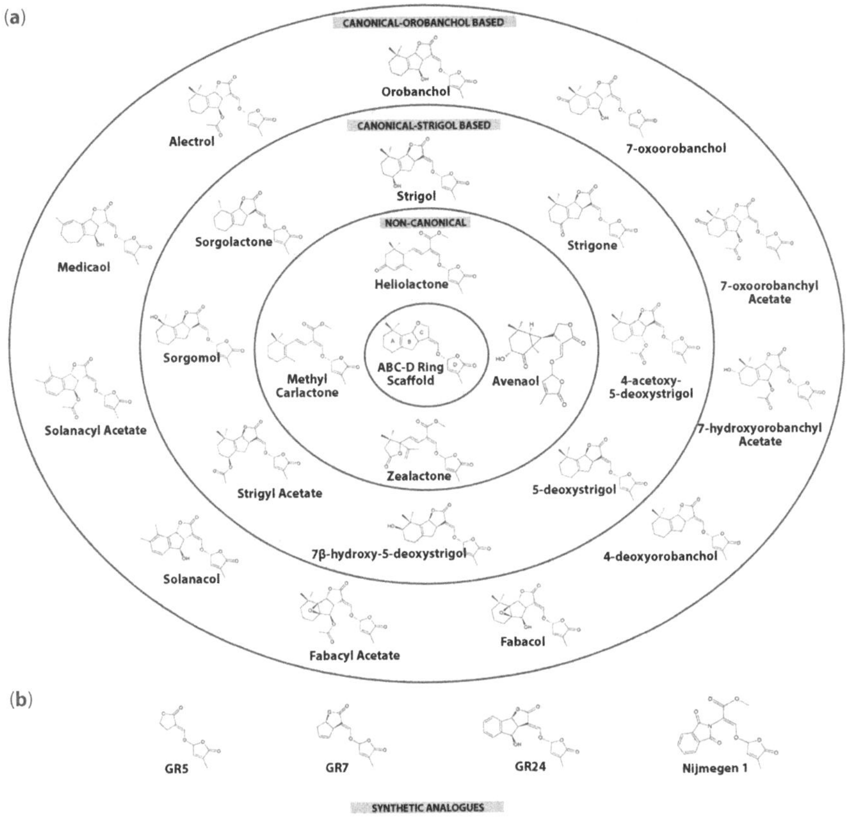


**Figure 3.2** (a) The central ring depicts the ABC-D ring scaffold base structure of canonical SLs. The second ring depicts non-canonical SLs, the third ring depicts canonical strigol SLs, and the final ring depicts canonical orobanchol SLs. (b) Four synthetic SL analogues, of which GR24 is the most commonly used in application (ChemSketch Freeware was used to draw the SL structures).

undergo conformational changes during ligand binding to interact with downstream signaling components [19]. The main difference to note between receptors is that the SL receptor contains a larger ligand-binding pocket than the KAR receptor, and this structural difference affects SL and KAR binding ability [21]. Crystallography has been used to show that the larger SLs cannot bind to the D14L pocket simply due to size, and thus this indicates an inability for SLs to bind to the D14L receptor—that said, the smaller size of KARs does not alone rule out the potential for them to bind to the D14 pocket [21]. Using isothermal titration calorimetry (ITC), GR24, and karrikinolide (KAR1—a phytoreactive karrikin-containing compound extracted from smoke water [25]), SL and KAR entropy-driven binding affinities were observable in the micromolar range for the first time between D14-GR24 and D14L-KAR1, respectively [21]. There was no observed binding between D14-KAR1 nor D14L-GR24, and so speculatively KARs may be able to enter and exit the D14 pocket without receptor interaction [21]. The structural differences paired with the ITC analysis confirm the different receptor usage by SL and KAR signaling pathways.

The conclusive differences between D14 and D14L were consistent with the analysis of the D14 ortholog in *Petunia hybrida*, DECREASED APICAL DOMINANCE2 (DAD2) [10]. Like in the D14-D14L study mentioned previously, X-ray crystallography identified the α/β-hydrolase structure of DAD2 and showed that the α/β-hydrolase fold consists of a seven-stranded β sheet domain core further surrounded by seven α helices. Meanwhile, the lid connects to the β sheet core via a short β hairpin that together forms the conserved catalytic triad Ser-His-Asp seen in homologs present in other species [10]. Crystallography further corroborated the understanding that the SL-receptor bind process induces an irreversible conformational change in the receptor [10]. Differential scanning fluorimetry proved the ligand/substrate suitability between DAD2 and GR24 by the observed melting point drop from 56.7°C to 47.7°C at a molar ratio ~2:1 GR24:DAD2 [10]. To test the credibility of the crystallography and fluorimetry observations, GR24 was applied to single, double, and triple *dad*-mutants, and only the triple mutant showed no phenotypic difference to the *dad2* mutant. Additional grafting experiments found that WT grafting reverted the phenotype in all single non-*dad2* mutants, but not the *dad2*-mutants [10]. The implication of this finding, along with other studies highlighting GR24 insensitivity in *d14* mutants, is that D14, DAD2, and their homologs are key players in SL signaling [10, 19].

Up to this point, this review has focused on pulling together studies that combine to form a strong foundation implicating D14 as the SL receptor, with a particular focus on the structure. However, a recent work by Seto *et al.* adds much more nuance to the discussion around how the receptor mechanism functions [26]. Rather than the conformational change inducing the SL active state, they found that it is actually the SL molecules that cause D14 to become activated, and then the hydrolyzing nature of D14 is what causes SLs to degrade and therefore become deactivated, following the signal transduction [26]. Before the SL can become degraded, the inactive D14 is able to interact with the downstream signaling components at the Asp location of the now disrupted Ser–His–Asp catalytic triad due to the SL-induced conformational change [26]. Once the negative downstream SL regulators are degraded, the conserved catalytic triad state is reconstructed, and the bioactive SLs are hydrolyzed and degraded [26]. This new nuanced view into the perception mechanism is an important focus for future studies to deduce.

## 3.3 Strigolactone Signal Transduction

Prior to the work of Seto *et al.*, the understanding was that the SL becomes hydrolyzed during the SL-D14 binding process, and the new SL-D14 structure together forms a covalently linked intermediate molecule (CLIM) that induces a conformational change to D14 [27]. The CLIM structure then interacts with MAX2. This "CLIM model" was widely accepted until 2018 when Carlsson *et al.* revisited the methods used to identify the SL receptor complexes and realized two important issues with the methods: (1) buffer and crystallization compounds are able to outcompete the modeled compounds and (2) crystallization gives preference to the apo enzyme subpopulation and so the modeled compounds may bind in quite low occupancies [27]. This, in general, means that the CLIM model cannot be accurately relied upon [27]. The study of Seto *et al.* came out the following year, and the current understanding is the dual-functionality role of D14 is to both perceive and deactivate SLs [28].

Key to the success of the SL signaling pathway is the SL dependency upon its receptor's interactions with the downstream F-box proteins MAX2/D3 [10]. Unlike what was found for the receptors, MAX2 is a shared component in both the KAR and SL signaling pathways [29]; however, the MAX2-mediated responses are able to distinguish the two different hormonal signals to respond appropriately—this was confirmed when SL-deficient *Arabidopsis* mutants (*max3* and *max4*) were subjected to KAR1, KAR2, and GR24, and only the GR24 treatment resulted in suppressed shoot branching, while *max2* mutants were insensitive to all treatments [30]. This suggests that, despite both SLs and KARs using MAX2 for signal transduction, MAX2 appears to be adapted to separately mediate responses based on the cue presented [30]. However, it is considered somewhat controversial to use *max2* mutants to determine SL function from KAR function during transduction due to observations of *max2* mutants producing phenotypes attributable to both SL and KAR insensitivity (aside from SL-deficient symptoms, a reduction in germination and photo-morphogenesis has been observed, which are KAR-regulated [31]) via the shared MAX2 pathway [32]. Conversely, receptor mutants are clearly distinguishable in SL and KAR insensitivity [33, 34]. D14-MAX2 ubiquitin proteasome-mediated protein degradation is a necessity in the

SL signaling pathway, but when questioning the function of SLs or KARs, only receptor mutants (D14/D14L) should be used to isolate the signaling pathway of interest.

As an F-box protein, MAX2 interacts with SKP and CULLIN proteins to form a type E3 Skp1, Cullin, F-Box containing (SCF) complex for the purpose of ubiquitination and degradation by the 26S proteasome via the ubiquitin-proteasome pathway (UPP) [29]. MAX2's role is to be an adaptor protein that acts in a SL concentration-dependent manner to bring together the SL SL-bound receptor (D14) and other SCF components to allow SL signal transferal to downstream signaling machinery, resulting in the ubiquitination and degradation of the downstream target proteins [35]. These target proteins normally repress SL signaling and, at present, three classes have been described as target candidates: (1) DELLA family transcriptional activators, (2) BES1 family of brassinosteroid response factors, and (3) SMAX1-LIKE (SMXL) family proteins [35].

No biological process occurs within a vacuum. The intricate complexities that must come together to create a living organism are only possible by constant communication and crosstalk. Phytohormones are the signaling molecules that allow plants to function by regulating the many processes involved in a plant's life cycle. Each class of phytohormone has its own specialized set of regulatory functions uncovered over the course of decades and diverse research systems. It is of no surprise that the ever changing needs a plant has over the course of its life cycle requires crosstalk to take place between the different phytohormone pathways so that a plant can regulate responses accordingly to internal and external signals and stressors. More distinctly, the proteolytic targets of SL signal transduction never bind to DNA; rather, they associate with DNA-bound TFs to modulate repressor gene expression, and it is by this approach that the different phytohormone pathways are characterized [34, 36]. To date, crosstalk has been described between SLs and ABA, IAAs, BRs, CKs, ETH, and KARs. The way in which SLs and these phytohormones interact to regulate the internal environment may be influenced by external environmental conditions [37] (see Table 3.1 for some instances in literature discussing crosstalk between SLs and other phytohormones). The following are brief descriptions highlighting significant aspects of the crosstalk relationships between SLs, ABA, IAAs, CKs, and ETH, while crosstalk with KARs is discussed elsewhere within the text.

**Crosstalk With Abscisic Acid**

ABA is a key phytohormone plants utilize for abiotic stress response regulation. Just like SLs, ABA derives by oxidative cleavage from the carotenoid

**Table 3.1** Some instances of SL crosstalk with other phytohormones in literature.

| Phytohormone | Relationship | Species | Effects | Experiment type | References |
|---|---|---|---|---|---|
| ABA | Synergistic | *Arabidopsis thaliana* | ABA import | *max2* mutant response | [38, 39] |
| | | *Oryza sativa* | Drought tolerance/ stomatal closure | GR24 application | [40, 41] |
| | | *Vitis vinifera* | Stomatal closure/ shoot branching | GR24 application | [42] |
| IAAs | Antagonistic | *Arabidopsis thaliana* | Combined IAA andGR24 additive root hair lengthening effect | SL/IAA mutants paired with GR24 and IAA application | [43, 44] |
| BRs | Antagonistic | *Oryza sativa* | BRs enhance tillering, SLs inhibit it (D53 and OsBZR1 interact to inhibit FC1 expression) | Double mutants *d14* + lines blocking BR signaling | [45] |
| CKs | Synergistic | *Arabidopsis thaliana* | Lateral root repression | GR24 application | [46] |
| | Antagonistic | *Oryza sativa* | IAA increase/ CK decrease | Insensitive mutants; low nitrogen/ phosphorus conditions | [47] |
| ETH | Synergistic | *Arabidopsis thaliana* | Root development | *max2*, *ein2*, and *etr1* mutants with GR24 application | [43] |
| KAR | Synergistic | *Arabidopsis thaliana* | Photosynthesis regulation (normal and osmotic stress conditions) | d*14* and *kai2* mutants | [48] |

biosynthesis pathway, which presents the potential scenario that SLs may also be used to regulate stress response. Notably, both ABA and SLs occur in higher concentrations in stressed mycorrhizal plant species. In *Lotus japonicus*, WT and SL-depleted plants were subjected to normal conditions alongside both osmotic and phosphate-depleted stress conditions to understand SL's role in drought resistance within the root system [49]. In shoots, SL-depleted plants responded to ABA with a slower stomatal closure time, verifying the role SLs play in drought resistance. In contrast, SL concentration in roots saw a drastic reduction that triggered the ABA biosynthesis gene, *LjNCED2*, to be upregulated when WT plants faced osmotic stress. As confirmation of this mechanism, an exogenous application of the synthetic SL GR24 was applied to WT plants prior to osmotic stress and the treated plants were unable to upregulate the *LjNCED2* gene for ABA biosynthesis [49].

**Crosstalk With Auxins**

So important are IAAs to plant growth and development that Charles Darwin, and later with his son Francis, had deduced that some "moving stimulus" must exist to control phototropism and elongation/etiolation [50]. Fritz W. Went built upon Darwin's work to show that the yet-to-be-named stimulus exuded from intact etiolated coleoptile tips of *Avena sativa* [51]. For the study, Went developed an "Avena Test" that could be used to identify natural and synthetic compounds that he, along with Thimann, went on to christen as "auxins," making them the first phytohormones to be described [51, 52]. Unlike Went and Thimann's original hypothesis that IAAs transport via simple diffusion [52], it is now understood that carrier proteins from the AUXIN RESISTANT1/LIKE AUX1 (AUX1/LAX) transporter family facilitate IAA uptake while PIN-FORMED (PIN) and ATP-BINDING CASSETTE SUBFAMILY B (ABCB) proteins facilitate IAA efflux [53].

Clathrin-mediated endocytosis (CME) is a crucial component in establishing PIN protein polarity [54]. These polarly localized PIN proteins regulate local IAA accumulation via directional transport [55]. In contrast to IAA-mediated growth, SLs act synergistically with IAA to inhibit axillary shoot branching through the D14-MAX2 mediated pathway specifically by regulating canalization of PIN proteins [56, 57]. More specifically, SL signaling prevents IAA feedback of PIN polar membrane localization that an activating bud would otherwise use to connect IAA flow to the main vasculature, which subsequently causes PIN protein levels to decrease in activating buds at a faster rate than can be inserted, thus resulting in axillary buds being prevented from branching out [56, 57]. Chabikwa *et al.*

tested IAA-mediated bud growth by applying the auxin transport inhibitors 1-N57 naphthylphthalamic acid (NPA—inhibits PIN protein efflux activity) and p-chlorophenoxyisobutyric acid (PCIB—blocks receptor function) to *P. sativum* axillary buds. Both NPA and PCIB prevented IAA from being transported out of the buds, causing an apparent buildup of IAA within the buds that triggered PIN1 upregulation 3 h following treatment, a response suggestive of IAA overproduction [58]. They additionally found that their applications did not impact IAA transport within the main stem nor did they impact initial bud outgrowth timing following decapitation—important additional findings as they indicate IAA transport in the main stem is not impacted by IAA transport during bud initiation, demarcating IAA transport between bud and stem [58]. Similarly in roots, SLs impact lateral root and root hair formation by regulating IAA distribution [59].

**Crosstalk With Cytokinins**

CKs are one of the best described classes of phytohormones that impact multiple processes, including root and shoot development, shoot branching, and cytokinesis, and its biosynthesis is downregulated under drought conditions to prevent undesirable growth. The type of CK impacts the way by which it is transported through a plant, e.g., xylem is used to transport trans-zeatin classed CKs from roots to shoots, while phloem is the transportation means by which isopentenyladenine (iP) CKs move—the isopentenyladenine CKs can move from root to shoot or be absorbed systemically [60].

Until 2012, only the antagonist relationship between IAA and CKs regulating bud outgrowth had been well explored, with just a few studies starting to scrutinize and acknowledge an impact on bud growth due to crosstalk between IAAs and SLs. Dun *et al.* recognized a lack of research investigating a potential crosstalk between CKs and SLs and so launched a study using the synthetic CK, 6-benzylaminopurine (BA) to determine the means by which local and long-distance CKs in the xylem sap and SLs come together to regulate bud outgrowth in the model species *P. sativum* [61]. *BRANCHED1* (*BRC1*) is an important gene impacting branching and will be discussed in greater detail in the following section. As it pertains to the investigation of Dun *et al.*, the *P. sativum* ortholog, *PsBRC1*, was a key focal point in understanding the antagonistic relationship between CKs and SLs as exogenously applied BA reduced *PsBRC1* expression and GR24 increased expression [61].

BCR1 expression prevents budding while repression results in greater branching, which can be seen in *brc1*-deficient mutants presenting a greater amount of branching as opposed to WT. Application of GR24 to *rms5* SL-deficient mutant buds inhibited expression of genes coding CK biosynthesis while simultaneously increasing expression of the CK dehydrogenase/oxidase enzymes (CKX)s CKX3, CKH5, and CKX7 [62]. Cao *et al.* further used a CK oxidase inhibitor on the *rms5* mutants to show that preventing endogenous CK degradation results in bud growth, confirming the antagonistic regulatory roles between CKs and SLs [62].

**Crosstalk With Ethylene**

Ethylene and SL work in an antagonistic relationship in some processes, while synergistically in other processes. *Striga hermonthica* germination is a synergistic example in which SLs trigger the germination cascade by first causing an upregulation of ethylene biosynthesis [63]. This was confirmed by subjecting *S. hermonthica* seeds first to GR24 to show that ethylene concentration increased prior to germination, then subjecting GR24-treated seeds to two ethylene biosynthesis inhibitors: aminoethoxyvinylglycine (AVG) and 1-methylcyclopropene, which caused germination to be inhibited. Then, finally, the treated seeds were subjected against ethylene biosynthesis with 1-aminocyclopropane-1-carboxylic acid (ACC) to trigger ethylene biosynthesis, which was found to reverse the AVG effects, indicating the role ethylene has in germination and the role SLs have in ethylene biosynthesis as part of the germination cascade [63].

## 3.4 Transcriptional Regulation

The SL signaling pathway is much less understood than the SL biosynthesis pathway, particularly the signaling pathway's downstream transcriptional regulation. The general framework by which SLs are perceived and interact with downstream signaling components, MAX2 and the SCF complex, have so far been laid out in this review. Of the three candidate classes of target proteins, the SMXL family has received the most attention. SMXLs are weakly similar to Class1 Hsp100/ClpB proteins, and evidence has converged around SMXL proteins via independent methods to identify them as genuine MAX2 targets [35]. *SUPPRESSOR OF MAX2 1* (*SMAX1*) is the MAX2 suppressor gene found in *A. thaliana*, while *DWARF 53* (*D53*) is nuclear-localized. SMXL repressor protein found in *O. sativa* and the co-orthologs SMXL6/7/8 (SMAX1-like 6/7/8) are found in *A. thaliana* [35]. D53 forms a repressor–corepressor–nucleosome complex with co-repressing TOPLESS (TPL) proteins by using ethylene-responsive element binding factor-associated amphiphilic repression (EAR) motifs to oligomerize the TPL proteins [64]. In the presence of SLs, D53 is ubiquitinated and degraded by the MAX2-mediated SCF complex [65]. By downregulating D53, the BRANCHED1 (BRC1) negative transcription factor (TF) expression is induced, as is the IAA efflux carrier PIN1 [64]. Though identifying D53 as a downstream target of the SL signaling pathway is a breakthrough, the effects D53 has on downstream TFs has only begun to be a less muddied area of understanding. The first known downstream TF target of D53 in the SL signaling pathway was identified as *IDEAL PLANT ARCHITECTURE1* (*IPA1*) in 2017 [66]. *IPA1* expression is regulated by the microRNAs: miRNA156 and miRNA529, of which point mutations in the latter correlate with an increase in both plant height and panicle branching while simultaneously causing a decrease in tillering in *O. sativa* [66]. *IPA1* encodes SPL14, a member of the SQUAMOSA PROMOTER BINDING PROTEIN-LIKE (SPL) TF family [67]. The connection to SL is that *ipa1*-loss-of-function mutants shared a similar phenotype consisting of high tillering numbers and dwarfism that has been frequently noted in SL-deficient and SL-signaling mutants [66]. Not only does D53 suppress SPL transcription activity, but it has also been implicated as a direct suppressor of CYTOKININ OXIDASE 9 (CKX9) in *O. sativa*, which is consistent with the known antagonistic relationship SLs and CKs have regarding lateral shoot development [68]. D53-mediated suppression of OsCKX9 directly results in an increase of tillering and CK levels, whereas GR24 application to *osckx9* mutants reduces tillering [69]. Due to GR24 not preventing bud growth in the *ipa1-10* mutant, it seems likely that IPA1 is an important regulator in tillering suppression [69].

Leading up to the discovery of IPA1 as a downstream TF target of D53 came studies focused on the negative regulator *BRANCHED1* (*BRC1*) in *A. thaliana*. *A. thaliana* has two BRC genes: *BRC1* and *BRC2*, which both encode a bud-outgrowth-inhibiting TF found in the TEOSINTE BRANCHED 1 (TB1)/CYCLOYDEA/PCF (TCP) protein family [70]. TCP genes make up an evolutionarily conserved family of proteins that regulate important biological processes that have been deduced across a number of species. These processes include reproduction (i.e., gametophyte development), germination, architecture (i.e., branching), and plant defense responses, among other processes [71]. A total of 24 TCP genes are present in *Arabidopsis*, of which nearly half play a role in axillary meristem development that go on to mature into either lateral shoots or flowers [70]. The phytohormone-driven regulation of TCP genes is done indirectly by either IAA or SL induction or by CK repression [61]. The first description of *BRC1* having a downstream role in the SL signaling pathway came about in *A. thaliana* and *P. sativum* mutant studies in which the *bcr1* mutant did not show sensitivity to an exogenously applied SL (GR24) despite BRC1

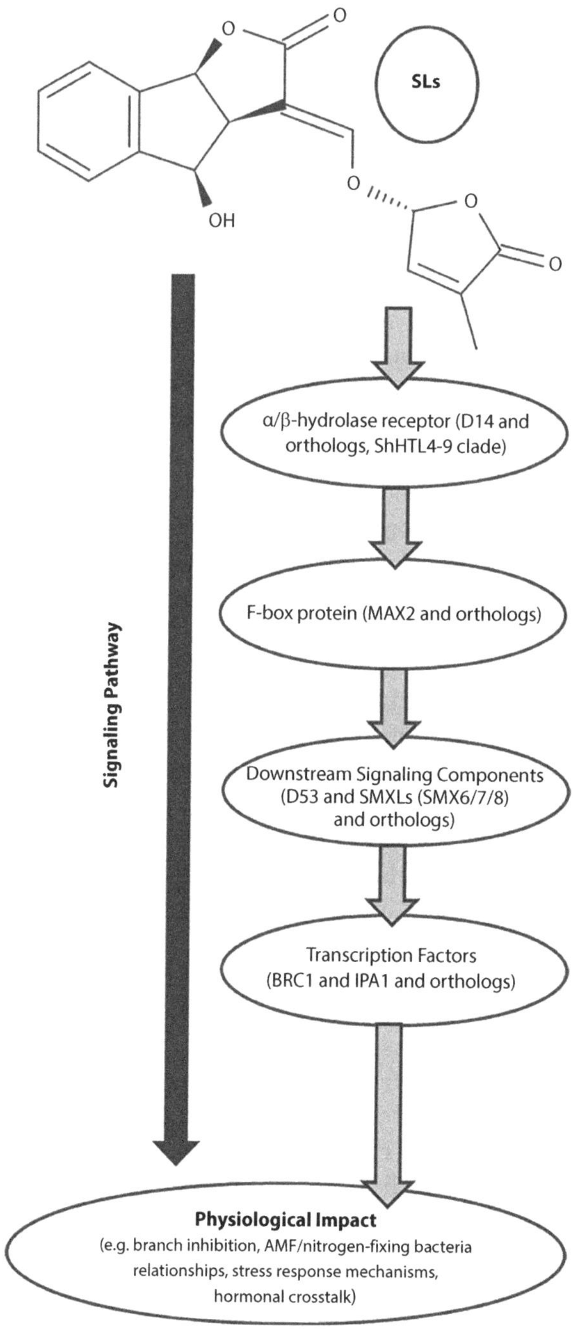


**Figure 3.3** A simplified flowchart of the current understanding of the SL signaling pathway.

being upregulated by SL [61]. Recalling from the previous paragraph, D53 is a SMXL repressor protein that is degraded by D14-mediated SL perception and D53-like deficient mutants have been noted to exhibit BRC1 upregulation. In *O. sativa*, IPA1 will directly bind to the BRC1 ortholog, OsTB1, to trigger OsTB1 upregulation [72]. However, if D53 is present, it interacts with IPA1 to inhibit OsTB1 upregulation, hence why D53-like deficiency results in BRC1 upregulation instead of repression [72]. Wang *et al.* have found that the IPA1 orthologs SQUAMOSA PROMOTOR BINDING PROTEIN-LIKE9 (SPL9) and SPL15 act in parallel pathways with SM6/7/8 for branching regulation within *A. thaliana* [73]. More so, they found that SMXL6 acts like a TF by directly binding to DNA, which is at odds with other known phytohormone signaling pathways [73]. This divergence from the more typical mechanisms that other phytohormone signaling pathways use opens an exciting new area of research by which future studies may continue to further investigate how these mechanistic variations impact plant physiology and architecture. Figure 3.3 depicts a simplified flowchart of our current understanding of the SL signaling pathway.

## 3.5 Environmental Factors Modulate Strigolactone Signaling

The stresses caused by the shifting of seasons over the course of a year and the many abiotic and biotic shifts that such shifts induce are the evolutionary drivers that push plants to evolve robust stress regulatory systems [74]. Though more on this will be discussed in the next section, here it is important to note that climate change is causing the once slow and steady shifts of seasonal change to become much more chaotic and intense, placing a higher pressure on species abilities to adapt and respond at faster rates than in the past [75]. Phytohormones regulate all aspects of plant growth and development, including their germination, growth/elongation, reproductive processes, dormancy (in non-annual species), senescence and everything in between, including stress response [76]. So far, this review has described the current knowledge of the SL signaling pathway along with some discussion crosstalk with other phytohormones. This section shifts focus to look more broadly at environmental factors that directly impact SL signaling. These environmental factors can be roughly divided into four categories: nutrient availability (phosphorus and nitrogen deficiencies), osmotic stress (drought and salinity), photo conditions (light availability/quality), and temperature fluctuations (heat/cold seasonal variations).

Nutritional deficiencies greatly hamper a plant's ability to grow and develop. Different nutrients are modulated by different pathways. Environmental deficiencies of nitrogen (N) and phosphorus (P) induce an increase in SL biosynthesis that correlates with an increase in mycorrhizal relationships with AMF and associated symbiotic bacteria [77]. In both *A. thaliana* and *Hordeum vulgare* L. (barley) studies, SL signaling mutants increased the density of lateral root formation but did not increase the lateral root length under normal nutrient conditions [78, 79]. Under nutrient-deficient conditions, however, the plants respond by decreasing the above-ground growth and instead focusing on the below-ground root architecture [77]—for example, *A. thaliana* increases both the density and length of lateral roots while decreasing the length of the primary roots under P-deficient conditions, which corresponds with the SL role in lateral root formation, but with no known participation in primary root development [80]. The crosstalk between SLs and IAAs is a closely tied relationship under P-deficiency—if you recall, SLs downregulate the expression of PIN proteins to modulate local IAA levels [81]. As far as N-deficiency goes, SL exudates appear to increase somewhat commonly across species [78]. Considering the well-described association between some species and N-fixing bacteria, it may be of no surprise that increased exudation is the response. In a study comparing *Glycine max* (soybean) and *G. soja* (wild soybean), *GsmlMaT2* was cloned from *G. soja* and overexpressed in the hairy root system of *G. max*, and it was found that the gene activates SL biosynthesis which work alongside the co-increase of nodulation production which together increase the ability to form associations with N-fixing *Rhizobia* [82]. Rather than relying upon the deficient environment, this provides a workaround for species to use symbiotic relationships to obtain N.

Drought conditions cause significant stress to plants, and climate change is intensifying and expanding the drought occurrences that farmers face, which can and does cause great costs to farmers both through yield loss and increased watering demands throughout the hot summer months [83]. Contrary to the response SLs induce under nutrient-deficient conditions, SLs do not appear to influence changes in root architecture under osmotic or salinity conditions [38]. *A. thaliana* WT and SL-deficient/loss-of-function mutants were exposed to drought and salinity conditions to deduce that SLs induce stomatal closure along with leaf senescence to improve resistance to stress [38]. An important finding in that study was that WT had improved drought tolerance upon foliar GR24 application [38]. Under drought and saline stress, SLs conduct a crosstalk with ABA and CKs in a coordinated response network [38].

SL levels positively correlate with light-associated processes. *A. thaliana* seedlings were treated with GR24, and within 90 min light signaling genes were upregulated [84]. *Solanum lycopersicum* (tomato) has been used for a number of studies investigating the impact of light on SLs to note the SL regulation of light processes. However, an important caveat pointed out by Koltai *et al.* is that it might be that increased light causes an increase in apically derived IAA biosynthesis pathway as described by Hayward *et al.*, which was found to induce *CCD7* and *CCD8* expression, which are key genes in SL biosynthesis [44, 85]. Koltai *et al.* used *S. lycopersium* in their study and found that transcription of the SL-biosynthesizing gene *SlCCD7* increased by 23.4% compared to those under 60% shade conditions, and it took a total of 5 days in full light for shade-treated plants to reach the same SL root concentration as their counterparts always in full light [85]. For the next few years, it could only be stated that light positively impacts SL biosynthesis and may be an important component to light adaptations, but no conclusive statements could be made regarding the chicken and the egg question about whether light directly or IAA indirectly stimulated SL biosynthesis [85]. Ge *et al.* recently took up this question in the context of AMF colonization and P-deficiency by using multiple techniques to understand how light impacts shoot to root mediation of AMF formation and P-uptake [86]. Phytochrome B (phyB), ELONGATED HYPOCOTYL5 (HY5), and SL biosynthesis genes were found to form a signaling cascade to regulate AMF symbiosis and P-utilization [86]. AMS and P uptake rely upon the red light spectrum, and in their study, they used *ccd7* mutants exposed to conditions with and without red light and found that the mutants did not significantly form AMF structures as opposed to WT while also reducing P-uptake and general plant size [86]. Though Ge *et al.* did provide further details on the importance of SLs in light-mediated reactions, it still seems likely that the SL increase is due to an increase in IAA. Red light also increases the biosynthesis of SA, which suggests future studies should consider potential light-mediated crosstalk between SLs and SA to further deduce the impact of light on SL signaling.

Temperature stress is regulated by an interacting network of ABA, GAs, JA, and SLs, but the relationship between ABA and SLs may be of particular importance considering that SL-deficient *S. lycopersicum* mutants have been found to have a lower ABA concentration as well as slower stomatal closure times [38]. Using the HEAT SHOCK PROTEIN (HSP) HSP70, Chi *et al.* recently found that *ccd7* mutants decreased HSP70 accumulation while the application of GR24 increased HSP70 accumulation, which makes it likely that HSP is responsive to SLs under stress and, possibly, under normal conditions [87]. In cold conditions, GR24 induced

C-REPEAT BINDING FACTOR 1 (CBF1) transcription, and interestingly, ABA biosynthesis deficiency caused the transcription of HSP70 and CBF1 to stop, which created an overall negative impact on GR24's ability to regulate temperature [87]. Thus, it can be stated that SLs do play an important role in regulating temperature-induced stress but require a crosstalk with ABA for modulation.

## 3.6 Applications and Future Perspectives

Climate change is the great challenge of this generation. Agriculture around the world is facing a daunting dilemma as extreme temperatures are breaking records annually around the globe [88]. Extreme weather events are becoming more prevalent, and crops are racing against time to become more resistant against spreading super pests and diseases, heat stress, drought stress, and continually increasing demands as the world population continues to climb [89]. Earth's carrying capacity has yet to be reached but is limited by the availability of adequate food and water resources [90]. By far, the agricultural industry has the heaviest demand for water globally. Per the Food and Agriculture Organization of the United Nations (FAO), 69% of global water withdrawal ratios are distinguished as agricultural (including crop irrigation, livestock, and aquaculture). However, the ratios are highly variable when divided by region (e.g., the African continent has an 82% agricultural use, while Europe is at 21%, and the Americas are at 51%) [91].

Among other factors such as war and poverty, droughts are already inducing more instances of famine across many regions of the globe [92]. Clearly, the demand for water is incredibly important to agricultural industries, and increasing population demands will create an even higher demand for water to feed the population. However, drawing down precious groundwater resources at a rate faster than replenishment is a recipe for future disaster in which there will not be enough water for growing crops, taking care of herds, nor drinking water for humans. It is crucial that agricultural practices adapt to the changing climate by focusing on practices that reduce water demands. Not only do more sustainable field methods need to be introduced on a widescale for farming but also the actual plants also do need to become more resistant to higher temperatures, stronger weather events, drought conditions, and pests and diseases [93].

Wheat, rice, maize, and soybeans are four of the top 10 major food crops grown globally. They were domesticated thousands of years ago and have since been bred into many different varieties. However, traditional breeding is a slow process that takes many years to produce plants with the ideal traits we know and see today [94]. The slow process of breeding is not fast enough to viably produce enough new varieties of plants that can withstand the impacts of climate change to match the increasingly chaotic nature of variable growing conditions found in the many ecosystems around the world. Additionally, breeding is not always a viable option as limited genetic diversity may mean that desired traits may simply not even be available [94]. This is where the biotechnology industry comes into play.

The process of making genetically modified (GM) crops is significantly faster than traditional breeding systems as desired traits can be stably incorporated into genomes at a much faster rate [94]. The miracle story of "Golden Rice" is just the beginning of what GM crops can do to tackle malnutrition in the world. Golden Rice is a genetically modified crop in which vitamin A was incorporated into rice endosperm to produce a fortified crop [95]. Vitamin A is a micronutrient that humans are incapable of producing and so must rely upon consuming enough of it through diet. Vitamin A deficiency (VAD) affects 190 million preschool-age children around the world, with most of those children concentrated across the African continent and Southeast Asia [96]. VAD has been implicated as the leading cause of childhood blindness, with half a million new cases occurring every year. Pregnant women are the second most affected population after young children. So extensive is the impact of VAD that it is now considered to be an immune deficiency syndrome, which caused the deaths of 1 to 2 million people alone in 2016 [96, 97]. The fortified Golden Rice has incorporated vitamin A in a readily bioavailable form from which consumers may obtain as much vitamin A as they do from other known sources, such as supplements, eggs, and milk [97].

Using Golden Rice as one example in which a major crop can be edited to make a major impact on global health, now the focus will shift back to SLs and biotechnological approaches that can be used for future crop improvement. The particular set of functions SLs have in plants are what makes them ideal targets for genetic modification. Returning to the original discovery of SLs as *Striga* and *Orobanche* germination stimulants, GM technology can be utilized to develop new crop varieties that produce modified SLs that prevent parasitic plants from recognizing hosts [98] or synthetic SLs may potentially be used in field applications to induce parasitic seed germination before conditions can support further growth and attachment to host plants. This was tried years ago when such an approach was simply not feasible on a wider scale due to the cost of synthesizing synthetic SLs, compound instability, and field delivery complications [99]. There is new technology available today to more cost-effectively synthesize synthetic SLs and produce better delivery methods and this is potentially a viable approach today [100], though synthetic SL analogues still remain relatively expensive. In either of these technologies, however, weed pressure decreases in fields alongside yield loss due to parasite infestation, though studies will need to determine feasibility and identify potential negative effects before wider usage.

Symbiotic relationships mediated by SLs are a particularly enticing area for biotechnology. Not only do mycorrhizal fungi improve nutrient uptake and water absorption in root systems but they also improve plant immunity toward soil-borne pathogens by inducing plant defense mechanisms to produce antibiotics that improve plant competition for space and resources within the rhizosphere [101]. To an extent, mycorrhizae are also capable of reducing susceptibility to pest infestation [102]. Using genetic editing techniques to increase SL biosynthesis within roots could be a potentially comprehensive method through which GM plants can bolster their symbiotic rhizosphere relationships, which may result in plants that have greater disease/pest resistance, improved nutrient uptake, and better environmental stress tolerance. These resulting GM plants should need fewer applications of fertilizer since increased mycorrhizal and nitrogen-fixing bacteria interactions will help draw in the required nutrients, without negative impact on plant growth and yield. Potentially, the downstream effects may cause a boost in growth and yield as soil health benefits from increased mycorrhizal networks [103]. Clearly, less need for fertilizer and better disease resistance result in an economic benefit to the farmer.

For many decades, agriculture has focused on growing crops with ever increasing size that correlates with increased yield. However, bigger is not always better, as crops such as maize (*Zea mays*) have been finding out. Maize yields can decrease anywhere from 3% to 25% of planted fields due to severe storm stress globally every year due to lodging and other storm effects, with potentially even greater losses depending on the weather event [104]. In August 2020, a derecho weather event occurred in the Midwest corridor of the United States that caused damage to ~4 million hectares of maize. Over 60% of all maize yield lost in 2020 in the United States came from that single weather event and the overall loss is estimated to be ~US$2.5 billion [104]. Bayer Crop Science was running a field trial at that time on a new short stature maize hybrid they were developing to improve lodging resistance against extreme weather events, such as the derecho, and they found that their new variety resulted in half the damage of taller hybrids while simultaneously having an improved yield by 13%. The models that Bayer Crop Science ran suggested that US$938 million of loss during the derecho could have been avoided if all of the maize affected by the storm had been their new short stature hybrid [104]. Not only does this particular case study depict so well why agriculture much focus on improving crops to better persist against increasing extreme weather events but also it highlights the time differential between traditionally bred and GM crops.

It can take 5+ years to traditionally breed a new maize variety, while GM maize can be bred much more quickly [105, 106]. Considering that farmers' needs are highly variable depending on their global location and specifics of their local environment, the cost, time, and resources it would take to create the number of varieties necessary to fill the many current niches are prohibitive and can be greatly mitigated through GM technology [107]. In the case of short stature maize, an inhibition within the GA pathway was the effective method [104]. Much like the way GA inhibition in maize is effective for creating a short stature crop, SLs hold a lot of promise that can be used via GM to alter plant architecture for desirable traits in the context of future climatic events, particularly with a focus on the SL role in root architecture and stress responses.

Compared to sustainable agricultural practices, rare, endangered plant conservation is often less discussed on wider scales, and yet just like sustainability, conservation is of great importance to preserve biodiversity and the integrity of ecosystems—if not for the sake of the individual plant, then for the health of the wider ecosystem and preservation of plant genetic resources for currently unknown future potential needs [108, 109]. As understudied as SLs are within agriculture, they are even more understudied in rare/endangered plant species. Orchidaceae, the second largest Angiosperm family following Asteraceae, has arguably the most intimate relationships with their mycorrhizal symbionts [110], so much so that they evolved to have dust-like seeds lacking almost all nutritional reserves and thus rely upon their mycorrhizal symbionts for germination and nutritional needs throughout early-stage growth and development [111]. Some species are mycoheterotrophic and rely fully upon their symbionts for all nutritional needs throughout their lives, whilst most develop photosynthetic capabilities over time [111]. However, cryptic mycoheterotrophy is a somewhat recently discovered phenomenon amongst some meadow species that would appear to rely upon photosynthesis for energy, but actually continue to rely fairly heavily upon the symbionts for nutrition [112]. Further exploration into the potential of SL-mediated mycorrhizal relationships may be able to answer crucial questions about the mechanisms driving the relationships to improve germination and culturing needs of individual species [113, 114].

## 3.7 Challenges and Unanswered Questions in Strigolactone Research

Over the course of this review, the discovery of SLs and an overview of our current understanding of the SL signaling pathway have been covered. Additionally, environmental factors impacting signaling along with some applications and future directions have been addressed. Despite the many advances that pertain to our understanding of the roles SLs hold in plant physiology and architecture over the past few decades, we continue to face elusive answers to a number of challenging questions they present to us—unknown aspects of their biosynthesis/signaling pathways, agricultural/biotechnological application possibilities, and in their potential for conservational use. The cost to produce and procure synthetic SLs is one significant barrier at this point of time [115]. A single milligram of GR24 from leading chemical suppliers on average costs US$100+, which makes studies using GR24 infeasible for many smaller labs with lower levels of funding available.

In 2023, Dun *et al.* pointed out that the most important question to currently answer about SLs is to determine the endogenous shoot branching SL structural identity, which continues to be prevented by the dual challenge of structural instability and synthesizing difficulty [116]. The reason why the structure is crucial to figure out is twofold—first, it can open the path to developing GM crops where shoot SL biosynthesis can be targeted separately from roots to induce and preserve ideal phenotypes combining the best shoot and root traits [116]. More SLs are being discovered and with that is the structural diversity they present [117]. Because structure plays a key role in which SLs regulate different processes, those processes cannot be better understood until the structures become known. Not only do different plant species produce different SL structures that cultivate different microbial populations within their respective rhizospheres, but so also do different genotypes, as found in *O. sativa* [118, 119]. Consider how the knowledge of SL structures exuded by specific genotypes would allow targeting crop varieties to soils containing desired microbial populations, how microbial treatments could be applied to soils to match desired genotypes, and how GM technology might be applied to bring together the best of both worlds.

Plant tissue culture is an important technique utilized in commercial horticulture, conservation, and throughout agriculture and the general biotech industry. Phytohormones are important media additives used to stimulate different stages of tissue culture (i.e. callus induction, shoot proliferation, and rooting). IAAs, CKs, and GAs (namely, $GA_3$) are the most commonly used phytohormones due to their well-described roles in plant growth. SLs, however, are still so newly understood in their phytohormone functions that they have yet to be adopted much within plant tissue culture media compositions. GR24 is a synthetic SL analogue available on the market and has been used to test plant responses outside of *in vitro* conditions, most often in bud growth experiments, but its use as a plant growth regulator (PGR) in the *in vitro* tissue culture environment is lacking. Mdodana observed *A. thaliana* callus cultures developed roots when moved from a low hormone medium containing 0.5 mg/L 2-4,D and 0.5 mg/L kinetin to a medium containing GR24, as opposed to callus coming from a higher hormone concentration medium [120]. A recent 2023 study on *Zingiber officinale* (ginger) grew plants in tissue culture and then treated them with GR24 after out-planting which resulted in a significant increase in height, stem diameter, and yield, while a decrease in overall shoot branching was observed when GR24 was used in a concentration of 7.5 um $L^{-1}$ [121]. This suggests agricultural potential for pairing plants grown in tissue culture to decrease disease potential with field SL application to increase plant health and yield. Fan *et al.* used GR24 *in vitro* with *Malus domestica* callus to highlight MdBRC1's inhibitory role on AR formation [37]. In general, however, the impact of endogenous production of SLs by *in vitro* tissues has been almost completely unexplored, and it is unknown how endogenous SLs may be influencing growth at different tissue culture stages along with the crosstalk influence of light, temperature, and media compositions. This is an exciting new aspect of the SL field of study that is wide open for future exploration.

## 3.8 Conclusion

The processes involved in the SL signaling pathway are highly conserved across plant families, being found in both monocotyledonous and dicotyledonous species [122]. The cascade of signaling events starts with the D14 receptor that interacts with the F-box protein MAX2 and the SCF complex to then be able to interact with downstream TF components, namely, IPA1 [66]. SLs provide a useful system by which to understand small-molecule receptor evolution [123]. Do not make the mistake of thinking SLs are unimportant in plant development simply because they are present at such low levels. As studies reviewed throughout the previous sections make clear, SLs do indeed significantly impact plant growth and development at every point of the life cycle. Despite all that has been deduced about the SL signaling pathway, there is still so much yet to be explored about this.

## References

1. Cook, C.E., Whichard, L.P., Turner, B., Wall, M.E., Egley, G.H., Germination of Witchweed (*Striga lutea* Lour.): Isolation and Properties of a Potent Stimulant. *Science*, 154, 3753, 1189–1190, 1966.
2. Chang, M., Netzly, D.H., Butler, L.G., Lynn, D.G., Chemical regulation of distance. Characterization of the first natural host germination stimulant for *Striga asiatica*. *J. Am. Chem. Soc.*, 108, 7858–7860, 1986.
3. Spallek, T., Mutuku, M., Shirasu, K., The genus *Striga*: A witch profile. *Mol. Plant Pathol.*, 14, 861–869, 2013.
4. Parker, C., Observations on the current status of orobanche and striga problems worldwide. *Pest. Manage. Sci.*, 65, 453–459, 2009.
5. Butler, L.G., Chemical Communication Between the Parasitic Weed *Striga* and Its Crop Host. *Allelopathy*, 582, 158–168, 1994.
6. Akiyama, K., Matsuzaki, K.I., Hayashi, H., Plant sesquiterpenes induce hyphal branching in arbuscular mycorrhizal fungi. *Nature*, 435, 824–827, 2005.
7. Corradi, N. and Bonfante, P., The Arbuscular mycorrhizal symbiosis: Origin and evolution of a beneficial plant infection. *PLoS Pathog.*, 8, 1–3, 2012.
8. Genre, A. and Bonfante, P., The making of symbiotic cells in arbuscular mycorrhizal roots, in: *Arbuscular Mycorrhizas: Physiology and Function*, pp. 57–71, Springer Netherlands, Dordrecht, Netherlands, 2010.
9. Gomez-Roldan, V. *et al.*, Strigolactone inhibition of shoot branching. *Nature*, 455, 189–194, 2008.
10. Hamiaux, C. *et al.*, DAD2 is an α/β hydrolase likely to be involved in the perception of the plant branching hormone, strigolactone. *Curr. Biol.*, 22, 2032–2036, 2012.
11. Yao, R. *et al.*, DWARF14 is a non-canonical hormone receptor for strigolactone. *Nature*, 536, 469–473, 2016.
12. De Saint Germain, A. *et al.*, An histidine covalent receptor and butenolide complex mediates strigolactone perception. *Nat. Chem. Biol.*, 12, 787–794, 2016.
13. Vaishnav, D. and Chowdhury, P., Plant Abiotic Stress Responses and Tolerance Mechanisms, Types and Function of Phytohormone and Their Role in Stress, in: *Plant Abiotic Stress Responses and Tolerance Mechanisms*, IntechOpen, Rijeka, Croatia, 2023.
14. Kende, H. and Zeevaart, J.A.D., The Five 'Classical' Plant Hormones. *Plant Cell*, 9, 1197–1210, 1997.
15. Sakakibara, H., Cytokinin biosynthesis and transport for systemic nitrogen signaling. *Plant J.*, 105, 421–430, 2021.
16. Hirose, N. *et al.*, Regulation of cytokinin biosynthesis, compartmentalization and translocation. *J. Exp. Bot.*, 59, 75–83, 2008.
17. Shan, X., Yan, J., Xie, D., Comparison of phytohormone signaling mechanisms. *Curr. Opin. Plant Biol.*, 15, 84–91, 2012.

18. Tal, L., Gil, M.X.A., Guercio, A.M., Shabek, N., Structural aspects of plant hormone signal perception and regulation by ubiquitin ligases. *Plant Physiol.*, 182, 1537–1554, 2020.
19. Arite, T. *et al.*, D14, a strigolactone-Insensitive mutant of rice, shows an accelerated outgrowth of tillers. *Plant Cell Physiol.*, 50, 1416–1424, 2009.
20. Miyakawa, T., Xu, Y., Tanokura, M., Molecular basis of strigolactone perception in root-parasitic plants: aiming to control its germination with strigolactone agonists/antagonists. *Cell. Mol. Life Sci.*, 77, 1103–1113, 2020.
21. Kagiyama, M. *et al.*, Structures of D14 and D14L in the strigolactone and karrikin signaling pathways. *Genes. Cells*, 18, 147–160, 2013.
22. Zwanenburg, B., Nayak, S.K., Charnikhova, T.V., Bouwmeester, H.J., New strigolactone mimics: Structure-activity relationship and mode of action as germinating stimulants for parasitic weeds. *Bioorg. Med. Chem. Lett.*, 23, 5182–5186, 2013.
23. Scaffidi, A. *et al.*, Exploring the molecular mechanism of karrikins and strigolactones. *Bioorg. Med. Chem. Lett.*, 22, 3743–3746, 2012.
24. Fukui, K. *et al.*, New branching inhibitors and their potential as strigolactone mimics in rice. *Bioorg. Med. Chem. Lett.*, 21, 4905–4908, 2011.
25. Dixon, K.W., Merritt, D.J., Flematti, G.R., Ghisalberti, E.L., Karrikinolide - A phytoreactive compound derived from smoke with applications in horticulture, ecological restoration and agriculture. *Acta Hortic.*, 813, 155–170, 2009.
26. Seto, Y. *et al.*, Strigolactone perception and deactivation by a hydrolase receptor DWARF14. *Nat. Commun.*, 10, 1–10, 2019.
27. Carlsson, G.H., Hasse, D., Cardinale, F., Prandi, C., Andersson, I., The elusive ligand complexes of the DWARF14 strigolactone receptor. *J. Exp. Bot.*, 69, 2345–2354, 2018.
28. Mashiguchi, K., Seto, Y., Yamaguchi, S., Strigolactone biosynthesis, transport and perception. *Plant J.*, 105, 335–350, 2021.
29. Zhang, X. *et al.*, F-box proteins in plants. *Annu. Plant Rev.*, 2, 307–327, 2019.
30. Nelson, D.C. *et al.*, F-box protein MAX2 has dual roles in karrikin and strigolactone signaling in Arabidopsis thaliana. *Proc. Natl. Acad. Sci. U.S.A.*, 108, 8897–8902, 2011.
31. Jia, K.-P., Li, C., Bouwmeester, H.J., Al-Babili, S., Strigolactone Biosynthesis and Signal Transduction, in: *Strigolactones-Biology and Applications*, 2019.
32. Waters, M.T. and Nelson, D.C., Karrikin perception and signalling. *New Phytol.*, 237, 1525–1541, 2023.
33. Machin, D.C., Hamon-Josse, M., Bennett, T., Fellowship of the rings: a saga of strigolactones and other small signals. *New Phytol.*, 225, 621–636, 2020.
34. Korek, M. and Marzec, M., An update on strigolactone signaling in plants, in: *Strigolactones*, pp. 53–73, 2024.
35. Soundappan, I. *et al.*, SMAX1-LIKE/D53 family members enable distinct MAX2-dependent responses to strigolactones and karrikins in *Arabidopsis*. *Plant Cell*, 27, 3143–3159, 2015.

36. Abd-Hamid, N.A., Ahmad-Fauzi, M.I., Zainal, Z., Ismail, I., Diverse and dynamic roles of F-box proteins in plant biology. *Planta*, 251, 1–31, 2020.
37. Cheng, X., Ruyter-Spira, C., Bouwmeester, H., The interaction between strigolactones and other plant hormones in the regulation of plant development. *Front. Plant Sci.*, 4, 1–16, 2013.
38. Ha, C.V. *et al.*, Positive regulatory role of strigolactone in plant responses to drought and salt stress. *Proc. Natl. Acad. Sci. U.S.A.*, 111, 851–856, 2014.
39. Ruiz-Lozano, J.M. *et al.*, Arbuscular mycorrhizal symbiosis induces strigolactone biosynthesis under drought and improves drought tolerance in lettuce and tomato. *Plant Cell Environ.*, 39, 441–452, 2016.
40. Ling, F. *et al.*, Effects of strigolactone on photosynthetic and physiological characteristics in salt-stressed rice seedlings. *Sci. Rep.*, 10, 1–8, 2020.
41. Haider, I. *et al.*, The interaction of strigolactones with abscisic acid during the drought response in rice. *J. Exp. Bot.*, 69, 2403–2414, 2018.
42. Min, Z. *et al.*, Transcriptome analysis revealed hormone signaling response of grapevine buds to strigolactones. *Sci. Hortic.*, 283, 1–12, 2021.
43. Kapulnik, Y. *et al.*, Strigolactones interact with ethylene and auxin in regulating root-hair elongation in *Arabidopsis*. *J. Exp. Bot.*, 62, 2915–2924, 2011.
44. Hayward, A., Stirnberg, P., Beveridge, C., Leyser, O., Interactions between auxin and strigolactone in shoot branching control. *Plant Physiol.*, 151, 400–412, 2009.
45. Fang, Z. *et al.*, Strigolactones and Brassinosteroids Antagonistically Regulate the Stability of the D53–OsBZR1 Complex to Determine FC1 Expression in Rice Tillering. *Mol. Plant*, 13, 586–597, 2020.
46. Jiang, L. *et al.*, Strigolactones spatially influence lateral root development through the cytokinin signaling network. *J. Exp. Bot.*, 67, 379–389, 2016.
47. Zha, M., Zhao, Y., Wang, Y., Chen, B., Tan, Z., Strigolactones and Cytokinin Interaction in Buds in the Control of Rice Tillering. *Front. Plant Sci.*, 13, 1–14, 2022.
48. Li, W. *et al.*, Different strategies of strigolactone and karrikin signals in regulating the resistance of *Arabidopsis thaliana* to water-deficit stress. *Plant Signal Behav., 15*, 9, 1789321, 2020.
49. Liu, J. *et al.*, Osmotic stress represses strigolactone biosynthesis in *Lotus japonicus* roots: exploring the interaction between strigolactones and ABA under abiotic stress. *Planta*, 241, 1435–1451, 2015.
50. Kutschera, U. and Niklas, K.J., Evolutionary plant physiology: Charles Darwin's forgotten synthesis. *Naturwissenschaften*, 96, 1339–1354, 2009.
51. Went, F.W., Wuchsstoff und Wachstum. *Recl. Trav. Bot. Néerl.*, 25, 1–116, 1928.
52. Went, F.W. and Thimann, K.V., *Phytohormones*, Macmillan, New York, 1937.
53. Enders, T.A. and Strader, L.C., Auxin activity: Past, present, and future. *Am. J. Bot.*, 102, 180–196, 2015.
54. Marhava, P., Recent developments in the understanding of PIN polarity. *New Phytol.*, 233, 624–630, 2022.

55. Petrášek, J. *et al.*, PIN proteins perform a rate-limiting function in cellular auxin efflux. *Science*, 312, 5775, 914–918, 2006.
56. Shinohara, N., Taylor, C., Leyser, O., Strigolactone Can Promote or Inhibit Shoot Branching by Triggering Rapid Depletion of the Auxin Efflux Protein PIN1 from the Plasma Membrane. *PLoS Biol.*, 11, 1–14, 2013.
57. Zhang, J. *et al.*, Strigolactones inhibit auxin feedback on PIN-dependent auxin transport canalization. *Nat. Commun.*, 11, 1–10, 2020.
58. Chabikwa, T.G., Brewer, P.B., Beveridge, C.A., Initial bud outgrowth occurs independent of auxin flow from out of buds. *Plant Physiol.*, 179, 55–65, 2019.
59. Haq, B.U.I. *et al.*, Functional characterization of soybean strigolactone biosynthesis and signaling genes in *Arabidopsis* MAX mutants and GmMAX3 in soybean nodulation. *BMC Plant Biol.*, 17, 1–20, 2017.
60. Durán-Medina, Y., Díaz-Ramírez, D., Marsch-Martínez, N., Cytokinins on the move. *Front. Plant Sci.*, 8, 1–7, 2017.
61. Dun, E.A., de Saint Germain, A., Rameau, C., Beveridge, C.A., Antagonistic action of strigolactone and cytokinin in bud outgrowth control. *Plant Physiol.*, 158, 487–498, 2012.
62. Cao, D. *et al.*, Auxin-independent effects of apical dominance induce changes in phytohormones correlated with bud outgrowth. *Plant Physiol.*, 192, 1420–1434, 2023.
63. Sugimoto, Y. *et al.*, Germination Strategy of *Striga Hermonthica* Involves Regulation of Ethylene Biosynthesis. *Physiol. Plant.*, 119, 1, 137–145, 2003.
64. Ma, H. *et al.*, A D53 Repression Motif Induces Oligomerization of TOPLESS Corepressors and Promotes Assembly of a Corepressor-Nucleosome Complex. *Sci. Adv.*, 3, 6, e1601217, 2017.
65. Jiang, L. *et al.*, DWARF 53 acts as a repressor of strigolactone signalling in rice. *Nature*, 504, 401–405, 2013.
66. Song, X. *et al.*, IPA1 functions as a downstream transcription factor repressed by D53 in strigolactone signaling in rice. *Cell Res.*, 27, 1128–1141, 2017.
67. Jiao, Y. *et al.*, Regulation of OsSPL14 by OsmiR156 defines ideal plant architecture in rice. *Nat. Genet.*, 42, 541–544, 2010.
68. Sun, H. *et al.*, Strigolactones interact with other phytohormones to modulate plant root growth and development. *J. Crop Sci.*, 10, 1517–1527, 2022.
69. Duan, J. *et al.*, Strigolactone promotes cytokinin degradation through transcriptional activation of CYTOKININ OXIDASE/ DEHYDROGENASE 9 in rice. *Proc. Natl. Acad. Sci. U.S.A.*, 116, 14319–14324, 2019.
70. Aguilar-Martínez, J.A., Poza-Carrión, C., Cubas, P., *Arabidopsis* Branched1 acts as an integrator of branching signals within axillary buds. *Plant Cell*, 19, 458–472, 2007.
71. Lan, J. *et al.*, *Arabidopsis* TCP4 transcription factor inhibits high temperature-induced homeotic conversion of ovules. *Nat. Commun.*, 14, 1–19, 2023.
72. Ishizaki, T., Ueda, Y., Takai, T., Maruyama, K., Tsujimoto, Y., In-frame mutation in rice TEOSINTE BRANCHED1 (OsTB1) improves productivity under phosphorus deficiency. *Plant Sci.*, 330, 1–12, 2023.

73. Wang, L. *et al.*, Transcriptional regulation of strigolactone signalling in *Arabidopsis*. *Nature*, 583, 277–281, 2020.
74. López-Maury, L., Marguerat, S., Bähler, J., Tuning gene expression to changing environments: From rapid responses to evolutionary adaptation. *Nat. Rev. Genet.*, 9, 583–593, 2008.
75. Litchman, E. and Thomas, M.K., Are we underestimating the ecological and evolutionary effects of warming? Interactions with other environmental drivers may increase species vulnerability to high temperatures. *Oikos*, 2023, 2, e09155, 2023.
76. Asif, R. *et al.*, Phytohormones as Plant Growth Regulators and Safe Protectors against Biotic and Abiotic Stress, in: *Plant Horm. Recent Adv. New Perspect. Appl.*, pp. 115–130, 2022.
77. Marzec, M. and Melzer, M., Regulation of root development and architecture by strigolactones under optimal and nutrient deficiency conditions. *Int. J. Mol. Sci.*, 19, 1–12, 2018.
78. Kapulnik, Y. *et al.*, Strigolactones affect lateral root formation and root-hair elongation in *Arabidopsis*. *Planta*, 233, 209–216, 2011.
79. Marzec, M., Gruszka, D., Tylec, P., Szarejko, I., Identification and functional analysis of the HvD14 gene involved in strigolactone signaling in *Hordeum vulgare*. *Physiol. Plant.*, 158, 341–355, 2016.
80. López-Bucio, J., Cruz-Ramírez, A., Herrera-Estrella, L., The role of nutrient availability in regulating root architecture. *Curr. Opin. Plant Biol.*, 6, 280–287, 2003.
81. Ruyter-Spira, C. *et al.*, Physiological effects of the synthetic strigolactone analog GR24 on root system architecture in *Arabidopsis*: Another below-ground role for strigolactones? *Plant Physiol.*, 155, 721–734, 2011.
82. Darwish, D.B.E. *et al.*, Constitutive overexpression of GsIMaT2 gene from wild soybean enhances *Rhizobia* interaction and increase nodulation in soybean (*Glycine max*). *BMC Plant Biol.*, 22, 1–18, 2022.
83. Naumann, G., Cammalleri, C., Mentaschi, L., Feyen, L., Increased economic drought impacts in Europe with anthropogenic warming. *Nat. Clim. Chang.*, 11, 485–491, 2021.
84. Mashiguchi, K. *et al.*, Feedback-regulation of strigolactone biosynthetic genes and strigolactone-regulated genes in *Arabidopsis*. *Biosci. Biotechnol., Biochem.*, 73, 2460–2465, 2009.
85. Koltai, H. *et al.*, Light is a positive regulator of strigolactone levels in tomato roots. *J. Plant Physiol.*, 168, 1993–1996, 2011.
86. Ge, S. *et al.*, Light-dependent activation of HY5 promotes mycorrhizal symbiosis in tomato by systemically regulating strigolactone biosynthesis. *New Phytol.*, 233, 1900–1914, 2022.
87. Chi, C. *et al.*, Strigolactones positively regulate abscisic acid-dependent heat and cold tolerance in tomato. *Hortic. Res.*, 8, 1–15, 2021.
88. Seneviratne, S.I. *et al.*, Weather and climate extreme events in a changing climate, in: *Climate Change 2021 – The Physical Science Basis: Working Group*

*I Contribution to the Sixth Assessment Report of the Intergovernmental Panel on Climate Change*, pp. 1513–1766, Cambridge University Press, 2023.
89. Roberts, D.P. and Mattoo, A.K., Sustainable agriculture—Enhancing environmental benefits, food nutritional quality and building crop resilience to abiotic and biotic stresses. *Agriculture (Switzerland)*, 8, 1–24, 2018.
90. Siegel, F.R., *The Earth's Human Carrying Capacity: Limitations Assessed, Solutions Proposed*, Springer Nature, Switzerland, 2021.
91. FAO, Water Use, AQUASTAT - FAO's Global Information System on Water and Agriculture, https://www.fao.org/aquastat/en/overview/methodology/water-use.
92. van Ginkel, M. and Biradar, C., Drought early warning in agri-food systems. *Climate*, 9, 1–23, 2021.
93. Rivero, R.M., Mittler, R., Blumwald, E., Zandalinas, S.I., Developing climate-resilient crops: improving plant tolerance to stress combination. *Plant J.*, 109, 373–389, 2022.
94. Wanga, M.A., Shimelis, H., Mashilo, J., Laing, M.D., Opportunities and challenges of speed breeding: A review. *Plant Breed.*, 140, 185–194, 2021.
95. Tang, G., Jian, Q., Dolnikowski, G.G., Russell, R.M., Grusak, M.A., Golden rice is an effective source of vitamin A. *Am. J. Clin. Nutr.*, 89, 1776–1783, 2009.
96. Song, P. *et al.*, The prevalence of vitamin A deficiency and its public health significance in children in low and middle-income countries: A systematic review and modelling analysis. *J. Glob. Health*, 13, 1–11, 2023.
97. Dubock, A., Golden Rice: To Combat Vitamin A Deficiency for Public Health, in: *Vitamin A*, IntechOpen, Rijeka, Croatia, 2019.
98. Aly, R., Matzrafi, M., Bari, V.K., Using biotechnological approaches to develop crop resistance to root parasitic weeds. *Planta*, 253, 1–11, 2021.
99. Vurro, M., Prandi, C., Baroccio, F., Strigolactones: how far is their commercial use for agricultural purposes? *Pest Manag. Sci.*, 72, 2026–2034, 2016.
100. Chesterfield, R.J., Vickers, C.E., Beveridge, C.A., Translation of Strigolactones from Plant Hormone to Agriculture: Achievements, Future Perspectives, and Challenges. *Trends Plant Sci.*, 25, 1087–1106, 2020.
101. Morgan, J.A.W., Bending, G.D., White, P.J., Biological costs and benefits to plant-microbe interactions in the rhizosphere. *J. Exp. Bot.*, 56, 1729–1739, 2005.
102. Sharma, Y.P., Watpade, S., Thakur, J.S., Role of mycorrhizae: a component of integrated disease management strategies. *J. Mycol. Plant Pathol.*, 44, 1, 12–20, 2014.
103. Busso, M.A. and Busso, M.B., Arbuscular mycorrhizal fungi and common mycorrhizal networks benefit plants through morphological, physiological and productive traits and soil quality. *Lilloa*, 59, 301–317, 2022.
104. Barten, T.J. *et al.*, Short-stature maize reduced wind damage during the 2020 midwestern derecho, improving yields and greenhouse gas outcomes. *Crop Sci.*, 62, 2439–2450, 2022.

105. Andorf, C. *et al.*, Technological advances in maize breeding: past, present and future. *Theor. Appl. Genet.*, 132, 817–849, 2019.
106. Glenn, K.C. *et al.*, Bringing new plant varieties to market: Plant breeding and selection practices advance beneficial characteristics while minimizing unintended changes. *Crop Sci.*, 57, 2906–2921, 2017.
107. Tester, M. and Langridge, P., Breeding technologies to increase crop production in a changing world. *Science*, 327, 818–822, 2010.
108. Schlaepfer, M.A. and Lawler, J.J., Conserving biodiversity in the face of rapid climate change requires a shift in priorities. *Wiley Interdiscip. Rev. Clim. Change*, 14, 11–14, 2023.
109. Salgotra, R.K. and Chauhan, B.S., Genetic Diversity, Conservation, and Utilization of Plant Genetic Resources. *Genes*, 14, 1–20, 2023.
110. Fay, M.F., Orchid conservation: how can we meet the challenges in the twenty-first century? *Bot. Stud.*, 59, 1–6, 2018.
111. Rasmussen, H.N., *Terrestrial Orchids: From Seed to Mycotrophic Plant*, Cambridge University Press, Cambridge, United Kingdom, 1995.
112. Schiebold, J.M.I., Bidartondo, M.I., Lenhard, F., Makiola, A., Gebauer, G., Exploiting mycorrhizas in broad daylight: Partial mycoheterotrophy is a common nutritional strategy in meadow orchids. *J. Ecol.*, 106, 168–178, 2018.
113. Hua, Z. *et al.*, The Armillaria response to *Gastrodia elata* is partially mediated by strigolactone-induced changes in reactive oxygen species. *Microbiol. Res.*, 278, 127536, 2024.
114. Leng, C., Hou, M., Xing, Y., Chen, J., Perspective and challenges of mycorrhizal symbiosis in orchid medicinal plants. *Chin. Herb. Med.*, 16, 172–179, 2024.
115. Wu, F., Gao, Y., Yang, W., Sui, N., Zhu, J., Biological Functions of Strigolactones and Their Crosstalk With Other Phytohormones. *Front. Plant Sci.*, 13, 1–10, 2022.
116. Dun, E.A., Brewer, P.B., Gillam, E.M.J., Beveridge, C.A., Strigolactones and Shoot Branching: What Is the Real Hormone and How Does It Work? *Plant Cell Physiol.*, 64, 967–983, 2023.
117. Mori, N. *et al.*, Chemical identification of 18-hydroxycarlactonoic acid as an LjMAX1 product and in planta conversion of its methyl ester to canonical and non-canonical strigolactones in *Lotus japonicus*. *Phytochem.*, 174, 1–16, 2020.
118. Kim, B. *et al.*, Effect of strigolactones on recruitment of the rice root-associated microbiome. *FEMS Microbiol. Ecol.*, 98, 1–12, 2022.
119. Janeeshma, E. *et al.*, Strigolactone and analogues: a new generation of plant hormones with multifactorial benefits in environmental sustainability. *Environ. Exp. Bot.*, 223, 105775, 2024.
120. Mdodana, N.T., The effects of the synthetic strigolactone GR24 on Arabidopsis thaliana callus culture, (Master's Thesis), University of Stellenbosch, Stellenbosch, South Africa, 2012.

121. Dong, X. *et al.*, Effects of exogenous strigolactone on growth, dry matter accumulation, and yield of tissue-cultured ginger. *Plant Prod. Sci.*, 26, 225–235, 2023.
122. Ur Rehman, N. *et al.*, Harmony but not uniformity: Role of strigolactone in plants. *Biomolecules*, 11, 1–24, 2021.
123. Arellano-Saab, *et al.*, Three mutations repurpose a plant karrikin receptor to a strigolactone receptor. *Proc. Natl. Acad. Sci.*, 118, 30, 1–8, 2021.

# 4

# Diverse Roles of Strigolactones in Plant Growth and Development: Shaping Above- and Below-Ground Architecture

**Alisha Hussain*, Kaiser Iqbal Wani and Shahla Faizan**

*Department of Botany, Faculty of Life Sciences, Aligarh Muslim University, Aligarh, India*

***Abstract***
Strigolactones (SLs), functioning as plant hormones, play diverse roles in regulating plant growth and architecture in both shoots and roots. Derived from carotenoids, this group of phytohormones is involved in a broad spectrum of physiological functions. In addition to regulating processes like photomorphogenesis, seed germination, nodulation, and the plant's reaction to diverse abiotic stimuli, these roles also encompass the regulation of plant architecture, such as preventing bud expansion and shoot branching. This chapter will focus on the growing body of evidence indicating the involvement of SLs in the regulation of shoot and root architecture, investigating their role in orchestrating the development of shoots and roots under different growth circumstances. Furthermore, the involvement of SLs in regulating shoot secondary growth and leaf senescence is also discussed.

***Keywords*:** Strigolactone, roots, shoot, senescence, *arabidopsis*, auxin, cytokinin

## 4.1 Introduction

Strigolactones (SLs) are a class of plant hormones that govern plant growth and development and the adaptation mechanisms of plants. SLs belong to the carotenoid derivative class, distinguished by a butenolide ring linked via an enol ether bridge to a less uniform secondary moiety. These signaling

**Corresponding author*: alishahussain302@gmail.com

Tariq Aftab and Kaiser Iqbal Wani (eds.) Strigolactones: Emerging Plant Hormones, (69–88) © 2025 Scrivener Publishing LLC

molecules were initially discovered for their role in the stimulation of seed germination of root parasitic weeds. Later on, their role in the regulation of shoot branching was discovered in 2008, making them eligible to be called hormones, but their influence extends to several other processes. These substances act as chemical signals when released into the soil, drawing in symbiotic arbuscular mycorrhizal fungus and promoting the development of seeds of root-parasitic weeds.

Strigolactones, named for their role in promoting the germination of root-parasitic plants (*Striga* and *Orobanche*), are captivating compounds with dual functions as plant hormones [1–3] and signals for inter-species communication [4, 5]. SLs play a regulatory role in several aspects of plant development, involving the inhibition of shoot branching (their most well-known hormonal function), modulation of root architecture, promotion of leaf senescence, and control of secondary growth [6–8].

Strigolactones play a dual role as signaling molecules within plants and in their surrounding environment. In the rhizosphere, SLs facilitate the interaction between plant roots and arbuscular mycorrhizal fungi (AMF), especially in environments deficient in nutrients [9]. AMF are symbiotic fungi found in the roots of terrestrial plants, forming mutually beneficial relationships with crop plants. They provide mineral nutrients to plants in exchange for carbon fixed during photosynthesis [10–12].

## 4.2 The Contribution of Strigolactones to Shoot Development

Plant architecture is greatly influenced by branching patterns, which are defined as the quantity, position, and size of lateral branches. By preventing the development of axillary buds along the stem, the main shoot apex controls branching (Figure 4.1). All leaf axils contain axillary meristems [13]; however, they typically only initiate a few leaf primordia before becoming dormant until the main shoot apex is removed (or inactivated), after which they become active. Additionally, they might spread out in response to either endogenous or external developmental cues. Apical dominance is the term used to describe this phenomenon.

The fate of axillary buds is tightly regulated throughout a plant's life cycle. Complicated signaling networks control whether a bud stays dormant or develops into a branch or tiller. These networks are impacted by both genetic and environmental factors. Even in genetically identical plants, the distinct control over these axillary buds adds to the diversity of plant shoot architecture. It is a crucial agronomic trait that contributes to a plant's overall shoot structure and is a potential target for optimizing yield in various crops. Therefore, understanding the regulation and genetic control of shoot branching is essential [14].

The observation of hyperbranching mutants led to the first identification of SLs as plant hormones. These mutants displayed changes in the quantity of a graft-transmissible signal that suppressed shoot branching. Since this phenotype could not be linked to the modifications in established plant hormones, it hinted at the presence of a novel signal associated with the observed effects [15]. Subsequently, this signal was identified as SL, serving as long-distance branching factors that prevent the growth of preformed axillary buds. This inhibition is dependent on the F-box protein MAX2 signaling [1, 2], inducing a state of active dormancy in metabolically active buds. However, it remains unclear whether axillary buds in SL mutant plants ever undergo dormancy or if they consistently proceed to grow. Despite the prevalence of non-growing axillary buds in SL mutant plants, it suggests the existence of additional dormancy mechanisms [16].

Strigolactones have been observed to regulate bud outgrowth both locally and over long distances. They have the ability to directly prevent young pea buds from growing [17, 18], potentially acting as a secondary messenger encouraged by auxin to suppress bud outgrowth [19–21]. Another proposed mechanism suggests that SLs reduce the ability of the main shoot to transport auxin from the apical meristem, which is crucial for activating buds. Consequently, this drop in auxin transport mediated by SLs may inhibit polar auxin transport from the buds, thus limiting their outgrowth [22–26]. *Arabidopsis* SL mutants, which exhibit hyperbranching and increased polar auxin transport, provide important evidence for the involvement of polar auxin transport in bud activation [26]. Systemic correlated inhibition of bud development, which is mediated by auxin and SL through a process known as auxin canalization, is the primary feature of apical dominance [27]. On the other hand, axillary bud activation is influenced by inducing signals like cytokinin and sugars [28]. Excess phosphate presence in the soil inhibits SL production in plants which in turn becomes a major determining factor for bud activation in plants (Figure 4.1).

### 4.2.1 Bud Activation Dynamics in *Arabidopsis* via Auxin–Strigolactone Interplay

The young, growing leaves at the primary apex produce high quantities of auxin, which are carried down the stem but do not enter buds, preventing bud activation. Two nonexclusive pathways, which involve signal integration, have been proposed to explain the indirect action of auxin.

First, there is a wide passive auxin flux from an auxin source to a sink that self-organizes into narrow files of cells with a high rate of auxin flux. This is based on Sachs' auxin transport canalization hypothesis, which claims that auxin flow orientates and reinforces itself by polarising auxin transport along the flow axis [29]. The canalization of auxin transport from the bud into the main stem appears to be linked with and necessary for sustained growth during axillary bud activation (Figure 4.1).

Auxin is transported via the main stem (acting as an auxin sink) in the polar auxin transport stream towards the root, while young growing leaves in the axillary buds and at the apex of the primary shoot act as auxin sources. The main stem turns into a poor auxin sink due to the high auxin content caused by the presence of an active apex which blocks the

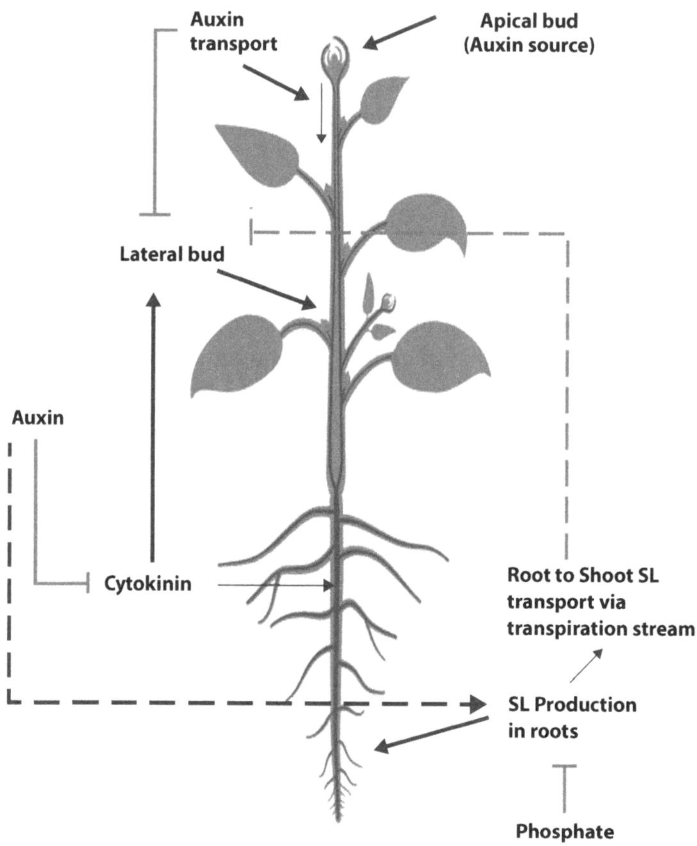


**Figure 4.1** Strigolactones interplay with auxins and cytokinins in the control of lateral shoot growth via regulation of lateral bud activation.

canalization of auxin transport from the buds into the stem [30]. This indirectly suppresses bud activity. Auxin levels in the main stem are decreased during decapitation, which makes the stem a potent auxin sink and permits auxin transport out of the bud, hence restoring high auxin levels in the main stem. As a result, apices compete over the use of a shared auxin sink in the primary stem [31]. Auxin induces a positive feedback loop on its own transport by stimulating the accumulation of polarised PIN-FORMED1 (*PIN1*) auxin exporters, lending credibility to this theory [32]. This might allow buds to transition from a non-canalized, inactive state to a canalised, active state [30].

In contrast to auxin transport, which is systemic, the transcription factor *BRANCHED1* (*BRC1*) controls shoot branching locally in buds [33]. Mutations in *BRC1* lead to branched plant morphology. Bud-activating cues, including decapitation, light quality, and basal cytokinin, cause *BRC1* to be rapidly downregulated in buds and accelerate their growth.

SL has an effect on both the auxin transport network and *BRC1*, which are two regulatory centres. *PIN1* and potentially *PIN7* are eliminated from the plasma membrane by SL [34] while increasing *BRC1* expression via D14 signaling [35]. This is expected to reduce the positive feedback of auxin flux on PIN accumulation at the plasma membrane, lowering bud activation potential [36]. This is anticipated to prevent or hinder the movement of auxin from the bud to the stem [25, 37].

Auxin in the main stem reduces bud activity by boosting the two mechanisms of SL action. Additionally, auxin stimulates SL biosynthesis [38], and strigolactone travels up the plant, most likely via the transpiration stream [39]. The auxin transporter ABCB19 may also serve as a link between these two centres for the regulation of shoot branching. According to van Rongen *et al.* [34], the ABCB19 mutation inhibits branching in both brc1 and strigolactone mutant buds. Auxin source strength in the bud may be increased by BRC1 adjusting the auxin loading that ABCB19 loads into the auxin transport stream in the bud.

## 4.3 Role of Strigolactones in Regulating Shoot Secondary Growth

The secondary growth of shoots engages lateral expansion of the shoot axes, resulting in increased girth. The vascular cambium, a structure resembling stem cells that produce secondary vascular tissues and encourage the growth of wood, is the driving force behind this process. Auxin and other hormones have an impact on the control of secondary growth from the vascular cambium [40].

Auxins and SLs engage in a distinctive interplay, with auxins governing SL production and influencing various developmental processes mediated by SLs [25, 41]. On the other hand, SLs increase auxin transport in plant stems, encouraging axillary branch and bud growth to compete with one another [26]. Acting as mediators, SLs interfere with auxin-induced secondary messengers, impeding bud outgrowth [42]. Additionally, auxins control the genes that produce SLs, and SLs are essential for auxin to prevent decapitation-induced shoot branching [43].

SLs play a crucial role in promoting secondary shoot growth by positively influencing cambial activity (Figure 4.2). This is accomplished via the local activation of the stem cell niche unique to the cambium and the development of vascular tissue. These findings were documented in pea, *Arabidopsis* and *Eucalyptus*, and the mechanism requires the same MAX2-dependent signaling pathway that prevents branching of the shoots [6]. The impact of SLs on secondary shoot growth is proposed to be localized and distinct from their influence on shoot branching [6]. Furthermore, experiments involving *max2-1* mutants expressing *MAX2* under the control of the cambium-specific *WUSCHEL-RELATED HOMEOBOX 4 (WOX4)* promoter illustrated that this expression was adequate to stimulate secondary growth at levels comparable to the wild type (WT).

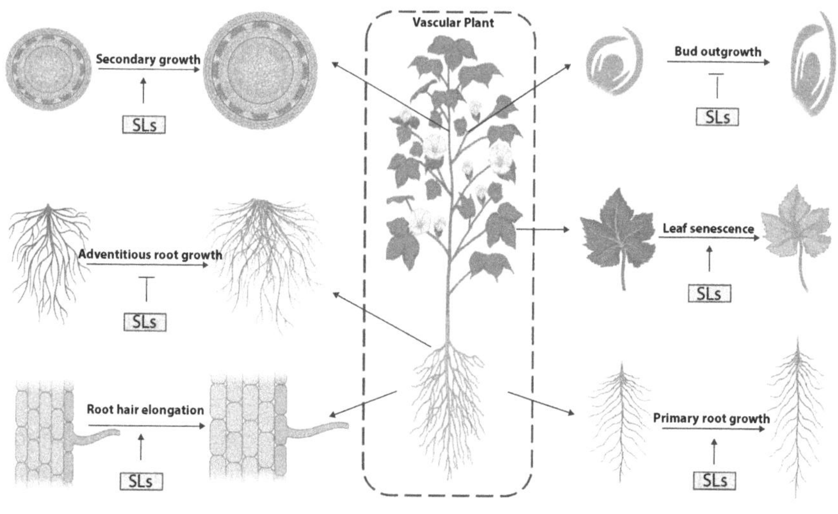


**Figure 4.2** Regulation of above- and below-ground architecture in plants by strigolactones.

This suggests a localized, cambium-specific, MAX2-dependent activity of SLs [6]. Therefore, Thus, SL signaling may operate as a secondary auxin messenger by controlling the cambium's secondary growth process in a cell-autonomous way [6].

## 4.4 Root Development in Plants: Potential Role of Strigolactones

According to a number of studies, there is increasing evidence that SLs are involved in controlling root growth in different plant species. Their main function is to regulate the movement of auxins within the main stem which may affect PR growth and LR initiation by influencing the lateral movement of these hormones in the elongation zone of roots [8, 25]. These chemicals promote the elongation of primary roots (PRs) as well as adventitious roots (ARs) while inhibiting lateral root (LR) formation simultaneously [8, 44]. In SL-deficient rice plants, crown root length is shorter than that one in wild-type (WT) plants [8] and externally applied auxin-mediated inhibition of root elongation was antagonized by SLs in tomato plant [45].

Studies on *Arabidopsis* mutants deficient in SLs have revealed that their PRs were shorter than those of typical plants because there were fewer meristem cells. However, the use of GR24, a synthetic SL counterpart, negated this effect [8, 46]. The meristem and elongation zones expanded as a result of the administration of GR24, which was linked to an increase in PR length and a decrease in lateral auxin transfer from the epidermis into the polar auxin transport (PAT) stream [47]. These findings suggest that SLs are involved in the regulation of lateral auxin reflux. The elongation zone of the root is a critical region for adjusting root growth in response to developmental stages and environmental cues, influenced not only by SLs but also by other hormones, making it a focal point for regulatory processes [8].

### 4.4.1 Primary Root Development

The root apical (RA) meristem, which is found at the root tip and is where actively dividing cells give rise to all PR tissues, controls the growth of the PR. The RA meristem, which is made up of a stem cell niche, a proximal and distal meristem, and controls cell division, elongation, and differentiation during root growth. Phytohormones, particularly auxin, play a central role in this process. Auxin is primarily imported from the shoot, but local biosynthesis also occurs in roots [48–51]. The auxin concentration gradient along the root meristem's longitudinal axis affects root growth. Auxin transporters, including as inflow carriers like AUX1 and LIKE-AUX1 and efflux carriers such *PIN*s and ABC transporters, work together to create this gradient [32, 35, 36]. Auxins can be more easily transported to the root tip of the PR due to the basal locations of *PIN1*, *PIN3*, and *PIN7* in the stele [52]. Within the root columella, *PIN3* and *PIN7* redirect auxin flow towards the epidermis, where *PIN2* aids in transporting auxins upwards into the elongation zone [52]. Additionally, a high auxin concentration is maintained at the root tip by *PIN2* in the cortex, which controls auxin flux both shootward and rootward [53].

The studies on the role of SLs in influencing PR length exhibit inconsistency, influenced by various environmental factors and sometimes showing species-specific effects [54]. When *rac*-GR24 is applied to SL-biosynthesis mutants (*max1, max3*, and *max4*), PR length rises under normal sugar and phosphate circumstances, but not in SL-signaling mutants (*max2*). This reaction is reliant on MAX2 [8, 55]. There is a correlation between the number of cortical cells in the PR meristem and transition zone and the increased PR length following GR24 therapy [8, 56].

Furthermore, GR24's stimulatory effect on PR length is evident under carbohydrate-limiting conditions, specifically in SL-biosynthesis mutants, confirming its MAX2-dependent role [57]. Under carbon-limiting conditions, both SL biosynthesis and signaling mutants in Arabidopsis exhibit shorter PRs compared to *WT* plants [57]. Similarly, under restricted phosphate and nitrate conditions, rice with SL biosynthetic and signaling mutants have short seminal roots; however, GR24 therapy only lengthens the roots of biosynthetic mutants—not those of signaling mutants [58]. Conversely, GR24 application increases PR length in SL-biosynthetic mutants of *Medicago truncatula* as well as tomato but has no impact on WT plants [45, 59]. However, in *Lotus japonicus*, silencing of the MAX3 ortholog causes a delay in PR development [60]. Overall, SLs have a positive regulatory effect on PR elongation (Figure 4.2); however, the extent of this effect differs between species and is dependent on growth circumstances.

### 4.4.2 Lateral Root Initiation and Development

LR formation initiates through auxin-controlled divisions in primed founder cells. SL-deficient *Arabidopsis* plants (*max1*, *max3*, *max4*, and *d14*) show longer LR lengths and larger LR counts under normal growth conditions than WT plants. Conversely, fewer LRs and shorter LR lengths are seen in mutant *smxl6,7,8* plants, suggesting that SLs have a negative regulatory role in LR formation and growth [8, 45, 56, 61–63]. Additionally, *d10* mutant rice plants exhibit the induction of secondary LRs (roots derived from primary LRs), whereas WT rice plants do not exhibit this phenomenon, indicating a potential negative control of secondary LR development by SL signaling in rice [49]. These findings suggest that SLs exert a negative regulatory influence on LR development in both grasses and eudicot plants under typical growth conditions [8, 45, 56, 61–63].

In *Arabidopsis*, the application of GR24 decreased the density of LRs in both WT plants and SL-deficient mutants *max1* and *max4* under conditions with sufficient phosphate (Pi) levels [45]. However, in contrast to SL-deficient *max1* and *max4* mutants, GR24 therapy enhanced LR development in WT *Arabidopsis* plants under Pi-deficient circumstances [45]. These results suggest that the impacts of SLs on LR formation in *Arabidopsis* are influenced by environmental Pi levels and may be related to the Pi-dependent auxin state.

Subsequent research in rice revealed that, in normal circumstances, there were no variations in LR density between WT plants and SL-related *d10* and *d27* mutants. However, in comparison to the *d10* and *d27* mutant plants, WT rice plants showed a much lower LR density under Pi-deficient circumstances [58, 64]. Moreover, LR density in SL-deficient *d10* and *d27* mutants was lowered to levels comparable to untreated WT plants by applying exogenous GR24. WT plants had somewhat lower LR densities than *d10* mutants under nitrogen (N) deficiency, and GR24 administration to SL-synthetic *d10* mutants produced a drop in LR density comparable to that seen in WT plants. These findings point to SLs playing a somewhat detrimental function in rice LR growth when Pi or N levels are low. Therefore, in the context of LR development under Pi-deficient conditions, SLs exhibit a positive regulatory role in *Arabidopsis* but a negative regulatory role in rice, indicating a dependence on nutrient conditions and plant species.

### 4.4.3 Root Hair Elongation

In both *Arabidopsis* and tomatoes, *rac*-GR24 has been found to stimulate the elongation of root hairs (RH) [7, 45, 65, 66]. However, in comparison to WT plants, neither SL-deficient nor SL-insensitive mutants show shorter RHs, indicating that variables other than SLs also affect total RH elongation [7, 67, 68]. Interestingly, under low phosphate (P) conditions, both *max2* and *max4* mutants showed a reduction in RH number and density [70]. SLs also function as positive regulators of RH elongation (Figure 4.2), as evidenced by the fact that high dosages of rac-GR24 could counteract this effect.

More investigation is still required to understand how SLs affect different facets of root formation completely. It is evident that root growth and development result from a complex interplay among multiple factors, with SLs being just one component. This complexity is highlighted by the varying responses of plants to SL treatments under different nutritional conditions. Additionally, other phytohormones play crucial roles in regulating root development [65, 89]. Therefore, nutrient status and interactions with other phytohormones must be taken into account in order to thoroughly comprehend the effects of SLs.

### 4.4.4 Adventitious Root Formation

ARs can emerge from any non-root plant tissue, either as part of normal development, like crown roots in rice, or in response to stresses like damage or nutrient scarcity. In *Arabidopsis* and pea, mutants lacking SLs or insensitive to them exhibit increased AR formation compared to wild-type plants [71, 72]. In wild-type and SL biosynthetic mutants, treatment with the synthetic SL analogue rac-GR24 decreases AR formation; however, this effect is not observed in perception mutants [69, 71, 72].

Rice plants possess two root types: seminal roots, which function in early seedling stages, and crown roots, which become the primary functional roots in mature plants. Crown roots, arising from stem nodes both above and below ground, are considered shoot-borne ARs. SL-deficient rice mutants show fewer and shorter crown roots than wild-type plants, in contrast to *Arabidopsis* and pea [58]. The application of GR24 can rescue this phenotype. This discrepancy in SL effects on roots might stem from the different roles of ARs across species. In rice, where crown roots are crucial for plant function, SLs influence them similarly to how they affect PRs in species with PR-based systems. Investigating SL effects on other forms of ARs, like nodal roots and prop roots, could shed further light on this. However, the effect of SLs on the start of roots from non-root tissues differs greatly throughout species; in *Arabidopsis* and pea, they function as negative regulators of adventitious rooting, but in rice, they function as positive regulators of crown root formation (Figure 4.2).

### 4.4.5 Root Development in Response to Nitrogen and Phosphate Deficiencies

The availability of nutrients in the soil is necessary for the typical growth and development of plants. Among the different nutrients, phosphorus (P) and nitrogen (N) are crucial to maintain normal plant growth and development. SLs play a significant role in controlling root development in response to nitrogen and phosphate deficiencies. In such conditions, plants exhibit increased SL levels in their roots, which trigger the branching of arbuscular mycorrhizal (AM) fungal hyphae and suppress bud outgrowth as an adaptation to nutrient deficiencies [2, 73–76]. Mutants of *Arabidopsis* deficient in SL biosynthesis and signaling display reduced responsiveness to phosphate limitation compared to wild-type plants [70]. However, this diminished response was restored when GR24, a synthetic SL analogue, was applied to SL biosynthesis mutants but not to signaling mutants. It has been proposed that SLs primarily regulate plant architecture, particularly under phosphate-deficient conditions. Under conditions of P deficiency, plants treated with SLs tend to produce more LRs, whereas SL-deficient plants exhibit fewer LRs. In *Medicago truncatula*, deficiencies in both P and N result in the upregulation of SL biosynthetic genes such as *CCD7*, *CCD8*, *D27*, and *MAX1* [77]. Yoneyama *et al.* studied six different plant species in order to investigate the effect of P and N scarcity on the biosynthesis and exudation of SLs [78]. According to their research, P deprivation increased the production of SLs in the other four plant species, but neither lucerne nor tomato showed any alterations in N-deficient conditions.

### 4.4.6 Strigolactone and Auxin Interplay During Root Development

The interaction and cross-talk between SLs and other phytohormones, particularly auxin, have been extensively studied. One prevailing hypothesis suggests that SLs exert an indirect influence by inhibiting the activity of PIN1, thereby altering the sink–source dynamics and reducing auxin export from buds [79]. Similarly, in a similar vein, SLs may control auxin flux in roots to control the elongation of root hair (RH). According to Koltai *et al.*, when exogenously supplied auxin is present, SLs may have a detrimental effect on the activity of auxin efflux carriers [45]. It has been noted that in the vascular tissue of the root tip, the administration of *rac*-GR24 reduces the expression of *PIN1*, *PIN3*, and *PIN7* [8]. RHs elongate independently of auxin transport in response to elevated auxin concentrations in epidermal cells [80]. This auxin transport through non-hair epidermal cells is facilitated by AUX1 [65, 81], suggesting that SL enhancement of root hair elongation may function via auxin [7].

Changes in LR density in *Arabidopsis* are also linked to the plant's auxin status [8]. When auxin concentrations are relatively low, treatments with *rac*-GR24 lead to a reduction in LR density. However, under conditions of higher auxin concentrations, the addition of *rac*-GR24 results in an increased LR density [8, 65]. The SL regulation of auxin flux through *PIN1* may account for this result. While the reduction of auxin levels by *rac*-GR24 may push the flux into the optimal range for LR development under high auxin conditions, it may also decline below the optimal concentration for LR development under reduced auxin conditions [8].

### 4.4.7 Strigolactone and Cytokinin Interplay During Root Development

CKs decrease root meristem size and activity by acting as negative regulators of PR elongation in *Arabidopsis* and LR development in *Arabidopsis* and rice [82]. It has been revealed that CKs and SLs interact to control root development and growth in rice and *Arabidopsis.* Lateral root formation in *Arabidopsis* plants has been reported to be insensitive to GR24 treatments due to mutations in cytokinin signaling components, such as *Arabidopsis* Histidine Kinase3 (ahk3), Arabidopsis Response Regulator1 (ARR1), and ARR12 [63]. This suggests that SL signaling negatively regulates LR formation by affecting cytokinin signaling. The transcriptional activation of cytokinin OXIDASE/DEHYDROGENASE 9 (CKX9) by GR24 has been demonstrated to enhance CK degradation in shoots of WT plants, but not in SL-specific d53 mutant shoots of rice, suggesting an SL–cytokinin metabolic interplay.

## 4.5 Role in Leaf Senescence

The process of leaf senescence, which denotes the end of leaf growth, is complex and impacted by a number of variables, such as age, flowering, dark treatment, nutrition shortage, and stressors [83]. Nutrients are transferred from older to younger tissues during leaf senescence [84]. A number of phytohormones are essential in controlling the senescence of leaves: cytokinins prevent the process, but ABA, salicylic acid, jasmonic acid, and ethylene can accelerate it [85]. Additionally, SLs have been identified as regulators of leaf senescence, as evidenced by the retardation of senescence observed in both SL-deficient and SL-insensitive mutants. Notably, SLs accelerate leaf senescence (Figure 4.2), as demonstrated by experiments involving the application of GR24 to the leaves of rice and *Arabidopsis* [86, 87]. In SL-deficient mutants of rice (*d27*, *d17*, and *d10*) and *Arabidopsis* (*max1*, *max2*, and *max3*), exogenous administration of GR24 induced leaf senescence [88].

## 4.6 Conclusion

With their diverse biological functions, SLs hold significant potential as a valuable tool for developing new agricultural methodologies and technologies in line with the evolving principles of sustainable agriculture. In the near future, it is expected that novel physiological impacts of SLs on root and shoot architectures will be discovered, providing insight into how they help plants become more resilient to environmental challenges, such as climate change. For instance, using SL inhibitors to encourage rooting has demonstrated potential in addressing the issues woody plants encounter with adventitious rooting and making it easier to propagate woody plants for use in industry and conservation [71]. Additionally, by modulating root branching, SLs have the potential to influence root architecture, which could be utilized to develop root systems with desired characteristics, such as hyperbranched root systems to enhance tolerance to nutrient deficiency or deeper roots to improve water use efficiency.

## References


1. Gomez-Roldan, V., Fermas, S., Brewer, P.B., Puech-Pagès, V., Dun, E.A., Pillot, J.P., Rochange, S.F., Strigolactone inhibition of shoot branching. *Nature*, *455*, 7210, 189–194, 2008.
2. Umehara, M., Hanada, A., Yoshida, S., Akiyama, K., Arite, T., Takeda-Kamiya, N., Yamaguchi, S., Inhibition of shoot branching by new terpenoid plant hormones. *Nature*, *455*, 7210, 195–200, 2008.
3. Xie, X., Yoneyama, K., Yoneyama, K., The strigolactone story. *Annu. Rev. Phytopathol.*, 48, 93–117, 2010.
4. Akiyama, K., K-i, M., Hayashi, H., Plant sesquiterpenes induce hyphal branching in arbuscular mycorrhizal fungi. *Nature*, 435, 824–827, 2005.
5. Cook, C.E., Whichard, L.P., Turner, B., Wall, M.E., Egley, G.H., Germination of witchweed (*Strigalutea* Lour.): isolation and properties of a potent stimulant. *Science*, 154, 1189–1190, 1966.
6. Agusti, J., Herold, S., Schwarz, M., Sanchez, P., Ljung, K., Dun, E.A., Brewer, P.B., Beveridge, C.A., Sieberer, T., Sehr, E.M., Greb, T., Strigolactone signaling is required for auxin-dependent stimulation of secondary growth in plants. *PNAS USA*, 108, 20242–20247, 2011.
7. Kapulnik, Y., Delaux, P.-M., Resnick, N., Mayzlish-Gati, E., Wininger, S., Bhattacharya, C. *et al.*, Strigolactones affect lateral root formation and root-hair elongation in *Arabidopsis*. *Planta*, 233, 209–216, 2011.
8. Ruyter-Spira, C., Kohlen, W., Charnikhova, T., van Zeijl, A., van Bezouwen, L., de Ruijter, N. *et al.*, Physiological effects of the synthetic strigolactone

analog GR24 on root system architecture in *Arabidopsis*: another below-ground role for strigolactones? *Plant Physiol.*, 155, 721–734, 2011.
9. Bouwmeester, H.J., Matusova, R., Zhongkui, S., Beale, M.H., Secondary metabolite signalling in host-parasitic plant interactions. *Curr. Opin. Plant Biol.*, 6, 358–364, 2003.
10. Jiang, Y., Wang, W., Xie, Q., Liu, N., Liu, L., Wang, D. *et al.*, Plants transfer lipids to sustain colonization by mutualistic mycorrhizal and parasitic fungi. *Science*, 356, 1172–1175, 2017.
11. Keymer, A., Pimprikar, P., Wewer, V., Huber, C., Brands, M., Bucerius, S.L. *et al.*, Lipid transfer from plants to arbuscular mycorrhiza fungi. *Elife*, 6, e29107, 2017.
12. Luginbuehl, L.H., Menard, G.N., Kurup, S., Van Erp, H., Radhakrishnan, G.V., Breakspear, A. *et al.*, Fatty acids in arbuscular mycorrhizal fungi are synthesized by the host plant. *Science*, 356, 1175–1178, 2017.
13. Wang, Q., Hasson, A., Rossmann, S., Theres, K., Divide et impera: boundaries shape the plant body and initiate new meristems. *New Phytol., 209*, 2, 485–498, 2016.
14. Kelly, J.H., Tucker, M.R., Brewer, P.B., The strigolactone pathway is a target for modifying crop shoot architecture and yield. *Biology*, 12, 1, 95, 2023.
15. Beveridge, C.A., Murfet, I.C., Kerhoas, L., Sotta, B., Miginiac, E., Rameau, C., The shoot controls zeatin riboside export from pea roots. Evidence from the branching mutant rms4. *Plant J.*, 11, 339–345, 1997.
16. Koltai, H. and Beveridge, C.A., Strigolactones and the coordinated development of shoot and root, in: *Long-distance systemic signaling and communication in plants*, pp. 189–204, Springer, Berlin Heidelberg, 2013.
17. Dun, E.A., de Saint, G.A., Rameau, C., Beveridge, C.A., Antagonistic action of strigolactone and cytokinin in bud outgrowth control. *Plant Physiol.*, 158, 487–498, 2012.
18. Dun, E.A., de Saint, G.A., Rameau, C., Beveridge, C.A., Dynamics of strigolactone function and shoot branching responses in *Pisum sativum. Mol. Plant*, 6, 128–140, 2013.
19. Brewer, P.B., Dun, E.A., Ferguson, B.J., Rameau, C., Beveridge, C.A., Strigolactone acts downstream of auxin to regulate bud outgrowth in pea and *Arabidopsis. Plant Physiol.*, 150, 482–493, 2009.
20. Ferguson, B.J. and Beveridge, C.A., Roles for auxin, cytokinin, and strigolactone in regulating shoot branching. *Plant Physiol.*, 149, 1929–1944, 2009.
21. Dun, E.A., Brewer, P.B., Beveridge, C.A., Strigolactones: discovery of the elusive shoot branching hormone. *Trends Plant Sci.*, 14, 364–372, 2009.
22. Bennett, T., Sieberer, T., Willett, B., Booker, J., Luschnig, C., Leyser, O., The *Arabidopsis* MAX pathway controls shoot branching by regulating auxin transport. *Curr. Biol.*, 16, 553–563, 2006.
23. Mouchel, C.F. and Leyser, O., Novel phytohormones involved in long-range signaling. *Curr. Opin. Plant Biol.*, 10, 473–476, 2007.

24. Ongaro, V. and Leyser, O., Hormonal control of shoot branching. *J. Exp. Bot.*, 59, 67–74, 2008.
25. Crawford, S., Shinohara, N., Sieberer, T., Williamson, L., George, G., Hepworth, J. *et al.*, Strigolactones enhance competition between shoot branches by dampening auxin transport. *Development*, 137, 2905–2913, 2010.
26. Domagalska, M.A. and Leyser, O., Signal integration in the control of shoot branching. *Nat. Rev. Mol. Cell Biol.*, 12, 211–221, 2011.
27. Zhang, J., Mazur, E., Balla, J., Gallei, M., Kalousek, P., Medveďová, Z. *et al.*, Strigolactones inhibit auxin feedback on PIN-dependent auxin transport canalization. *Nat. Commun.*, 11, 1, 3508, 2020.
28. Barbier, F.F., Dun, E.A., Kerr, S.C., Chabikwa, T.G., Beveridge, C.A., An update on the signals controlling shoot branching. *Trends Plant Sci.*, 24, 3, 220–236, 2019.
29. Ravichandran, S.J., Linh, N.M., Scarpella, E., The canalization hypothesis–challenges and alternatives. *New Phytol.*, *227*, 4, 1051–1059, 2020.
30. Prusinkiewicz, P., Crawford, S., Smith, R.S., Ljung, K., Bennett, T., Ongaro, V. *et al.*, Control of bud activation by an auxin transport switch. *PNAS U.S.A.*, 106, 17431–17436, 2009.
31. Bennett, T., Hines, G., van Rongen, M., Waldie, T., Sawchuk, M.G., Scarpella, E. *et al.*, Connective auxin transport in the shoot facilitates communication between shoot apices. *PLoS Biol.*, 14, 4, e1002446, 2016.
32. Sauer, M., Balla, J., Luschnig, C., Wisniewska, J., Reinohl, V., Friml, J. *et al.*, Canalization of auxin flow by Aux/IAA-ARF-dependent feedback regulation of PIN polarity. *Genes Dev.*, 20, 2902–2911, 2006.
33. Wang, M., Le Moigne, M.A., Bertheloot, J., Crespel, L., Perez-Garcia, M.D., Ogé, L., Sakr, S., BRANCHED1: a key hub of shoot branching. *Front. Plant Sci.*, *10*, 430480, 2019.
34. van Rongen, M., Bennett, T., Ticchiarelli, F., Leyser, O., Connective auxin transport contributes to strigolactone-mediated shoot branching control independent of the transcription factor BRC1. *PLos Genet.*, 15, 3, e1008023, 2019.
35. Braun, N., de Saint Germain, A., Pillot, J.P., Boutet-Mercey, S., Dalmais, M. *et al.*, The pea TCP transcription factor PsBRC1 acts downstream of strigolactones to control shoot branching. *Plant Physiol.*, 158, 225–238, 2012.
36. Seale, M., Bennett, T., Leyser, O., BRC1 expression regulates bud activation potential but is not necessary or sufficient for bud growth inhibition in Arabidopsis. *Development*, 144, 9, 1661–1673, 2017.
37. Shinohara, N., Taylor, C., Leyser, O., Strigolactone can promote or inhibit shoot branching by triggering rapid depletion of the auxin efflux protein PIN1 from the plasma membrane. *PLoS Biol.*, 11, e1001474, 2013.
38. Arite, T., Iwata, H., Ohshima, K., Maekawa, M., Nakajima, M., Kojima, M. *et al.*, DWARF10, an RMS1/MAX4/DAD1 ortholog, controls lateral bud outgrowth in rice. *Plant J.*, 51, 1019–1029, 2007.

39. Foo, E., Turnbull, C.G., Beveridge, C.A., Long-distance signaling and the control of branching in the rms1 mutant of pea. *Plant Physiol.*, 126, 203–209, 2001.
40. Miyashima, S., Sebastian, J., Lee, J.-Y., Helariutta, Y., Stem cell function during plant vascular development. *EMBO J.*, 32, 178–193, 2012.
41. Hayward, A., Stirnberg, P., Beveridge, C., Leyser, O., Interactions between auxin and strigolactone in shoot branching control. *Plant Physiol.*, 151, 1, 400–412, 2009.
42. Johnson, X., Brcich, T., Dun, E.A., Goussot, M., Haurogne, K., Beveridge, C.A. *et al.*, Branching genes are conserved across species: genes controlling a novel signal in pea are coregulated by other long-distance signals. *Plant Physiol.*, 142, 1014–1026, 2006.
43. Beveridge, C.A., Symons, G.M., Turnbull, C.G., Auxin inhibition of decapitation-induced branching is dependent on graft transmissible signals regulated by genes Rms1 and Rms2. *Plant Physiol.*, 123, 689–698, 2000.
44. Sun, H., Tao, J., Hou, M., Huang, S., Chen, S., Liang, Z., Xie, T., Wei, Y., Xie, X., Yoneyama, K. *et al.*, A strigolactone signal is required for adventitious root formation in rice. *Ann. Bot.*, 115, 1155–1162, 2015.
45. Koltai, H., Dor, E., Hershenhorn, J., Joel, D.M., Weininger, S., Lekalla, S., Shealtiel, H., Bhattacharya, C., Eliahu, E., Resnick, N., Barg, R., Kapulnik, Y., Strigolactones' effect on root growth and root-hair elongation may be mediated by auxin-efflux carriers. *J. Plant Growth Regul.*, 29, 129–136, 2010.
46. Koltai, H., Strigolactones are regulators of root development. *New Phytol.*, 190, 545–549, 2011.
47. Grieneisen, V.A., Xu, J., Maree, A.F.M., Hogeweg, P., Scheres, B., Auxin transport is sufficient to generate a maximum and gradient guiding root growth. *Nature*, 449, 1008–1013, 2007.
48. Chen, H. and Xiong, L., Localized auxin biosynthesis and postembryonic root development in Arabidopsis. *Plant Signal. Behav.*, 4, 752–754, 2009.
49. Petersson, S.V., Johansson, A.I., Kowalczyk, M., Makoveychuk, A., Wang, J.Y., Moritz, T. *et al.*, An auxin gradient and maximum in the *Arabidopsis* root apex shown by high-resolution cell-specific analysis of IAA distribution and synthesis. *Plant Cell*, 21, 1659–1668, 2009.
50. Kleine-Vehn, J., Dhonukshe, P., Swarup, R., Bennett, M., Friml, J., Subcellular trafficking of the *Arabidopsis* auxin influx carrier AUX1 uses a novel pathway distinct from PIN1. *Plant Cell*, 18, 3171–3181, 2006.
51. Zazimalova, E., Murphy, A.S., Yang, H.B., Hoyerova, K., Hosek, P., Auxin transporters—why so many? *Cold Spring Harb. Perspect. Biol.*, 2, a001552, 2010.
52. Petrasek, J. and Friml, J., Auxin transport routes in plant development. *Development*, 136, 2675–2688, 2009.
53. Rahman, A., Takahashi, M., Shibasaki, K., Wu, S., Inaba, T., Tsurumi, S. *et al.*, Gravitropism of *Arabidopsis thaliana* roots require the polarization of PIN2

toward the root tip in meristematic cortical cells. *Plant Cell*, 22, 1762–1776, 2010.

54. Marzec, M. and Melzer, M., Regulation of root development and architecture by strigolactones under optimal and nutrient deficiency conditions. *Int. J. Mol. Sci.*, 19, 7, 1887, 2018.
55. Jain, A., Poling, M.D., Karthikeyan, A.S., Blakeslee, J.J., Peer, W.A., Titapiwatanakun, B., Murphy, A.S., Raghothama, K.G., Differential effects of sucrose and auxin on localized phosphate deficiency-induced modulation of different traits of root system architecture in *Arabidopsis*. *Plant Physiol.*, 144, 232–247, 2007.
56. Koren, D., Resnick, N., Gati, E.M., Belausov, E., Weininger, S., Kapulnik, Y. *et al.*, Strigolactone signaling in the endodermis is sufficient to restore root responses and involves SHORT HYPOCOTYL 2 (SHY2) activity. *New Phytol.*, 198, 866–874, 2013.
57. Claassens, A.P. and Hills, P.N., Role of strigolactones in adventitious rooting of Arabidopsis thaliana callus *in vitro*. *S. Afr. J. Bot.*, 100, 109, 329, 2017.
58. Sun, H., Tao, J., Liu, S., Huang, S., Chen, S., Xie, X., Xu, G., Strigolactones are involved in phosphate-and nitrate-deficiency-induced root development and auxin transport in rice. *J. Exp. Bot.*, 65, 22, 6735–6746, 2014.
59. De Cuyper, C., Fromentin, J., Yocgo, R.E., DeKeyser, A., Guillotin, B., Kunert, K., Boyer, F.-D., Goormachtig, S., From lateral root density to nodule number, the strigolactone analogue GR24 shapes the root architecture of *Medicago truncatula*. *J. Exp. Bot.*, 66, 137–146, 2015.
60. Liu, J., He, H., Vitali, M., Visentin, I., Charnikhova, T., Haider, I., Cardinale, F., Osmotic stress represses strigolactone biosynthesis in Lotus japonicus roots: exploring the interaction between strigolactones and ABA under abiotic stress. *Planta*, 241, 1435–1451, 2015.
61. Soundappan, I., Bennett, T., Morffy, N., Liang, Y., Stanga, J., Abbas, A., Leyser, O., Nelson, D., SMAX1-LIKE/D53 family members enable distinct MAX2-dependent responses to strigolactones and karrikins in Arabidopsis. *Plant Cell*, 27, 3143–3159, 2015.
62. Villaecija-Aguilar, J., Hamon-Josse, M., Carbonnel, S., Kretschmar, A., Schmidt, C., Dawid, C., Bennett, T., Gutjahr, C., SMAX1/SMXL2 regulate root and root hair development downstream of KAI2-mediated signalling in Arabidopsis. *PLos Genet.*, 15, e1008327, 2019.
63. Jiang, L., Matthys, C., Marquez-Garcia, B., De Cuyper, C., Smet, L., De Keyser, A., Boyer, F., Beeckman, T., Depuydt, S., Goormachtig, S., Strigolactones spatially influence lateral root development through the cytokinin signaling network. *J. Exp. Bot.*, 67, 379–389, 2016.
64. Sun, H., Xu, F., Guo, X., Wu, D., Zhang, X., Lou, M., Luo, F., Zhao, Q., Xu, G., Zhang, Y., A strigolactone signal inhibits secondary lateral root development in rice. *Front. Plant Sci.*, 10, 1527, 2019.

65. Koltai, H. and Prandi, C., Strigolactones: biosynthesis, synthesis and functions in plant growth and stress responses, in: *Phytohormones: A window to metabolism, signaling and biotechnological applications*, pp. 265–288, 2014.
66. Wani, K.I., Chaudhary, S., Zehra, A., Naeem, M., Aftab, T., Precise Role of Strigolactones and Its Crosstalk Mechanisms in Root Development, in: *Rhizobiology: Molecular Physiology of Plant Roots*, pp. 253–270, 2021.
67. Kohlen, W., Charnikhova, T., Bours, R., López-Ráez, J.A., Bouwmeester, H., Tomato strigolactones: a more detailed look. *Plant Signal. Behav.*, 8, e22785, 2013.
68. Pandya-Kumar, N., Shema, R., Kumar, M., Mayzlish-Gati, E., Levy, D., Zemach, H., Belausov, E., Wininger, S., Abu-Abied, M., Kapulnik, Y., Koltai, H., Strigolactone analog GR24 triggers changes in PIN2 polarity, vesicle trafficking and actin filament architecture. *New Phytol.*, 202, 1184–1196, 2014.
69. Claassens, A.P. and Hills, P.N., Effects of strigolactones on plant roots, in: *Root Biology*, pp. 43–63, 2018.
70. Mayzlish-Gati, E., De-Cuyper, C., Goormachtig, S., Beeckman, T., Vuylsteke, M., Brewer, P.B. *et al.*, Strigolactones are involved in root response to low phosphate conditions in Arabidopsis. *Plant Physiol.*, 160, 1329–1341, 2012.
71. Rasmussen, A., Beveridge, C.A., Geelen, D., Inhibition of strigolactones promotes adventitious root formation. *Plant Signal. Behav.*, 7, 694–697, 2012a.
72. Rasmussen, A., Mason, M.G., De Cuyper, C., Brewer, P.B., Herold, S., Agusti, J., Geelen, D., Greb, T., Goormachtig, S., Beeckman, T., Beveridge, C.A., Strigolactones suppress adventitious rooting in Arabidopsis and Pea. *Plant Physiol.*, 158, 1976–1987, 2012b.
73. Arite, T., Kameoka, H., Kyozuka, J., Strigolactone positively controls crown root elongation in rice. *J. Plant Growth Regul.*, 31, 165–172, 2012.
74. Czarnecki, O., Yang, J., Weston, D., Tuskan, G., Chen, J.-G., A dual role of strigolactones in phosphate acquisition and utilization in plants. *Int. J. Mol. Sci.*, 14, 7681–7701, 2013.
75. De Jong, M., George, G., Ongaro, V., Williamson, L., Willetts, B., Ljung, K., McCulloch, H., Leyer, O., Auxin and strigolactone signaling are required for modulation of *Arabidopsis* shoot branching by N supply. *Plant Physiol.*, 166, 1, 384–395, 2014.
76. Xu, J., Zha, M., Li, Y., Ding, Y., Chen, L., Ding, C., Wang, S., The interaction between nitrogen availability and auxin, cytokinin, and strigolactone in the control of shoot branching in rice (*Oryza sativa* L.). *Plant Cell Rep.*, 34, 1647–1662, 2015.
77. Bonneau, L., Huguet, S., Wipf, D., Pauly, N., Truong, H.N., Combined phosphate and nitrogen limitation generates a nutrient stress transcriptome favorable for arbuscular mycorrhizal symbiosis in *Medicago truncatula*. *New Phytol.*, 199, 188–202, 2013.
78. Yoneyama, K., Xie, X., Kim, H.I., Kisugi, T., Nomura, T., Sekimoto, H. *et al.*, How do nitrogen and phosphorus deficiencies affect strigolactone production and exudation? *Planta*, 235, 1197–1207, 2012.

79. Bennett, T., Hines, G., Leyser, O., Canalization: what the flux? *Trends Genet.*, 30, 41–48, 2014.
80. Pitts, R.J., Cernac, A., Estelle, M., Auxin and ethylene promote root hair elongation in Arabidopsis. *Plant J.*, 16, 553–560, 1998.
81. Jones, A.R., Kramer, E.M., Knox, K., Swarup, R., Malcolm, J., Lazarus, C.M., Leyser, H.M.O., Grierson, C.S., Auxin transport through non-hair cells sustain root-hair development. *Plant Cell*, 11, 78–84, 2009.
82. Ioio, R.D., Linhares, F.S., Sabatini, S., Emerging role of cytokinin as a regulator of cellular differentiation. *Curr. Opin. Plant Biol.*, *11*, 1, 23–27, 2008.
83. Nooden, L.D., The phenomena of senescence and aging, in: *Senescence and aging in plants*, A.C. Leopold and L.D. Nooden (Eds.), pp. 1–50, Academic Press, San Diego, CA, USA, 1988.
84. Brundrett, M.C. and Tedersoo, L., Evolutionary history of mycorrhizal symbioses and global host plant diversity. *New Phytol.*, *220*, 4, 1108–1115, 2018.
85. Jibran, R., Hunter, D.A., Dijkwel, P.P., Hormonal regulation of leaf senescence through integration of developmental and stress signals. *Plant Mol. Biol.*, 82, 547–561, 2013.
86. Ueda, H. and Kusaba, M., Strigolactone regulates leaf senescence in concert with ethylene in *Arabidopsis*. *Plant Physiol.*, 169, 1, 138–47, 2015.
87. Tsuchiya, Y., Vidaurre, D., Toh, S., Hanada, A., Nambara, E., Kamiya, Y. *et al.*, A small-molecule screen identifies new functions for the plant hormone strigolactone. *Nat. Chem. Biol.*, 6, 741–749, 2010.
88. Yamada, Y. and Umehara, M., Possible Roles of Strigolactones during Leaf Senescence. *Plants*, 4, 664–677, 2015.

5

# Regulation of Phosphorus Nutrition in Tomato Plants: Unveiled Roles of Strigolactones

**Veronica Santoro[1], Michela Schiavon[1]*, Cristina Prandi[2] and Luisella Celi[1]**

[1]*Department of Agricultural, Forest and Food Sciences, University of Turin, Turin, Italy*

[2]*Department of Chemistry, University of Turin, Turin, Italy*

***Abstract***

Phosphorus (P) stands as a soil macronutrient indispensable for plant growth. As an integral component within diverse biomolecules, it assumes pivotal roles in both primary and secondary metabolic processes, contributing significantly to essential cellular functions in plants. However, abiotic soil processes strongly reduce P bioavailability, consequently constraining P nutrition and limiting plant growth. In attempts to address this challenge, farmers resort to the excessive use of P fertilizers, leading to adverse environmental impacts and the gradual exhaustion of phosphate rock reservoirs. In order to overcome P limitation in the soil, plants deploy a diverse array of molecular, physiological, and morphological responses. These encompass reconfiguration of their root system architecture, enhanced production and release of exudates containing phosphatases and phytases, modifications of rhizosphere pH, and release of substances that foster symbiosis with beneficial microorganisms, particularly mycorrhizal associations. The newly discovered class of phytohormones, strigolactones, plays a pivotal role in the regulation of several of these responses. Recent studies conducted in the model plant tomato (*Solanum lycopersicum* L.) have brought to light the modulation of strigolactone synthesis as a strategic mechanism to optimize plant growth under P limitation. In particular, strigolactone depletion is associated with distorted plant shoot and root phenotype, along with an altered perception of external P. This chapter is dedicated to elucidating the synthesis and functions of strigolactones in plants, placing particular emphasis on their pivotal roles within the intricate signaling pathways that regulate P nutrition in tomato plants.

**Corresponding author*: michela.schiavon@unito.it

Tariq Aftab and Kaiser Iqbal Wani (eds.) Strigolactones: Emerging Plant Hormones, (89–108) © 2025 Scrivener Publishing LLC

***Keywords*:** Strigolactone, tomato, roots, shoot, phosphorous, signaling, canonical, non-canonical

## 5.1 Introduction

Various physiological, morphological, and biochemical processes are essential for plant adaptation to environmental stressors. These processes necessitate synchronized molecular-level changes within an intricate plant growth network, demanding cross-talk among hormonal signals [1].

Within the realm of plant phytohormones, strigolactones (SLs) stand out as pivotal contributors to plant communication with the external environment [2 and references therein]. Functioning as signaling molecules, they enable the early perception of changes in nutrient availability in the rhizosphere exerting influence over the plant phenotype and nutrient use efficiency [3–6]. In addition, SLs appear to play a significant role in endowing plants with tolerance to other suboptimal growth conditions, particularly in mitigating the impacts of drought, salinity, low light, heat and cold stress, and certain pathogens [2 and references therein]. Therefore, understanding the role of SLs in plants can aid in optimizing the use of fertilizers in agriculture by implementing plant resilience to multiple types of stress [4, 8].

Remarkably, SLs are crucial in coordinating the harmonized development of both roots and shoots, especially in conditions characterized by low phosphorus (P) availability [4, 7]. Phosphorus is a vital element for plants, is integral to cell structure, and has a central role in fundamental biological processes like photosynthesis, respiration, and the synthesis of nucleic acids and membranes [9]. Moreover, it acts as a master regulator of several enzymes.

Despite its critical importance for plants, P in soil is typically poorly available due to processes such as adsorption, immobilization, precipitation, and coprecipitation, which render it either immobile or not accessible [10]. Consequently, plants frequently experience P deficiency, which results in reduced yield and productivity. To enhance P availability in soil and improve its acquisition rate and use efficiency, plants employ a plethora of strategies, encompassing physiological, biochemical and molecular adaptations, most of which are under hormonal control [11]. In particular, SLs, in synergy with other phytohormones, exert regulatory control over plant morphogenesis during P starvation, intricately governing processes such as shoot branching, leaf senescence, reproductive development, the formation of adventitious roots, and the density of root hairs [12, 13]. Moreover, SLs have the capacity to influence the expression of key regulatory genes of P uptake and, consequently, of P transporters, thereby fine-tuning a number of metabolic pathways to enhance the plant resilience against P limitation [14].

The roles of SLs in plant morphogenesis and mitigation of abiotic stress, whether nutritional or osmotic, and their regulation and perception by plants have been investigated across a number of species, particularly in tomato (*Solanum lycopersicum* L.) [7, 15, 16]. This is due to the

**Table 5.1** Main root trait differences between tomato wild-type and strigolactone-depleted plants in response to different P conditions.

| Aspect | Wild-type plants | Strigolactone-depleted plants |
|---|---|---|
| Response to P deprivation | - Impaired typical stress responses | - Drastically reduced growth compared to WT |
| | - Limited root capacity to respond to P stress | - Hypersensitivity to P deprivation stress in root tips<br>- Inability to respond to P stress in the long term |
| Response to sub-optimal P supply | - Induced responses to P stress | - Some features of P-stress attenuated |
| Response to P stress after acclimation (from high P to no P) | - Emphasized P-stress trait modifications | - Significant repression of P-stress trait modifications |
| Root traits affected | - Primary and total root elongation | - Impaired primary and total root elongation |
| | - Lateral root number | - Reduced lateral root number |
| | - Root volume and diameter | - Reduced root volume and diameter |
| Phenotypic differences | - Differences in root morphology and tip anatomy levels | - Hypersensitivity to P deprivation |

emergence of tomato as a model crop for research on SLs, facilitated by the generation of mutants exhibiting either modified perception or synthesis of these compounds. A number of SLs produced by tomato have also been identified, which include orobanchol, solanacol, and the two isomers of didehydro-orobanchol. Additionally, various other SLs have been recently identified in root exudates and extracts, encompassing strigol-type SLs among them [17]. The production of SLs in tomato has been recently reported to be markedly upregulated under P deficiency, resulting in the activation of P starvation responses (PSRs) aimed to improve plant P acquisition and P internal recycling [4, 18]. These responses have been thoroughly investigated under multiple P conditions, encompassing differences in P concentration and sources applied, as well as changes in P availability over time (see Table 5.1 in [7]). This comprehensive approach has ameliorated the understanding of the regulatory network underlying the activation of plant responses to P limitation and holds promise for devising innovative strategies to enhance the phosphorus use efficiency (PUE) of other crops beyond tomato.

## 5.2 Strigolactone Chemistry and Biosynthesis

### 5.2.1 Structural Features and Classifications of Natural SLs

Strigolactones can be divided into two main classes based on their chemical structure: canonical and non-canonical SLs (Figure 5.1). Canonical SLs have similar sesquiterpene carbon skeleton structures, featuring a tricyclic ABC core lactone linked to a 5-hydroxybutenolide (D-ring) through an enol ether bridge. They are further subdivided into two groups depending on their stereochemistry, namely strigol and orobanchol (1 and 5, Figure 5.1) [19].

It was found that the D-ring and the enol ether bridge are essential to retain bioactivity, and the R configuration at the C-2' of the D-ring was crucial to maintain SL-like activities [20]. The presence of various functional groups such as hydroxyl, alkyl, or acetoxy on the A and B rings contributes to the wide structural diversity of natural SLs [21]. Non-canonical SLs were discovered more recently and are characterized by a conserved enol ether and D-rings, but the classical ABC structure is missing and replaced by alkyl chains or partial cyclic structures (C, or BC rings, 7–10, Figure 5.1).

All natural SLs, comprising more than 30 derivatives identified so far [22], contain 5-hydroxyfuranone (D-ring) as a structurally crucial fragment responsible for their biological activity. This role extends beyond their

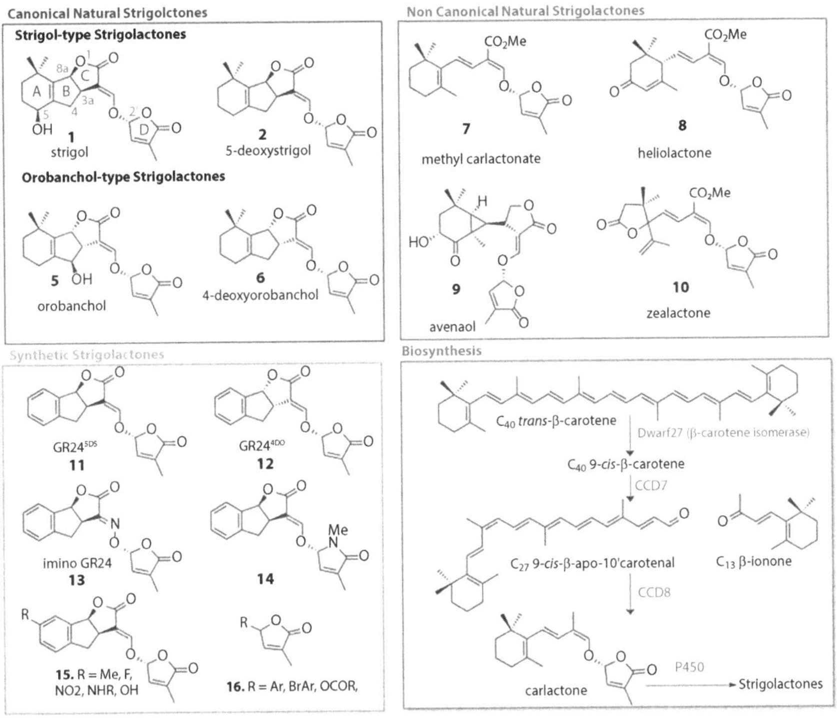


**Figure 5.1** Canonical and non-canonical natural SLs.

function as plant hormones to include other effects such as anti-tumoral and anti-inflammatory properties, which are of pharmaceutical interest [23, 24]. Stereochemistry plays a crucial role, and all natural SLs can be categorized into the strigol family or the orobanchol family depending on the orientation of the BC ring junction with respect to the stereocenter at the C-2', which in natural SLs is always R.

### 5.2.2 Synthetic Strigolactones

Structure–activity studies have identified key regions in the SL molecule responsible for receptor interactions. Targeted structural modifications have led to the synthesis of analogues with enhanced or altered biological activities, contributing to a deeper understanding of the molecular interactions involved in plant responses to SLs [25].

GR24 is the most representative synthetic analogue of indane lactones in which the A ring is replaced with a benzene (see 11 and 12, Figure 5.1). It has been widely used in various studies as a reference compound;

however, in many cases, it has been used as a mixture of stereoisomers, leading to results that might be attributable not only to SLs but, in some cases, to karrikin signaling pathways [26]. GR24 has four stereoisomers identified as GR245DS, GR24ent-5DS, GR244DO, and GR24ent-4DO (see 11 and 12, Figure 5.1). These stereoisomers have been shown to exhibit different response specificities, emphasizing the importance of using the correct stereoisomer in experiments. Modifications of the original structure have been targeted at the enol ether bridge (13, Figure 5.1) [27, 28], the D-ring (14, Figure 5.1) [29, 30], or the A/B rings (15, Figure 5.1) [31]. Most modifications at the enol ether bridge and at the D-ring led to diminished activity, confirming that these are both key conserved features to maintain phytohormone activity. A special mention among SLs analogues is deserved by aryl furanone ethers and esters, very simplified structures retaining biological activity, commonly known as SLs mimics (16, Figure 5.1). In this case, stereochemical complexity is reduced, and the mimics can be produced in bulk for applicative purposes.

### 5.2.3 Biosynthetic Pathway

SLs are biosynthesized via the carotenoid pathway (see Figure 5.1) [32]. In 2012, [33] found that Dwarf 27 (D27 in rice), CCD7 (carotenoid cleavage dioxygenase 7), and CCD8 (carotenoid cleavage dioxygenase 8) sequentially convert all-trans-β-carotene to 9-cis-β-apo-10'-carotenal and generate a non-canonical SL intermediate, CL (carlactone). A subfamily of cytochrome P450 (CYP) enzymes has been found to be crucial in converting CL into canonical or non-canonical SLs in vascular plants. CL is a substrate for the *Arabidopsis* SLs mutant MAX1 protein, converting CL to carlactonoic acid (CLA) and then to 4DO (4-deoxyorobanchol) and orobanchol (ORO) [34].

## 5.3 The SL Signaling Pathway

The activation of PSRs relies on the integration of local and systemic signaling pathways to maintain P homeostasis, though changes in root development are mainly governed locally based on how much P is available in the external environment [35]. Sensing of external P concentrations mainly takes place at the root tip, prompting adjustments in root system architecture [36–38]. This process is dependent on the auxin receptor TRANSPORT INHIBITOR RESPONSE1 (TIR1) and MORE AXILLARY

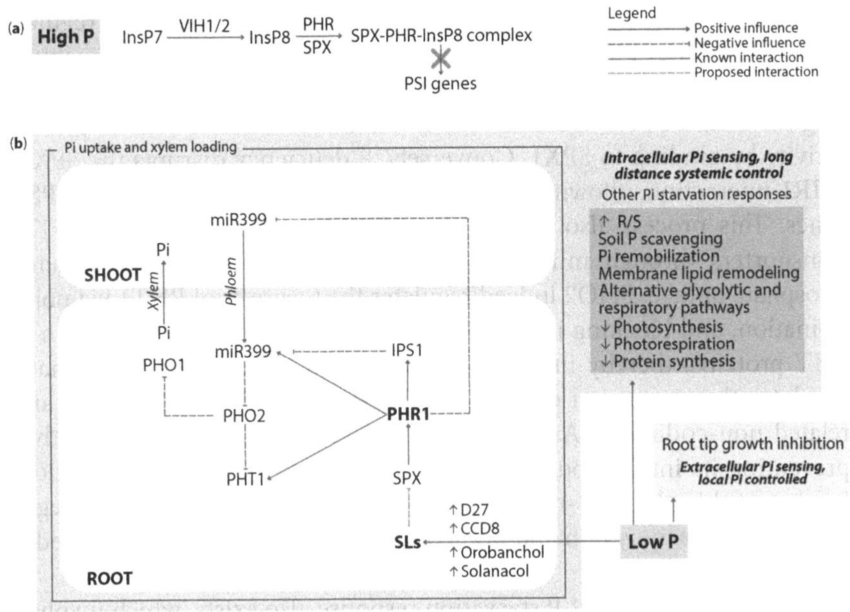


**Figure 5.2** Schematic model of the main players involved in the regulation of P signaling pathway in plants. This is constituted of a branch centered around PHR1 and related transcription factors, which is primarily dependent on intracellular Pi sensing and is controlled by overall plant Pi status (systemically controlled) and a branch dependent on extracellular Pi sensing and controlled by local Pi, which influences root tip growth and other root system architecture traits. (a) A high level of intracellular Pi promotes the synthesis of inositol octaphosphate (InsP8), which binds to the SPX domain in SPX1 and promotes the binding of SPX1 to PHR1, blocking PHR1 transcriptional activity and the activation of P starvation-induced (PSI) genes. (b) Under Pi deficiency, strigolactone (SL) biosynthesis is promoted, which affects the complex SPX–PHR1. The complex becomes unstable and releases the regulator PHR1, which, in turn, promotes the expression of P transporters from the PHT1 family in the roots, favoring Pi uptake. In addition, by inducing the expression of microRNA miR399, PHR1 indirectly inhibits PHO2, which mediates the posttranslational negative control of PHT1 and/or PHO1 (involved in xylem loading of Pi). On the other hand, IPS1 can interact and block miR399 transcripts, preventing miR399-PHO2 binding and degradation of PHO2. InsP7, inositol pyropolyphosphate. Based on [4, 18, 46].

GROWTH 2 (MAX2), a F-box protein that also serves as a crucial signaling component in the SL pathway [39–42].

The P starvation pathway in *Arabidopsis* is regulated by PHOSPHATE STARVATION RESPONSE 1 (PHR1) and associated transcription factors, with F-box genes being among the main targets of PHR1 [43, 44]. SPX domain-containing proteins like SPX1 and SPX2 play critical roles in sensing both external and internal phosphate levels, negatively regulating PHR1 activity [45]. Inositol octaphosphate (InsP8) serves as the intracellular phosphate signal, binding to SPX1 and controlling P homeostasis (Figure 5.2). Under high P levels, InsP8 inhibits PHR1 transcriptional activity by binding to SPX1. Conversely, P deficiency disrupts the SPX1-PHR1 interaction, allowing PHR1 to activate P starvation-induced (PSI) genes. This process also relieves P-mediated negative control of PHT1 transporters through miR399-induced inhibition of PHO2, enhancing phosphate uptake. PHO2 indeed regulates the turnover of PHT1 via ubiquitination, and blocking its activity reduces the proteolytic breakdown of PHT proteins, thereby increasing phosphate uptake. PHR1 additionally regulates the expression of genes encoding IPS1 (TPSl1 in tomato) and related non-coding RNAs, which sequester and modulate miR399, thus preventing its interaction with PHO2 transcripts and their degradation. In view of this, the IPS1-miR399-PHO2 triad plays a central role in regulating plant P acquisition and homeostasis, especially during prolonged P deprivation.

A PHR1-independent P starvation response also exists, which involves root tip inhibition and relies on extracellular P sensing and local P levels. This pathway differs from the PHR1-dependent pathway, which is influenced by the overall plant P status. Despite their distinct roles, these pathways are however tightly integrated.

Recent research suggests that SLs act as signals triggering plant responses to P deficiency, as indicated by changes in the IPS1-miR399-PHO2 triad following exogenous SL application or in SL-biosynthesis mutants [4, 14, 18, 47]. In tomato plants, brief exposure to 2'-*epi*-GR24 increased *PHO2*, *IPS1*, and *miR399* gene expression, mirroring responses observed under P starvation [18]. Exogenous SL application induced *IPS1* and *miR399* gene expression even under optimal P conditions, partially resembling effects occurring during P deficiency. Notably, the SL-deficient line SlCCD8-RNAi showed altered levels of key regulatory elements, especially PHO2, and displayed reduced sensitivity to P starvation [18].

The response to P deficiency can vary depending on its severity, especially when SL production is hindered or suppressed—for instance, the dysregulation of the miR399-PHO2 module was observed in SL-deficient tomato plants at 80 μM $P_i$ (inorganic phosphate) [14]. This effect was less pronounced in other studies, where higher Pi concentration was used to simulate a P-sufficient condition [4, 18]. However, while PHO2 transcripts were observed to exhibit an inverse correlation with Pi uptake and shoot P content in both wild-type and SL mutants supplied with Pi, this correlation was disrupted when phosphate was provided in alternative forms, such as organic P compounds or sparingly-available P forms [14, 47].

Beyond being implied in P nutrition, SLs also control a number of aspects related to shoot morphogenesis, particularly axillary bud outgrowth. In SL-insensitive dwarf 14 (d14) rice mutants, an enzyme from the α/β-fold hydrolase family was identified as playing a role for the control of this process [48, 49]. It has been reported that this enzyme operates in concert with MAX2 and that D14 serves as a SL receptor implied not only in shoot branching but also in controlling lateral root density and epidermal cell length [50–52]. Notably, plants possess a D14 homolog (KARRIKIN INSENSITIVE2, KAI2), which binds karrikins (KARs), smoke-derived signaling compounds belonging to the class of butenolides. Therefore, KARs share chemical similarity with SLs. Despite this, the roles of KARs and SLs in plant growth regulation differ significantly. Karrikins, for instance, are not involved in hyphal branching of arbuscular mycorrhizal fungi (AMF) or in the inhibition of shoot branching [53, 54]. In *Arabidopsis*, it has been discovered that KAI2-mediated signaling plays a predominant role in governing both root hair density and length [55]. This finding stands in contrast to previous beliefs that root hair formation and elongation were primarily regulated by the crosstalk between SLs and auxin [56–59]. KAI2-mediated signaling also regulates root skewing, straightness, and diameter, seed germination, and stress response. In contrast, both KAI2 and the SL receptor D14 pathways jointly regulate shoot branching, lateral root density, and epidermal cell length [50–52].

## 5.4 Function of Strigolactones in Tomato Plants: Emphasis on Phosphorus Nutrition

The roles of SLs in plant morphogenesis and alleviation of nutritional stresses have been extensively explored across multiple plant species. However, in recent years, there has been a notable increase in studies focusing on the model plant tomato, which have underscored the SL-dependent P-deficiency responses in this crop (Figure 5.3). This heightened attention is due to the economic significance of tomato as a horticultural crop worldwide, its ability to produce SLs, and the accessibility of SL mutants for research purposes [7]—for instance, tomato mutants carrying an antisense *SlCCD7* construct display increased branching, with root extracts substantially unable to induce the seed germination of the parasitic plant *Orobanche ramosa* compared to their wild-type counterparts [60, 61]. Tomato plants with RNAi-silenced

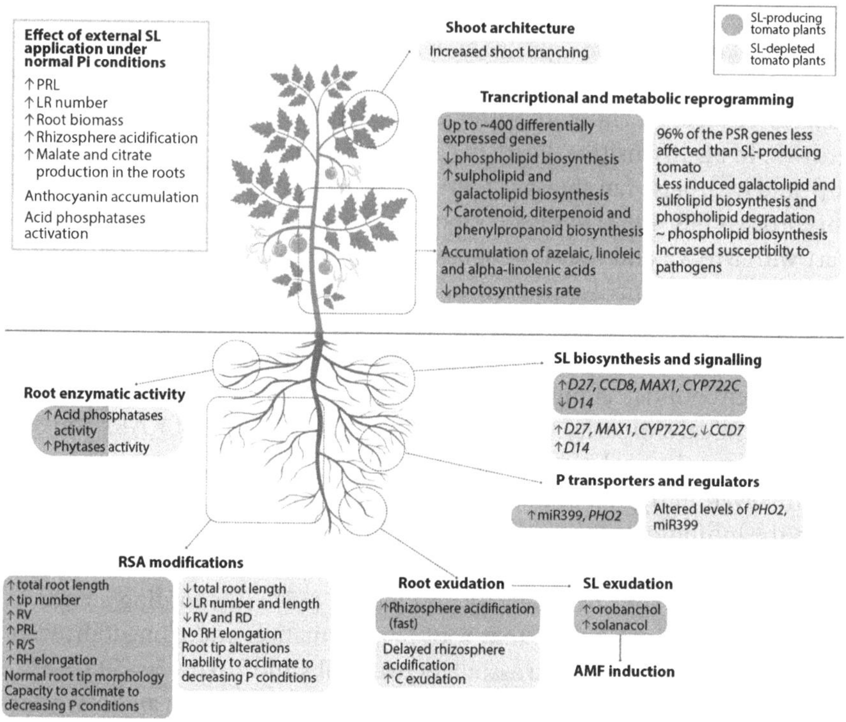


**Figure 5.3** Conceptual summary of how strigolactones (SLs) are involved in P-deficiency responses in tomato plants. Belowground, they influence root architecture, root exudation, and enzymatic activity. Strigolactone exudation can indirectly increase P uptake through the instauration of symbiosis with arbuscular mycorrhizal fungi (AMF). In the shoots, SLs have an effect on P remobilization and on transcriptional and metabolic reprogramming. InsP6, inositol hexaphosphate; LR, lateral root; PRL, primary root length; R/S, root to shoot ratio; RD, root diameter; RH, root hair; RSA, root system architecture; RV, root volume.

*SlCCD8*, instead, exhibit diminished plant height and decreased exudation of SLs accompanied by augmented shoot branching. Research employing these mutants has shed light on the SL-regulated responses in tomato, particularly under conditions of P deficiency [42].

When P in soil is limiting, plants respond by activating complex regulatory mechanisms and *ad hoc* responses, which aim to optimize P use efficiency through genetic, physiological, developmental and morphological changes. These PSR responses include strategies for P acquisition, translocation, allocation, utilization, and remobilization within the plant and primarily involve root architecture adjustments, symbiotic interactions with mycorrhizae, increased expression of high-affinity P transporters, secretion of P-solubilizing enzymes and low molecular weight organic acids, acidification of the rhizosphere, internal P remobilization via recycling enzymes, adjustments in carbon metabolism, utilization of alternative respiratory pathways, and the synthesis of low-P membranes [7, 62].

Recently, a study aimed to determine the SL-dependent transcriptional changes in tomato as a function of P availability revealed that SLs are not only the end-signals secreted by plants into the rhizosphere for inducing several PSRs in plants but also crucial regulators of PSRs [42]. Indeed molecular changes typically triggered by P starvation and referred to glyco- and galactolipid metabolism, carbohydrate metabolism, secondary metabolite biosynthesis (including terpenoids and polyketides), glycan biosynthesis and metabolism, and amino acid metabolism exhibited a diminished magnitude in the *SlCCD8* RNAi line compared to the wild type. In addition, the metabolism of the plant hormones ABA, GA, ethylene, and auxin was less affected in *SlCCD8* mutants in response to P starvation, thus indicating that SLs play a pivotal role in orchestrating these hormonal changes in plants.

In the context of PSRs, SLs also play a crucial role in enhancing root traits associated with improved root soil exploration (Table 5.1). Their effect can manifest via the promotion of primary root elongation and lateral root proliferation and the redistribution of biomass towards the root system. In particular, species like maize and *Arabidopsis* adapt to P deficiency through lateral root modification [63], but tomato plants behave differently, as increased P levels induce lateral root proliferation [64]. Nevertheless, when facing P deficiency, tomato plants, like many other species, increase root surface area while reducing total root weight and average root diameter.

In Santoro *et al.* [5], root characteristics of wild type tomato seedlings were examined under high *versus* low $P_i$ conditions, with or without the addition of exogenous strigolactones (GR24). The application of GR24 enhanced the development of primary and lateral roots under low $P_i$ condition, thus confirming a key role for SLs in driving root architecture of tomato plants under $P_i$ limitation. Additionally, root traits were analyzed in both wild-type and SL-deficient plants (*SlCCD7* mutants) initially subjected to high versus no $P_i$ conditions, followed by a transition to high versus no $P_i$ after adaptation to low $P_i$ levels. The optimal synthesis of SLs was revealed to be indispensable for the overall root growth, which was suppressed in SL-depleted mutants subjected to persistent $P_i$ deprivation. Furthermore, SL synthesis played a critical role in regulating a range of root parameters, including total root length, root hair length, lateral root number, and root tip morphology. These effects were particularly pronounced in plants cultivated under low $P_i$ conditions or transferred from environments with low to no $P_i$ availability, suggesting that SLs are essential for allowing plants to acclimatize to P deficiency and orchestrate typical responses to low P. However, under sufficient P supply, no significant differences in root traits were evident between SL-mutants and wild-type plants, which means SLs do not significantly impact root traits when plants receive adequate P supplementation.

Strigolactone exudation by tomato is enhanced by the lack of external P [65]. In particular, the presence of orobanchol (2 $\mu g\ L^{-1}$) and solanacol (60 $\mu g\ L^{-1}$) was detected in the exudates of P-starved tomato plants by Rial *et al.* [66] and Marro *et al.* [4], but not in the exudates or root extracts of P-sufficient plants. Similarly, Gamir *et al.* [18] observed a significant increase in the quantities of orobanchol and solanacol in tomato roots after 1 week of P starvation. The increased production of SLs in the roots of tomato plants is thought to be essential for the induction of several P-starvation responses primarily within this organ, with secondary effects extending to the entire plant. On the other hand, the exudation of these compounds from the roots is likely associated with the plant requirement to modify the rhizosphere environment, facilitating crucial processes such as the establishment of symbiosis with AMF, that may occur in the soil and has a pivotal role in enhancing P mobilization and supply to the plant [67].

It must be noted that insights gleaned from the same series of experiments, conducted with varying P concentrations and timings of P deficiency initiation, underscore the nuanced nature of plant responses to P starvation. These responses are markedly influenced by the degree of P stress applied and the presence of an acclimation period before applying complete P deprivation. It is noteworthy that constant P provision and abrupt P depletion, scenarios less common in open field environments, have been largely explored in several studies. However, in soils, P availability tends to gradually decline due to plant uptake or fixation processes, leading to increasing P deficiency over time. Consequently, the synthesis of SLs becomes particularly significant as it impacts the capability of tomato roots to effectively respond to a gradual P decrease when plants are afforded the opportunity to acclimate before experiencing complete P deprivation [68].

According to Santoro *et al.* [5], SL-depleted plants experienced a reduction in total root length during P starvation compared to those grown with sufficient P levels. This decrease was not attributed to inhibited primary root elongation, but to a decline in both the number and length of lateral roots. The observed anatomical changes in the root tips of P-starved SL-depleted plants suggested that SL synthesis plays a critical role in regulating lateral root formation and overall root development in tomato plants. These findings align with research on *Arabidopsis* and rice mutants, underscoring the compensatory capacity of other hormones to counter alterations in root morphology in the absence of SLs.

In addition to regulating a large array of PSRs in plants, SLs have been recently reported to have a key role in the nitrogen (N)-P signaling interplay. Similarly to P, N acts as a modulator of downstream N responses, which are vital for plants to adapt to variations in nutrient availability. In particular, the nitrate transporter NRT1.1, endowed with dual affinity (high and low) for nitrate depending on external N supply, regulates the nitrate-induced PSR responses depending on PHR1, while the levels of PHO2 are reduced by P starvation in plants that receive nitrate [69]. NRT1.1 can also interact with specific SPX proteins causing its degradation and promoting PHR1 activation. SPX, in turn, can interact with the family of transcription factors NINE-LIKE PROTEIN (NLP), which stand as master regulators of the nitrate-related responses, consequently triggering NLP activation under high nitrate condition [70]. However, nitrate uptake has been reported to be repressed under P starvation via PHR1 [71].

Recently, the responses of wild-type and SL-deficient tomato plants (line *SlCCD8-RNAi L09*) were analyzed under varying P and N supplies [4], and in plants treated with a short-term pulse of the synthetic SL analogue 2'-epi-GR24. The main outcomes indicate that plants prioritize N over P status via modulation of SL biosynthesis and that SLs modulate the expression of key regulatory genes of phosphate and nitrate signaling pathways.

## 5.5 Conclusions

In summary, SLs emerge as critical regulators of PSRs in plants, including tomato, supported by their profound influence on various responses during P starvation. A disruption in either their biosynthesis or sensing can result in plants with altered or diminished capacity to initiate PSR-related changes, rendering them less responsive to P limitation, and in turn to N fluctuations. This highlights the crucial role SLs play in ensuring plant adaptability to variable environmental conditions.

Moreover, delving into the chemistry of natural and synthetic SLs provides insights into their molecular interactions and potential applications. With the application of rational drug design and Artificial Intelligence in drug research, it is reasonable to expect that more and more SLs analogues will be designed. These advancements are crucial for the development of targeted agricultural strategies and a comprehensive understanding of the roles of these molecules in plant physiological processes.

One significant concern regarding the utilization of SLs in agriculture, however, pertains to their classification within specific regulatory frameworks. This classification may indeed involve categorizing SLs either as products used in plant protection or as plant strengtheners or biostimulants during the registration process. This would pose a challenge in imposing regulations on their usage, given the complexity inherent in determining their precise regulatory status [72].

## References

1. Weyers, J.D.B. and Paterson, N.W., Plant hormones and the control of physiological processes. *New Phytol.*, 152, 3, 375–407, 2001.
2. Aliche, E.B., Screpanti, C., De Mesmaeker, A., Munnik, T., Bouwmeester, H.J., Science and application of strigolactones. *New Phytol.*, 227, 4, 1001–1011, 2020.
3. Aquino, B., Bradley, J.M., Lumba, S., On the outside looking in: roles of endogenous and exogenous strigolactones. *Plant J.*, 105, 2, 322–334, 2021.
4. Marro, N., Lidoy, J., Chico, M.Á., Rial, C., García, J., Varela, R.M., Macías, F.A., Pozo, M.J., Janoušková, M., López-Ráez, J.A., Strigolactones: New players in the nitrogen–phosphorus signalling interplay. *Plant Cell Environ.*, 45, 512–527, 2022.
5. Santoro, V., Schiavon, M., Gresta, F., Ertani, A., Cardinale, F., Sturrock, C.J., Celi, L., Schubert, A., Strigolactones control root system architecture and tip anatomy in *Solanum lycopersicum* L. plants under P starvation. *Plants*, 9, 612, 2020.
6. Yuan, K., Zhang, H., Yu, C. *et al.*, Low phosphorus promotes NSP1-NSP2 heterodimerization to enhance strigolactone biosynthesis and regulate shoot and root architecture in rice. *Mol. Plant*, 16, 11, 1811–1831, 2023.
7. Santoro, V., Schiavon, M., Celi, L., Role of soil abiotic processes on phosphorus availability and plant responses with a focus on strigolactones in tomato plants. *Plant Soil*, 494, 1–49, 2023.
8. Tariq, A., Ullah, I., Sardans, J. *et al.*, Strigolactones can be a potential tool to fight environmental stresses in arid lands. *Environ. Res.*, 229, 115966, 2023.
9. Prathap, V., Kumar, A., Maheshwari, C., Tyagi, A., Phosphorus homeostasis: acquisition, sensing, and long-distance signaling in plants. *Mol. Biol. Rep.*, 49, 8, 8071–8086, 2022.
10. Celi, L., Martin, M., Barberis, E., Phosphorus in soil, in: *Encyclopedia of Soils in the Environment*, vol. 2, pp. 152–156, Elsevier, Academic Press, 2023.
11. Chiou, T.J. and Lin, S.I., Signaling network in sensing phosphate availability in plants. *Annu. Rev. Plant Biol.*, 62, 185–206, 2011.

12. Koltai, H., Strigolactones activate different hormonal pathways for regulation of root development in response to phosphate growth conditions. *Ann. Bot.*, 112, 409–441, 2013.
13. Wu, F., Gao, Y., Yang, W., Sui, N., Zhu, J., Biological Functions of Strigolactones and Their Crosstalk With Other Phytohormones. *Front. Plant Sci.*, 13, 821563, 2022.
14. Santoro, V., Schiavon, M., Visentin, I., Constán-Aguilar, C., Cardinale, F., Celi, L., Strigolactones affect phosphorus acquisition strategies in tomato plants. *Plant Cell Environ.*, 44, 3628–3642, 2021.
15. Chi, C., Xu, X., Wang, M. *et al.*, Strigolactones positively regulate abscisic acid-dependent heat and cold tolerance in tomato. *Hortic. Res.*, 8, 1, 237, 2021.
16. López-Ráez, J.A., Charnikhova, T., Gómez-Roldán, V. *et al.*, Tomato strigolactones are derived from carotenoids and their biosynthesis is promoted by phosphate starvation. *New Phytol.*, 178, 4, 863–874, 2008.
17. Wakabayashi, T., Moriyama, D., Miyamoto, A. *et al.*, Identification of novel canonical strigolactones produced by tomato. *Front. Plant Sci.*, 13, 1064378, 2022.
18. Gamir, J., Torres-Vera, R., Rial, C., Berrio, E., de Souza Campos, P.M., Varela, R., Macías, F.A., Pozo, M.J., Flors, V., López-Ráez, J.A., Exogenous strigolactones impact metabolic profiles and phosphate starvation signalling in roots. *Plant Cell. Environ.*, 43, 1655–1668, 2020.
19. Mashiguchi, K., Seto, Y., Yamaguchi, S., Strigolactone biosynthesis, transport and perception. *Plant J.*, 105, 2, 335–350, 2021.
20. Yoneyama, K., Xie, X., Yoneyama, K., Takeuchi, Y., Strigolactones: structures and biological activities. *Pest Manag. Sci.*, 65, 5, 467–470, 2009.
21. Humphrey, A.J., Galster, A.M., Beale, M.H., Strigolactones in chemical ecology: waste products or vital allelochemicals? *Nat. Prod. Rep.*, 23, 4, 592–614, 2006.
22. Zhaoyong, K., Xiaoqi, D., Shengnan, L., Qingzhi, G., Strigolactone and Its Novel Derivatives. *Prog. Chem.*, 35, 9, 1341, 2023.
23. Dell'Oste, V., Spyrakis, F., Prandi, C., Strigolactones, from plants to human health: achievements and challenges. *Molecules*, 26, 15, 4579, 2021.
24. Mayzlish-Gati, E., Laufer, D., Grivas, C.F., Shaknof, J., Sananes, A., Bier, A., Ben-Harosh, S. *et al.*, Strigolactone analogs act as new anti-cancer agents in inhibition of breast cancer in xenograft model. *Cancer Biol. Ther.*, 16, 11, 1682–1688, 2015.
25. Xie, X., Kusumoto, D., Takeuchi, Y., Yoneyama, K., Yamada, Y., Yoneyama, K., 2'-Epi-orobanchol and solanacol, two unique strigolactones, germination stimulants for root parasitic weeds, produced by tobacco. *J. Agric. Food Chem.*, 55, 20, 8067–8072, 2007.
26. Scaffidi, A., Waters, M.T., Sun, Y.K., Skelton, B.W., Dixon, K.W., Ghisalberti, E.L., Flematti, G.R., Smith, S.M., Strigolactone hormones and their stereoisomers signal through two related receptor proteins to induce different

physiological responses in Arabidopsis. *Plant Physiol.*, 165, 3, 1221–1232, 2014.
27. Boyer, F.D., de Saint Germain, A., Pillot, J.P., Pouvreau, J.B., Xiao Chen, V., Ramos, S., Stévenin, A. *et al.*, Structure-activity relationship studies of strigolactone-related molecules for branching inhibition in garden pea: molecule design for shoot branching. *Plant Physiol.*, 159, 4, 1524–1544, 2012.
28. Zwanenburg, B., Mwakaboko, A.S., Reizelman, A., Anilkumar, G., Sethumadhavan, D., Structure and function of natural and synthetic signalling molecules in parasitic weed germination. *Pest Manag. Sci.*, 65, 5, 478–491, 2009.
29. Lombardi, C., Artuso, A., Grandi, E., Lolli, M., Spirakys, F., Priola, E., Prandi, C., Recent advances in the synthesis of analogues of phytohormones strigolactones with ring-closing metathesis as a key step. *Org. Biomol. Chem.*, 15, 38, 8218–8231, 2017.
30. Yoneyama, K., Mori, N., Sato, T., Yoda, A., Xie, X., Okamoto, M., Iwanaga, M. *et al.*, Conversion of carlactone to carlactonoic acid is a conserved function of MAX 1 homologs in strigolactone biosynthesis. *New Phytol.*, 218, 4, 1522–1533, 2018.
31. Malik, H., Rutjes, F.P.J.T., Zwanenburg, B., A new efficient synthesis of GR24 and dimethyl A-ring analogues, germinating agents for seeds of the parasitic weeds *Striga* and *Orobanche* spp. *Tetrahedron*, 66, 35, 7198–7203, 2010.
32. Matusova, R., Rani, K., Verstappen, F.W.A., Franssen, M.C.R., Beale, M.H., Bouwmeester, H.J., The strigolactone germination stimulants of the plant-parasitic *Striga* and *Orobanche* spp. are derived from the carotenoid pathway. *Plant Physiol.*, 139, 920–934, 2005.
33. Alder, A., Jamil, M., Marzorati, M., Bruno, M., Vermathen, M., Bigler, P., Ghisla, S., Bouwmeester, H., Beyer, P., Al-Babili, S., The path from β-carotene to carlactone, a strigolactone-like plant hormone. *Science*, 335, 1348–1351, 2012.
34. Seto, Y., Sado, A., Asami, K., Hanada, A., Umehara, M., Akiyama, K., Yamaguchi, S., Carlactone is an endogenous biosynthetic precursor for strigolactones. *Proc. Natl. Acad. Sci.*, 111, 4, 1640–1645, 2014.
35. Péret, B., Clément, M., Nussaume, L., Desnos, T., Root developmental adaptation to phosphate starvation: better safe than sorry. *Trends Plant Sci.*, 16, 442–450, 2011.
36. Müller, J., Toev, T., Heisters, M., Teller, J., Moore, K.L., Hause, G., Dinesh, D.C., Bürstenbinder, K., Abel, S., Iron-dependent callose deposition adjusts root meristem maintenance to phosphate availability. *Dev. Cell*, 33, 216–230, 2015.
37. Ticconi, C. and Abel, S., Short on phosphate: plant surveillance and countermeasures. *Trend Plant Sci.*, 9, 548–555, 2004.
38. Svistoonoff, S., Creff, A., Reymond, M., Sigoillot-Claude, C., Ricaud, L., Blanchet, A., Nussaume, L., Root tip contact with low-phosphate media reprograms plant root architecture. *Nat. Genet.*, 39, 792–796, 2007.

39. Al-Babili, S. and Bouwmeester, H.J., Strigolactones, a novel carotenoid-derived plant hormone. *Annu. Rev. Plant Biol.*, 66, 161–186, 2015.
40. Mayzlish-Gati, E., De-Cuyper, C., Goormachtig, S., Beeckman, T., Vuylsteke, M., Brewer, P.B., Beveridge, C.A., Yermiyahu, U., Kaplan, Y., Enzer, Y., Wininger, S., Resnick, N., Cohen, M., Kapulnik, Y., Koltai, H., Strigolactones are involved in root response to low phosphate conditions in Arabidopsis. *Plant Physiol.*, 160, 1329–1341, 2012.
41. Wang, L. and Smith, S.M., Strigolactones redefine plant hormones. *Sci. China Life Sci.*, 59, 1083–1085, 2016.
42. Wang, Y., Duran, H.G.S., van Haarst, J.C., Schijlen, E.G.W.M., Ruyter-Spira, C., Medema, M.H., Dong, L., Bouwmeester, H.J., The role of strigolactones in P deficiency induced transcriptional changes in tomato roots. *BMC Plant Biol.*, 21, 349, 2021.
43. Akash, P.A.P., Srivastava, A., Mathur, S., Sharma, A.K., Kumar, R., Identification, evolutionary profiling, and expression analysis of F-box superfamily genes under phosphate deficiency in tomato. *Plant Physiol. Biochem.*, 162, 349–362, 2021.
44. Srivastava, R., Sirohi, P., Chauhan, H., Kumar, R., The enhanced phosphorus use efficiency in phosphate deficient and mycorrhiza-inoculated barley seedlings involves activation of different sets of PHT1 transporters in roots. *Planta*, 254, 38, 2021.
45. Zhou, Z., Wang, Z., Lv, Q., Shi, J., Zhong, Y., Wu, P., Mao, C., SPX proteins regulate Pi homeostasis and signaling in different subcellular level. *Plant Signal. Behav.*, 10, 9, e1061163, 2015.
46. Puga, M.I., Rojas-Triana, M., de Lorenzo, L., Leyva, A., Rubio, V., Paz-Ares, J., Novel signals in the regulation of pi starvation responses in plants: Facts and promises. *Curr. Opin. Plant Biol.*, 39, 40–49, 2017.
47. Santoro, V., Schiavon, M., Visentin, I., Martin, M., Said-Pullicino, D., Cardinale, F., Celi, L., Tomato plant responses induced by sparingly available inorganic and organic phosphorus forms are modulated by strigolactones. *Plant Soil*, 474, 355–372, 2022.
48. Arite, T., Umehara, M., Ishikawa, S., Hanada, A., Maekawa, M., Yamaguchi, S., Kyozukaet, J., *d14*, a Strigolactone insensitive mutant of rice, shows an accelerated outgrowth of tillers. *Plant Cell Physiol.*, 50, 1416–1424, 2009.
49. Ruyter-Spira, C., Al-Babili, S., van der Krol, S., Bouwmeester, H., The biology of strigolactones. *Trends Plant Sci.*, 18, 72–83, 2013.
50. Kramna, B., Prerostova, S., Vankova, R., Strigolactones in an experimental context. *Plant Growth Regul.*, 88, 113–128, 2019.
51. Kyozuka, J., Nomura, T., Shimamura, M., Origins and evolution of the dual functions of strigolactones as rhizosphere signaling molecules and plant hormones. *Curr. Opin. Plant Biol.*, 65, 102154, 2022.
52. Villaecija-Aguilar, J.A., Hamon-Josse, M., Carbonnel, S., Kretschmar, A., Ljung, K., Bennett, T., Gutjahr, C., KAI2 regulates root and root hair development by modulating auxin distribution, bioRxiv, 539734, 2019.

53. Xie, X., Yoneyama, K., Yoneyama, K., The strigolactone story. *Annu. Rev. Phytopathol.*, 48, 93–117, 2010.
54. Yang, T., Lian, Y., Wang, C., Comparing and contrasting the multiple roles of butenolide plant growth regulators: strigolactones and karrikins in plant development and adaptation to abiotic stresses. *Int. J. Mol. Sci.*, 20, 6270, 2019.
55. Carbonnel, S., Das, D., Varshney, K., Gutjahr, C., The karrikin signaling regulator SMAX1 controls *Lotus japonicus* root and root hair development by suppressing ethylene biosynthesis. *PNAS*, 117, 35, 21757–21765, 2020.
56. Czarnecki, O., Yang, J., Weston, D., Tuksan, G.A., Chen, J.-G., A dual role of strigolactones in phosphate acquisition and utilization in plants. *Int. J. Mol. Sci.*, 14, 7681–7701, 2013.
57. López-Bucio, J., Cruz-Ramírez, A., Herrera-Estrella, L., The role of nutrient availability in regulating root architecture. *Curr. Opin. Plant Biol.*, 6, 280–287, 2003.
58. Koltai, H., Dor, E., Hershenhorn, J., Joel, D.M., Weininger, S., Lekalla, S., Shealtiel, H., Bhattacharya, C., Eliahu, E., Resnick, N., Barg, R., Kapulnik, Y., Strigolactones' effect on root growth and root-hair elongation may be mediated by auxin-efflux carriers. *J. Plant Growth Regul.*, 29, 129–136, 2010a.
59. Kapulnik, Y., Delaux, P.-M., Resnick, N., Mayzlish-Gati, E., Wininger, S., Bhattacharya, C., Séjalon-Delmas, N., Combier, J.-P., Bécard, G., Belausov, E., Beeckman, T., Dor, E., Hershenhorn, J., Koltai, H., Strigolactones affect lateral root formation and root-hair elongation in Arabidopsis. *Planta*, 233, 209–216, 2011.
60. Koltai, H., LekKala, S.P., Bhattacharya, C., Mayzlish-Gati, E., Resnick, N., Wininger, S., Dor, E., Yoneyama, K., Yoneyama, K., Hershenhorn, J., Joel, D.M., Kapulnik, Y., A tomato strigolactone-impaired mutant displays aberrant shoot morphology and plant interactions. *J. Exp. Bot.*, 61, 1739–1749, 2010b.
61. Vogel, J.T., Walter, M.H., Giavalisco, P., Lytovchenko, A., Kohlen, W., Charnikhova, T., Simkin, A.J., Goulet, C., Strack, D., Bouwmeester, H.J., Fernie, A.R., Klee, H.J., SlCCD7 controls strigolactone biosynthesis, shoot branching and mycorrhiza-induced apocarotenoid formation in tomato. *Plant J.*, 61, 300–311, 2010.
62. Lambers, H., de Britto Costa, P., Cawthray, G.R. *et al.*, Strategies to acquire and use phosphorus in phosphorus-impoverished and fire-prone environments. *Plant Soil*, 476, 133–160, 2022.
63. Dixon, M., Simonne, E., Obreza, T., Liu, G., Crop response to low phosphorus bioavailability with a focus on tomato. *Agronomy*, 10, 617, 2020.
64. Jiang, X., Chen, W., Xu, C., Zhu, H.-H., Yao, Q., Influences of arbuscular mycorrhizal fungus and phosphorus level on the lateral root formation of tomato seedlings. *J. Appl. Ecol.*, 26, 1186–1192, 2015.
65. Yoneyama, K., Xie, X., Kim, H.I., Kisugi, T., Nomura, T., Sekimoto, H., Yokota, T., Yoneyama, K., How do nitrogen and phosphorus deficiencies

affect strigolactone production and exudation? *Planta*, 235, 1197–1207, 2012.

66. Rial, C., Varela, R.M., Molinillo, J.M.G., López-Ráez, J.A., Macías, F.A., A new UHPLC-MS/MS method for the direct determination of strigolactones in root exudates and extracts. *Phytochem. Anal.*, 30, 110–116, 2019.
67. Lanfranco, L., Fiorilli, V., Venice, F., Bonfante, P., Strigolactones cross the kingdoms: plants, fungi, and bacteria in the arbuscular mycorrhizal symbiosis. *J. Exp. Bot.*, 69, 9, 2175–2188, 2018.
68. Garcia, M. and Ascencio, J., Root morphology and acid phosphatase activity in tomato plants during development of and recovery from phosphorus stress. *J. Plant Nutr.*, 15, 2491–2503, 1992.
69. Medici, A., Szponarski, W., Dangeville, P., Safi, A., Dissanayake, I.M., Saenchai, C., Emanuel, A., Rubio, V., Lacombe, B., Ruffel, S., Tanurdzic, M., Rouached, H., Krouk, G., Identification of molecular integrators shows that nitrogen actively controls the phosphate starvation response in plants. *Plant Cell*, 31, 1171–1184, 2019.
70. Hu, B. and Chu, C., Nitrogen–phosphorus interplay: Old story with molecular tale. *New Phytol.*, 225, 1455–1460, 2020.
71. Maeda, Y., Konishi, M., Kiba, T., Sakuraba, Y., Sawaki, N., Kurai, T., Ueda, Y., Sakakibara, H., Yanagisawa, S., A NIGT1-centred transcriptional cascade regulates nitrate signalling and incorporates phosphorus starvation signals in Arabidopsis. *Nat. Commun.*, 9, 1376, 2018.
72. Vurro, M., Prandi, C., Baroccio, F., Strigolactones: how far is their commercial use for agricultural purposes? *Pest Manag. Sci.*, 72, 11, 2026–2034, 2016.

6

# Strigolactones' Role in Heat and Saline Stress Tolerance in Horticultural and Field Crops

**Juan Pablo Rodriguez[1]*, Jose Delatorre Herrera[2], Luisa Bascuñán[3] and Enrique Ostria[4]**

*[1]Julius Kuhn-Institut (JKI), Federal Research Centre for Cultivated Plants, Institute for Plant Protection in Horticulture and Urban Green, Braunschweig, Germany*

*[2]Doctoral Program in Agriculture for Arid-Desert Environments, Faculty of Renewable Natural Resources, Desert Agriculture Area, Universidad Arturo Prat, Iquique, Chile*

*[3]Laboratory of Plant Physiology, Department of Botany, Universidad de Concepción, Concepción, Chile*

*[4]Department of Botany, Faculty of Natural Sciences and Oceanography, Universidad de Concepcion, Concepcion, Chile*

## Abstract

Salinity and heat are the primary constraints on crop production. Salinity is affecting a significant area of land worldwide, and agriculture is becoming increasingly challenging. Climate is also affecting crops through cold and heat. Globally, the temperatures have been reported to be increasing, and the forecast is that the global temperature will increase by 1.7°C. This means that new forms of crop production and stress-tolerant mechanisms must be understood and utilized to enhance food production. This chapter will present the role and use of endogenous and exogenous strigolactones in horticultural and field crops as well as the mechanisms by which plants cope with stress due to salinity and heat.

***Keywords:*** Endogenous, exogenous, strigolactones, abiotic stress, heat, salinity

**Corresponding author*: jprodriguezcalle@yahoo.fr

Tariq Aftab and Kaiser Iqbal Wani (eds.) Strigolactones: Emerging Plant Hormones, (109–136) © 2025 Scrivener Publishing LLC

## 6.1 What are Strigolactones

Plant roots secrete a diverse array of compounds into the rhizosphere, the soil immediately surrounding the roots. These exudates influence the organisms that inhabit this niche. Among these exudates are the strigolactones (SLs). These were initially known as host-derived germination stimulants for parasitic weeds from the genera *Striga* and *Orobanche*, where they play a crucial role in synchronizing the life cycles of the parasitic plant with its host. Subsequently, SLs were identified as a signal for symbiosis between plants and arbuscular mycorrhizal (AM) fungi. In this symbiotic association, SLs are exuded by the roots, promoting the hyphal branching of AM fungi and facilitating their growth toward the plant's roots [1, 2]. However, recent research has expanded our understanding of their involvement in various plant physiological processes, which are pivotal functions in fostering plant growth, developmental processes, and communication within the rhizosphere. This includes the discovery of a new class of branch-inhibiting phytohormones, which regulate the overall architecture of land plants [3, 4]. To date, a significant body of genetic research has identified the molecular pathways associated with the production of SLs as a compound essential for suppressing lateral (axillary) branching across a diverse range of plant species.

From a chemical perspective, SLs represent a category of plant hormones derived from carotenoids, consisting of a butenolide ring (D-ring) connected via an enol-ether bridge to a tricyclic lactone (ABC-ring). Over 20 natural SLs have been identified, differing in the structure of the ABC-ring or in the substituents on the D-ring, which significantly impacts their biological activity [5].

The synthesis of SLs commences with the transformation of all-trans-β-carotene into 9-cis-β-carotene, a reaction that is catalyzed by the enzyme DWARF27. Subsequent cleavage by carotenoid cleavage dioxygenases (CCDs), specifically CCD7 and CCD8, leads to the production of carlactone (CL), a direct precursor of various SLs. SLs are classified into two distinct categories: canonical and non-canonical. While canonical SLs are characterized by the presence of an ABCD ring structure (Figure 6.1), non-canonical SLs lack one or more of the A, B, or C rings but include an enol ether-D ring component that is crucial for their biological functions [6].

Several modifications of CL occur under the action of cytochrome P450 monooxygenases, notably more axillary growth 1 (MAX1) from the CYP711A family, among other enzymes. These modifications contribute to the wide array of over 30 recognized natural canonical and non-canonical

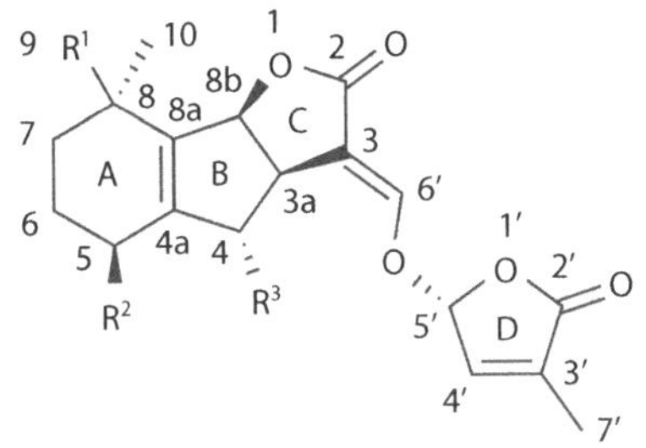

**Figure 6.1** Strigolactones' general chemical structure (source: Wikimedia Commons, the free media repository).

structure SLs [7, 8]. SLs exert their effects through a receptor designated as DWARF14 (D14), an α/β hydrolase. Upon perception of SLs, D14 interacts with the F-box protein D3 (MAX2 in *Arabidopsis*), leading to the ubiquitination and subsequent degradation of suppressor proteins (SMXLs), which are involved in the inhibition of shoot branching [8]. This pathway is intricately linked with the biosynthesis of SLs and environmental and developmental signals that modulate plant growth [9].

SLs are generated and secreted in minute amounts, with an effective range of pico- to nanomolar concentrations. They have been reported in a variety of plant species, including those studied by Andreo-Jimenez *et al.* [10], Delaux *et al.* [11], and Xie [7]. It has been demonstrated that individual plant species can produce multiple forms of SLs, with distinct varieties within the same species generating a diverse mixture of SL compounds in varying proportions [12, 13].

## 6.1.1 Endogenous Production Under Abiotic Stress

While roots are typically the primary source of SLs, research indicates that these compounds act as long-distance signals that travel to shoots, influencing shoot tillering and branching suppression, regulating secondary stem growth, plant height, and leaf shape [14, 15]. Such strategic adaptations enable plants to modify root and shoot architectures, optimizing resource allocation and reducing energy use during stress [16]. SLs play a pivotal role in assisting plants with various abiotic stresses. They contribute to oxidative stress responses by enhancing osmolyte production to maintain cellular balance, influencing the expression of a wide array of genes, including those involved in hormone signaling pathways, modulating the activity of ion transporters, and enhancing the activity of antioxidant enzymes [17, 18].

Pi deficiency: Under phosphorus (Pi) deficiency, there is a surge in SL production within roots that leads to improved development of root architecture [19], enabling the plant to explore a larger soil area for nutrient uptake (Table 6.1). This systemic mechanism allows SLs to rapidly adjust the aboveground architecture by altering PIN accumulation, while SL biosynthesis in the shoot is also increased, influencing branching patterns [19]. This is because SLs' activity is finely tuned through interactions with other hormones, particularly auxin and abscisic acid (ABA), highlighting their significant role in root architecture modulation under low Pi conditions. Additionally, it leads to the upregulation of the SL exporter PDR1, located in the plasma membrane of root sub-epidermal cells, facilitating SL solubilization into the rhizosphere and promoting symbiotic interactions with arbuscular mycorrhizal fungi (Table 6.1), which enhances Pi absorption by the plant [20].

In response to salinity stress, plants experience an imbalance in ions, particularly elevated external $Na^+$ concentrations. SLs enhance plant tolerance by modulating the expression and activity of ion transporters. SLs

**Table 6.1** Endogenous generation of SLs under abiotic stresses in plants.

| **Endogenous** | **Characteristic** | **References** |
|---|---|---|
| Pi deficiency | SL production within roots improves root development.<br>Interaction with auxin and abscisic acid (ABA) | [19]<br>[20] |
| In response to salinity stress | SLs influence root architecture and promote the deeper growth of roots.<br>SLs enhance antioxidant defence systems like SOD and CAT and enhance stress tolerance. | [12]<br>[22, 23] |
| Drought | SL with ABA facilitates stomatal closure. | [24, 25] |

regulate the expression of genes for sodium/potassium transporters, which are crucial for maintaining ion balance by controlling $Na^+$ exclusion and $K^+$ retention. These processes are essential for plant survival under salt stress [21]. SLs significantly influence root architecture, promoting the growth of deeper root systems that can access water from less saline soil layers. This is achieved by altering the expression of genes involved in root development and by affecting the hormone signaling pathways that control root growth and branching [12]. Salinity stress leads to the generation of reactive oxygen species (ROS), which are detrimental to cellular structures. SLs enhance the antioxidant defense systems by upregulating the expression of antioxidant enzymes such as superoxide dismutase (SOD) and catalase (CAT), thereby reducing oxidative damage and enhancing stress tolerance [22, 23] (Table 6.1).

Drought: Stomatal closure is a rapid and effective plant response to minimize water loss under stress conditions, including high salinity. Research indicates that SLs engage with ABA signaling pathways to facilitate stomatal closure. This interaction involves the upregulation of ABA-responsive genes and can be mediated through SL-induced changes [24, 25] (Table 6.1). However, the effects of drought stress on the production of endogenous SLs differ between mycorrhizal and non-mycorrhizal plants—for example, increased drought stress led to a reduction in the synthesis of SLs, DDH, and solanacol in non-mycorrhizal tomato plants, whereas their production was augmented in mycorrhizal tomato plants subjected to similar drought conditions [26].

### 6.1.2 Exogenous Use to Alleviate Abiotic Stress

Research has shown that applying SLs externally enhances plant resilience against various abiotic stresses—for example, using synthetic SL analogues such as GR24 has proven effective in boosting tolerance to nutrient deficiency, drought, heat, and heavy metals in diverse plant species, thereby improving their growth and survival rates. This phenomenon is mainly linked to modifications in root development and architectural changes [16, 18].

Pi availability: The impact of SLs on lateral root development can be either inhibitory or promotive, contingent on the availability of phosphorus (Pi). The application of the synthetic SL GR24 on *Arabidopsis* in an auxin-containing medium has been demonstrated to accelerate lateral root elongation in comparison to controls without GR24, suggesting a synergistic

interaction with auxin [12] (Table 6.2). Moreover, when GR24 is applied to plants with adequate Pi, it results in a significant reduction in lateral root numbers compared to plants experiencing Pi deficiency without GR24 treatment [12]. Furthermore, GR24 has been observed to reduce auxin concentrations in foliage, indicating that SLs' overall effects under low Pi scenarios are contingent upon the plant's auxin levels, which are susceptible to environmental conditions and Pi availability [27].
Drought: Considerable research efforts aim to unravel how exogenous SLs influence drought stress mechanisms in plants. Exogenous SLs have been identified as positive regulators that enhance drought resistance in both wild-type and SL-deficient *max* mutant *Arabidopsis* plants (Table 6.2) [24]. Furthermore, plants with compromised SL biosynthesis and signaling pathways exhibit heightened sensitivity to water deficits [24]. SLs have been found to improve drought tolerance in wheat varieties

**Table 6.2** Exogenous application of SLs under abiotic stresses in plants.

| Exogenous | Characteristic | References |
|---|---|---|
| Pi availability | Application of synthetic SL GR24, e.g., on *Arabidopsis* can accelerate root elongation. | [12] |
| | Can reduce auxin concentration in foliage | [27] |
| Heavy metal stress | SLs have a main role in the management of heavy metal contamination, e.g., Cd. | [36, 37] |
| | Involved with shoot formation and facilitates interaction with microbes | [38] |
| Drought | SLs have a positive role, e.g., wild-type *Arabidopsis*. | [24] |
| | SL with salicylic acid enhances drought tolerance in wheat cultivation. | [28] |
| | SLs have a beneficial impact on plant growth in drought conditions. | [29, 30] |
| *Heat stress* | SL GR24 could benefit the elongation of crown roots, e.g., *Festuca arundinaea*. | [32, 33] |
| | Mitigates stress and enhances germination using GR24, e.g., *Lupinus polyphyllus* | [33] |

by reducing electrolyte leakage and boosting the activity of antioxidant enzymes. The extent of improvement varies with the drought tolerance inherent to each cultivar. Additionally, SLs have been observed to act synergistically with salicylic acid, enhancing drought tolerance in wheat grown under water-deficient conditions [28]. SLs applied externally have uniformly demonstrated beneficial impacts on plant growth metrics under drought conditions, enhancing transpiration rate, stomatal conductance, water use efficiency, and net photosynthesis rate [29, 30]. Furthermore, these treatments have been demonstrated to mitigate the adverse effects of drought on both the absorption-based performance index and the maximum photochemical efficiency of photosystem II [30]. It is notable that while SL treatment initially enhances the activity of photosynthetic enzymes, it subsequently lowers the ABA content in *Pennisetum purpureum* after 120 h [30]. Furthermore, external SL applications regulate stomatal closure through mechanisms involving reactive oxygen species (ROS) or ABA, modulate chlorophyll components, and activate antioxidant capabilities to diminish the detrimental impacts of drought, thereby promoting an interaction between SL and ABA pathways. Such treatments have been reported to decrease the malondialdehyde content and increase the relative water content under drought stress, further enhancing leaf ascorbic acid, total phenolics, and antioxidant activities in maize, thereby improving water use efficiency, stomatal conductance, chlorophyll content, gas exchange traits, net $CO_2$ assimilation rate, photosynthetic pigments, and overall plant growth under drought conditions [31]. Despite the numerous studies that have been conducted, further refinement is required to develop more effective application protocols concerning the timing, dosage, and methods of SL treatments to alleviate the effects of drought stress.

Heat stress: A critical constraint that impedes root development in cold-season plants is their requirement for considerably lower temperatures, which renders them exceptionally vulnerable to increased external temperatures. Studies have demonstrated the beneficial effects of GR24 on the elongation of crown roots and the number of root cells in heat-stressed *Festuca arundinacea*. This is evidenced by alterations in the expression patterns of key genes involved in cell division and the cell cycle within the root system, including *Proliferating Cell Nuclear Antigen* (*PCNA*), *Cyclin-D2* (*CycD2*), and *Cyclin-Dependent Kinase B* (*CDKB*) [32]. Furthermore, temperature has a significant influence on seed germination, with high temperatures generally impeding this process. The GR24 treatment has been demonstrated to mitigate the adverse effects of heat stress on the germination of *Lupinus polyphyllus* seeds [33]. Furthermore, GR24 enhances antioxidant activity, which facilitates the scavenging of ROS during germination. It has been observed that SLs can counteract temperature-induced dormancy in *Pasteuria ramosa* seeds. ABA is known to be a strong inhibitor of seed germination, while gibberellins (GA) promote it. The administration of SLs has been demonstrated to reduce the ABA/GA ratio, thereby increasing the germination rates of *Lactuca sativa* seeds at elevated temperatures [34]. Furthermore, the exogenous application of GR24 stimulated the expression of the ABA catabolic gene (*CYP707A1*), resulting in the breakdown of ABA and facilitating seed germination even under conditions of heat stress [35].

Enhancement of metal chelation: SLs play a pivotal role in the management of heavy metal contamination, such as cadmium stress (Table 6.2). The synthetic analogue GR24 is of paramount importance for regulating growth, root development, and ROS scavenging [36, 37].

SLs are involved in shoot formation in higher plants under varied environmental and nutritional conditions, facilitating interactions with microbes [38]. They are particularly effective in mitigating cadmium toxicity, as evidenced by the alleviation of phytotoxicity in switchgrass following the application of GR24. Furthermore, residual GR24 in the soil enhances the levels of iron and zinc, which interact with other heavy metals to enhance plant antioxidative enzymes, including peroxidase (POD), superoxide dismutase (SOD), and catalase (CAT) [39].

The action of SLs on stomatal closure is notably fast, and they enhance nitric oxide (NO) production. This is counteracted by cPTIO (widely used as a NO scavenger), which serves as an effective counteragent [40]. In areas subjected to the stress of heavy metals, deficiencies in nitrogen and phosphate stimulate root elongation [18]. Phosphorus, in its inorganic form as phosphate (Pi), plays a pivotal role in stimulating root architecture, thereby reducing heavy metal uptake through enhanced lateral root formation when adequate Pi is available. This adjustment in root architecture, including increased root hair length and density, equips plants to manage heavy metal stress more effectively [12, 41, 42].

In plants, SLs facilitate the binding of heavy metals. This process of chelation prevents the metals from catalysing the production of harmful free radicals, thereby protecting cellular structures from damage. This mechanism has been demonstrated in studies where plants exposed to metals such as cadmium and lead showed improved resilience when pretreated with SLs.

## 6.2 Production, Role, and Understanding of Strigolactones in Alleviating Heat Stress in Horticultural and Field Crops

Temperature is a significant factor that controls the distribution, productivity, and physiological activity of plants both spatially and temporally [43]. One of the most significant threats associated with climate change and global warming is the increase in frequency, intensity, and duration of extreme climatic anomalies such as heatwaves and high temperature events [44]. Anomalous high-temperature events and heatwaves not only bring high temperatures but also an increase in soil water evaporation [45]. These conditions can have significant impacts and cause long-term ecological changes [46, 47].

In the context of plant biology, heat stress is defined as temperatures that exceed a plant's optimal threshold. Consequently, heat stress affects the physiological, biochemical, and molecular responses of plants [48]. These impacts are manifested primarily through disruptions in photosynthesis, nutrient uptake, and overall plant metabolism, which ultimately affect plant growth and productivity [44, 49–52]. Nevertheless, despite the growing number of studies, we are still far from a comprehensive understanding of the physiological mechanisms of plants in response to heat stress [44]. The primary physiological consequence of heat stress is the inhibition of photosynthesis. This is not solely due to direct thermal damage to the photosynthetic apparatus but also due to the induced stomatal closure that reduces $CO_2$ availability [53–55]. Elevated temperatures have been shown to affect carboxylation efficiency and induce the deterioration of components involved in the electron transport rate [56]. Furthermore, heat stress affects the water relations of plants. High temperatures increase the water vapor pressure deficit between the leaf and the atmosphere, thereby augmenting transpiration rates. This can result in rapid dehydration if not counterbalanced by adequate water uptake from the roots, which itself can be impeded by heat-induced reductions in root growth and function [54, 57]. Consequently, regardless of the climate change scenario that we face, the increasing temperatures pose a significant threat to the protection of ecosystems, the functioning of agricultural systems, and the security of food supplies.

Rather than representing a discouraging outlook, plant biologists are presented with a significant opportunity to conduct research involving holistic analyses with multiple tools, with the aim of better understanding plant responses to extreme temperature events. Consequently, investigating the consequences of heat stress at a physiological and metabolic level serves as a cornerstone for two key purposes: firstly, to discover mechanisms of resistance and/or tolerance and, secondly, to develop and implement crops that are resistant and suited to the prevailing new climatic conditions [58–60]. To mitigate the effects of heat stress, several strategies have been employed. A primary focus of breeding for heat tolerance is the incorporation of genetic traits that enhance cellular thermotolerance, improve water retention, and stabilize photosynthesis during heat events [61]. Agronomic practices, such as altering planting dates, increasing plant density, and using shading or evaporative cooling, are also effective in managing the microenvironment of crops, thereby reducing the heat load [62].

Strigolactones are multifunctional plant metabolites that act as allelochemicals in the rhizosphere and as a novel class of hormones regulating growth and development in plants as well as their response to various abiotic stresses, including heat stress. Studies have indicated that SLs may play a role in enhancing plant resilience to heat stress by modulating physiological and molecular responses [24]. They are involved in the regulation of stomatal closure, which is crucial under heat stress conditions to reduce water loss while maintaining the carbon dioxide influx for photosynthesis [25]. Furthermore, SLs have been demonstrated to upregulate heat-shock proteins (HSPs) (Figure 6.2). The figure on heat stress and the role of strigolactones, which play a pivotal role in protecting plants against heat stress by ensuring protein stability and preventing the aggregation of denatured proteins [63]. Moreover, these hormones facilitate the enhancement of antioxidant defence systems in plants, thereby reducing oxidative damage triggered by elevated temperatures [64]. The enhancement of lateral root growth by SLs facilitates the uptake of water and nutrients, thereby ensuring the maintenance of turgor and cellular functions during

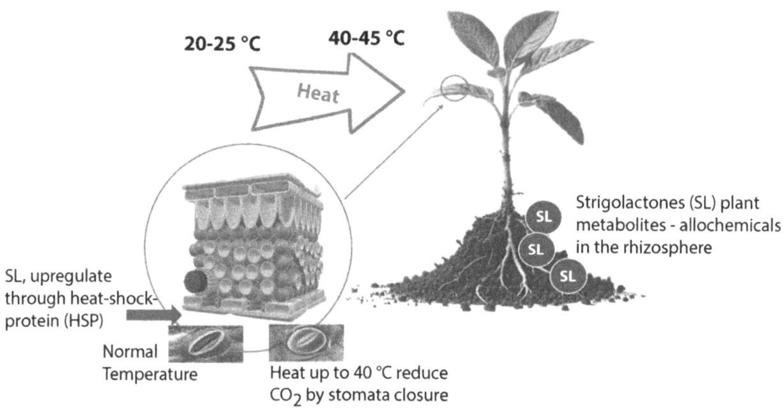


**Figure 6.2** Role, and understanding of strigolactones in alleviating heat stress in horticultural and field crops.

periods of elevated heat stress [17]. This enhanced water uptake facilitated by improved root systems can be crucial during periods of high temperature, as it reduces the evaporation of moisture from the soil.

Considering these functions, SLs present a promising avenue for the development of heat-tolerant crop varieties. This section will examine the morpho-physiological, biochemical, and molecular mechanisms in which SLs are involved in plant biology, with a particular focus on their potential role in mitigating the effects of heat stress.

### 6.2.1 Morpho-Physiological Responses of Plants to Strigolactones

It has been demonstrated that SLs exert their control over physiological and morphological processes through a direct or indirect influence on the activities of other hormones and/or integrants of signaling cascades of other growth regulators [33]. SLs regulate hypocotyl elongation, thereby exerting regulatory influence on the photomorphogenic response [65]. They have been demonstrated to stimulate internode elongation and secondary growth in stems. Additionally, SLs modulate leaf angle and orientation, reducing the surface area exposed to solar radiation. This reduces leaf temperature and the plant's overall heat load. SLs influence shoot architecture by regulating branching. This is achieved by modulating polar auxin transport to interrupt auxin canalization and independently of auxin by acting directly within buds by promoting the key branching inhibitor TEOSINTE BRANCHED1. In optimal conditions, buds will grow out. However, in suboptimal conditions, SL levels increase to restrict branching. This can be problematic in agricultural applications, as reductions in branching can have adverse effects on crop yield. A recent study has identified variations in the promoter elements of key SL-related genes, such as *IDEAL PLANT ARCHITECTURE1*, which promote a phenotype with enhanced yield performance [66].

Conversely, SLs can modulate root architecture by stimulating adventitious root formation, which enhances water uptake and is crucial for maintaining turgor pressure and overall plant hydration during heat stress [4]. Additionally, SLs facilitate the acquisition of defensive characteristics against drought stress by initiating physiological responses and mediating the interaction with soil microorganisms. Nutrient deficiency represents a significant abiotic stress factor, prompting plants to deploy a range of strategies to ensure their survival. These include enhancing the efficiency of nutrient uptake and forming beneficial relationships with microorganisms. In this sense, SLs act as a signal molecule to attract various microorganisms and provide the roots with essential elements, including nitrogen and phosphorus. Among these beneficial microorganisms are AMF, which regulate plant metabolic activities through the provision of phosphorus to the roots [67].

The biosynthesis of SLs in response to stress, such as heat stress, confers several morpho-physiological adaptive advantages to plants, enhancing their ability to cope with thermal stress and maintain physiological homeostasis. These hormones are involved in the enhancement of leaf epidermal features, including increased trichome density and thicker cuticles, which provide a physical barrier against heat stress. These features assist in reflecting solar radiation and reducing water loss, further contributing to heat tolerance [68]. One of the most significant morpho-physiological effects of SLs under heat stress is the regulation of stomatal aperture. SLs enhance the plant's responsiveness to ABA, resulting in more efficient stomatal closure. This action reduces water loss through transpiration, a critical adaptation during heat stress that helps to prevent dehydration and maintain cellular turgor, which is crucial for sustaining photosynthetic and metabolic activities during high temperatures [24]. Furthermore, SLs play a pivotal role in maintaining photosynthetic performance under heat stress conditions by promoting the expression of genes associated with photosynthesis, chloroplast development, and carbon fixation. By enhancing photosynthetic capacity and minimising photooxidative damage, SLs enable plants to maintain carbon assimilation rates and sustain growth even in the face of thermal stress—for instance, the application of the synthetic SL analog GR24 to *Lupinus angustifolius* conferred resistance to heat stress-induced damage to the oxygen evolution complex and prevented the inactivation of PSII reaction centers, thus ensuring PSII thermotolerance [33].

Morpho-physiological responses are of critical importance for maintaining growth and productivity in horticultural and field crops subjected to episodic or prolonged heat stress. Although considerable research has been conducted to elucidate the effects of SLs interactions with other hormones at the morpho-physiological level, a more comprehensive understanding of the signaling and perception of SLs may provide insight into the sustainability of plants under heat stress. The SLs may be regarded as a gateway to sustainable agriculture.

### 6.2.2 Biochemical Attributes of Strigolactones in Alleviating Heat Stress in Horticultural and Field Crops

Strigolactones facilitate a range of biochemical modifications that improve the responses of plants to elevated temperatures. One of the primary biochemical roles of SLs under heat stress is the enhancement of the plant's antioxidant defence system. SLs stimulate the activity of key antioxidant enzymes, including SOD, CAT, and POD. These enzymes play a pivotal role in the detoxification of ROS, which are copiously generated under stressful conditions. By mitigating oxidative stress, SLs help to preserve cellular integrity and functionality [24]. This action is of vital importance for maintaining photosynthesis and other metabolic processes during periods of thermal stress.

SLs also play a role in regulating the synthesis of heat shock proteins (HSPs), which are crucial for protein homeostasis under stress conditions. HSPs act as molecular chaperones, stabilising proteins, preventing their aggregation, and facilitating their refolding if denatured by heat. This process is essential for the maintenance of enzymatic activities that are critical for plant growth and stress mitigation [38]. The upregulation of HSPs by SLs thus provides a protective buffer against the damaging effects of high temperatures on the proteome.

Furthermore, SLs influence the biosynthesis of secondary metabolites that are involved in stress defence mechanisms. These include flavonoids and other phenolic compounds, which possess antioxidant properties and contribute to cellular protection against thermal stress. The biochemical roles of SLs in modulating the pathways that lead to the accumulation of these compounds can enhance the plant's overall resilience to environmental challenges [13].

### 6.2.3 Molecular Response of Strigolactones in Alleviating Heat Stress in Horticultural and Field Crops

At the molecular level, SLs orchestrate a complex network of responses that enable plants to cope with heat stress more effectively. SLs modulate gene expression, signaling pathways, and protein function, thereby providing plants with the necessary tools to adapt to increased temperatures.

At the level of gene expression regulation, SLs exert a significant influence on the expression of genes encoding HSPs. These proteins protect plants from the detrimental effects of heat by preventing protein denaturation and aggregation. SLs also enhance the expression of genes involved in the biosynthesis and signaling of ABA, a key hormone that mediates stomatal closure during heat stress, thus reducing water loss [38]. SLs exert their influence on the activity of specific transcription factors that regulate stress-responsive genes. Moreover, they are involved in epigenetic modifications that alter gene expression in response to heat stress. These modifications may include methylation and acetylation processes that alter the chromatin structure and subsequently affect gene transcription levels, allowing plants to quickly adapt to changing environmental conditions in a more rapid manner [13].

Regarding the modulation of signaling pathways, it is established that SLs modulate the ABA signaling pathway, thereby enhancing its sensitivity and effectiveness during heat stress conditions. Furthermore, SLs are involved in the crosstalk with other signaling molecules, such as salicylic acid and jasmonic acid, which are crucial for the plant's defence against both biotic and abiotic stresses. This crosstalk enables a coordinated response that optimises the plant's survival strategies under heat stress [24].

In addition to alleviating immediate stress, SLs also prepare plants, for future stress encounters. This priming effect is mediated by epigenetic modifications, which facilitate a more rapid and robust transcriptional response to subsequent stress events. Conversely, heat-tolerant crops, including some legumes and drought-adapted species, may exhibit constitutive or primed responses to heat stress. This is due to the basal levels of SLs contributing to their resilience under challenging conditions. It is of great importance to gain an understanding of the specific dynamics of heat-induced SL production in crops, as this knowledge is essential for the development of tailored agronomic strategies and breeding approaches aimed at enhancing heat stress tolerance and productivity in diverse cropping systems. Such memory effects are of crucial importance for long-term adaptation and survival under recurring stress conditions [13].

## 6.3 Production, Role, and Understanding of Strigolactones to Alleviate Salinity Stress in Horticultural and Field Crops

### 6.3.1 Morpho-Physiological Responses of Plants

The presence of excess salts in the soil can alter the porosity and hydraulic conductivity of the soil, which, in turn, affect the water-holding capacity of the soil. This can also result in a low water potential, which can lead to osmotic stress. It has been demonstrated that salinity can result in a range of physiological challenges for plants, including reduced growth and development [69]. This is due to the destabilisation of cell membranes and an increase in protein degradation, which are both the result of the toxic effects of sodium ions ($Na^+$). Furthermore, intracellular osmotic pressure is also increased due to the large accumulation of sodium, which can reach toxic levels [70]. In general terms, salinity causes osmotic and ionic alterations in the physiology of plants [71–73], which will influence the following:

- ❖ The reduction in leaf area is a consequence of the inhibition of cell division and growth.
- ❖ This results in a reduction in the growth of tissues, including roots, stems, leaves, and ultimately the death of the plant.
- ❖ The plant may also become intoxicated due to the accumulation of sodium in the cells.
- ❖ Furthermore, there is a reduction in photosynthetic activity.
- ❖ This results in a reduction in water uptake, which in turn impairs water nutrition and metabolism.

Saline–alkaline soils are distinguished by a high salt content and high pH values, which concurrently inflict combined damage due to osmotic and ionic stress, ionic toxicity, high pH and $HCO_3/CO_3^{2-}$ stress. These factors collectively induce oxidative stress [74–77].

## 6.3.2 Biochemical Responses at the Plant Level

### *6.3.2.1 Ion Imbalance*

Ion stress occurs when toxic ions have a detrimental effect on plants, and an imbalance in ion homeostasis disrupts the uptake of other ions. These two stresses result in oxidative stress in plants, which is the generation of excess ROS (hydroxyl radicals, hydrogen peroxide, and superoxide anions) [78–80].

Reactive oxygen species induced by salt-alkali stress, including superoxide anions ($O_2^-$), hydrogen peroxide ($H_2O_2$), and singlet oxygen, cause oxidative stress that can lead to increased permeability of the plant cell membrane and increased ionic permeability [81–83]. The alkaline pH affects the absorption of inorganic elements, thus altering the intracellular ionic balance [84]—for example, in apple trees, the absorption of K, Mg, and Zn is inhibited, while the absorption of Fe, Cu, or Mn increases under salt-alkali stress [85].

These essential mineral elements, including $K^+$, $Ca^{2+}$, $Fe^{2+}$, and $Cu^{2+}$, participate in numerous metabolic pathways, such as signal transduction and carbon assimilation [86–88]—for instance, $Ca^{2+}$ plays a pivotal role in signaling by modulating $Na^+$ influx directly and maintaining $Na^+/K^+$ homeostasis via the *SOS* gene pathway in the context of salt stress [86, 89].

In contrast, $Fe^{2+}$ is an essential nutrient for plants, participating in photosynthesis in the ferredoxin complex and also involved in the synthesis of chlorophyll and electron transport [90].

#### *6.3.2.2 Osmotic Imbalance*

Metabolic acclimatization through the accumulation of compatible solutes is frequently regarded as a fundamental strategy for the protection and survival of plants under abiotic stress [91, 92]. A considerable number of plant species have been observed to accumulate considerable quantities of glycine betaine, proline, polyols, ABA and other metabolites in response to high salinity [93]. A number of functions have been proposed for these compounds, with suggestions that they act as low-molecular-weight chaperones, stabilising the photosystem II complex and protecting the structure of enzymes and proteins, maintaining membrane integrity and eliminating ROS [94]. The water deficit resulting from osmotic stress induces stomatal closure, thereby reducing the internal $CO_2$ concentration (Ci). This in turn causes a decrease in plant photosynthesis [22].

### 6.3.3 Molecular Response Level

#### *6.3.3.1 Strigolactones and Salinity*

A factor that is rarely considered in conditions of salinity and nutrient deprivation is the synthesis of SLs. Under salt stress conditions, plants reduce their nutrient absorption; to compensate, plants must activate survival mechanisms, one of which is related to SL synthesis. High concentrations of SLs lead to the suppression of the branching of aerial shoots, modification of the root structure and stimulation of symbiosis with AM fungi [3, 4].

SLs play a pivotal role in alleviating nitrogen and phosphorus deficiencies by modifying root and shoot architecture [95]. They facilitate the symbiotic relationship with bacteria and/or fungi, ensuring the supply of water and nutrients (particularly phosphate and nitrogen). An experiment conducted by Ruyter-Spira *et al.* [12] examined the effects of a synthetic analogue of SL GR24 in wild type (WT) Arabidopsis and a mutant insensitive to SL, grown with low phosphorus concentrations. The results demonstrated that WT plants treated with GR24 exhibited a greater number of lateral roots in comparison to the mutants. Conversely, the application of GR24 in the presence of high phosphate concentrations results in a reduction in lateral root density, which is attributed to the suppression of both outgrowth and lateral root formation potential.

In addition to their role in regulating phosphate homeostasis, SLs have also been identified as a potential pathway for regulating plant growth in response to nitrogen supply [96]. Ravazzolo *et al.* [97] demonstrate that N deficiency can strongly induce SL biosynthesis in maize roots. Zealactone represents the SL of corn and manifests itself in N deficiency. However, the current results indicate that carlactonoic acid (CLA) could also mark this condition in the roots. CLA is obtained by oxidation of carlactone and is the putative intermediate of strigol-type and orobanchol-type SLs [98, 99]. Moreover, CLA is the precursor of the non-canonical SL called zealactone [100].

Tests on apple trees demonstrated that the application of 100 μM of exogenous SL analog (GR24) effectively alleviates salinity-alkalinity stress, as evidenced by an increase in chlorophyll content, which indicates a higher photosynthetic rate. Conversely, the exogenous application of GR24 regulates the expression of $Na^+/K^+$ transporter genes and decreases the $Na^+/K^+$ ratio in the cytoplasm, thereby maintaining ion homeostasis. Secondly, the application of exogenous GR24 resulted in an increase in the enzymatic activities of SOD, POD, and CAT, thereby reducing the production of ROS. Thirdly, the exogenous application of GR24 alleviated high pH stress by regulating the expression of $H^+$-ATPase genes and inducing organic acid production. Finally, the exogenous application of GR24 resulted in an increase in the endogenous contents of acetic acid, ABA, zeatin riboside and GA3, which indirectly responded to salinity-alkalinity stress [21] (Figure 6.3).

In a study conducted by Ma *et al.* [101] on rapeseed (*Brassica napus* L.), a crop that is significantly affected by salinity, treatments with GR24 were tested. The results demonstrated that salinity inhibited shoot and root growth in plants that did not receive SLs, while applications of GR24 enhanced growth under salt stress. Furthermore, leaf chlorophyll contents and gas exchange parameters (net photosynthetic rates, stomatal conductance, intercellular $CO_2$ concentration, and transpiration rate) were also significantly reduced with increasing salinity. The GR24 treatment significantly increased and decreased the quantum yield of photosystem II and

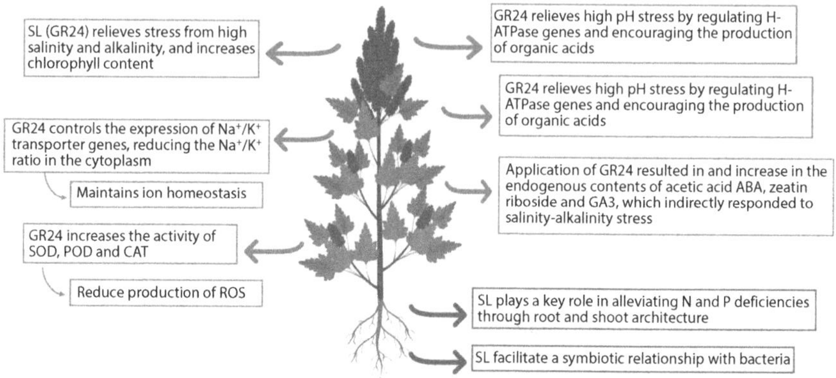


**Figure 6.3** Role of strigolactones in salt stress conditions in horticultural and field crops.

non-photochemical quenching, respectively, under salinity stress conditions. The application of GR24 resulted in an increase in POD and SOD activities and a reduction in malondialdehyde levels, indicating a decrease in lipid peroxidation. Quantitative PCR analysis demonstrated that the observed stress relief was primarily associated with the expression of genes involved in tryptophan metabolism, plant hormone signal transduction, and photosynthesis.

Studies on chilli (*Capsicum annuum* L) have demonstrated that 20 μM SL increases height by 21% and dry weight by 33%. Furthermore, fruit length and circumference reach values of 19.24% and 35.37% respectively, resulting in an increase in fruit yield of 60.74% compared to the salt-stressed control. These results are accompanied by an enhancement in chlorophyll a (18.65%) and chlorophyll b (43.52%) content compared to the control. Furthermore, the improvement in nitrogen, phosphorus and potassium concentration in the leaves substantiates the efficacious performance of 20 μM SL in comparison to other concentrations under salt stress. The study concluded that 20 μM SL is the most effective treatment for mitigating salt stress in chilli plants [102].

## 6.4 Conclusions and Future Perspectives of Strigolactones' Use for Alleviating Stress in Plants

- The role of strigolactones (SLs) in mitigating heat stress in plants is multifaceted, encompassing various physiological, biochemical, and molecular mechanisms that enhance plant resilience. SLs act as pivotal modulators of plant responses to elevated temperatures by influencing stomatal conductance, promoting heat-shock protein (HSP) synthesis, and enhancing antioxidant defences. These hormones facilitate the maintenance of photosynthetic efficiency and cellular homeostasis under stress conditions.
- Furthermore, SLs influence root architecture, enhancing water and nutrient uptake efficiency, which is of vital importance during periods of elevated soil moisture evaporation rates, such as those experienced during heat stress. This enhanced root system functionality is instrumental in supporting plant turgor and hydration, which are of critical importance for the sustenance of metabolic activities.
- The application of SLs or their analogues represents a promising strategy for the development of heat-tolerant horticultural and crop varieties. The synthesis of SLs under stress conditions confers adaptive advantages, including enhanced leaf epidermal features that reduce water loss and reflect solar radiation, thereby further contributing to heat tolerance. An understanding of the role of SLs' in these processes could lead to the development of innovative approaches to sustainable agriculture, with the potential to optimise crop performance under the increasingly frequent and severe heat stress events anticipated with climate change.
- In summary, SLs represent a crucial avenue for enhancing crop resilience to heat stress, with potential applications in breeding programmes and agronomic practices aimed at ensuring food security in the face of global warming. By harnessing the protective mechanisms mediated by SLs, crops can be better equipped to withstand the challenges posed by rising temperatures, thereby sustaining productivity and ecosystem stability.

## References

1. Akiyama, K., Matsuzaki, K.I., Hayashi, H., Plant sesquiterpenes induce hyphal branching in arbuscular mycorrhizal fungi. *Nat.*, *435*, 7043, 824–827, 2005.
2. Besserer, A., Puech-Pagès, V., Kiefer, P., Gomez-Roldan, V., Jauneau, A., Roy, S., Séjalon-Delmas, N., Strigolactones stimulate arbuscular mycorrhizal fungi by activating mitochondria. *PLoS Biol.*, *4*, 7, e226, 2006.
3. Gomez-Roldan, V., Fermas, S., Brewer, P.B., Puech-Pagès, V., Dun, E.A., Pillot, J.P., Rochange, S.F., Strigolactone inhibition of shoot branching. *Nat.*, *455*, 7210, 189–194, 2008.

4. Umehara, M., Hanada, A., Yoshida, S., Akiyama, K., Arite, T., Takeda-Kamiya, N., Yamaguchi, S., Inhibition of shoot branching by new terpenoid plant hormones. *Nat.*, *455*, 7210, 195–200, 2008.
5. López-Ráez, J.A., Shirasu, K., Foo, E., Strigolactones in plant interactions with beneficial and detrimental organisms: The Yin and Yang. *Trends Plant Sci.*, *22*, 6, 527–537, 2017.
6. Al-Babili, S. and Bouwmeester, H.J., Strigolactones, a novel carotenoid-derived plant hormone. *Annu. Rev. Plant Biol.*, *66*, 1, 161–186, 2015.
7. Xie, X., Structural diversity of strigolactones and their distribution in the plant kingdom. *J. Pestic. Sci.*, *41*, 4, 175–180, 2016.
8. Ito, S., Braguy, J., Wang, J.Y., Yoda, A., Fiorilli, V., Takahashi, I., Al-Babili, S., Canonical strigolactones are not the major determinant of tillering but important rhizospheric signals in rice. *Sci. Adv.*, *8*, 44, eadd1278, 2022.
9. Bouwmeester, H.J., Roux, C., Lopez-Raez, J.A., Becard, G., Rhizosphere communication of plants, parasitic plants and AM fungi. *Trends Plant Sci.*, *12*, 5, 224–230, 2007.
10. Andreo-Jimenez, B., Ruyter-Spira, C., Bouwmeester, H.J., Lopez-Raez, J.A., Ecological relevance of strigolactones in nutrient uptake and other abiotic stresses, and in plant-microbe interactions below-ground. *Plant Soil, 394*, 1–19, 2015.
11. Delaux, P.M., Séjalon-Delmas, N., Bécard, G., Ané, J.M., Evolution of the plant-microbe symbiotic 'toolkit'. *Trends Plant Sci.*, *18*, 6, 298–304, 2013.
12. Ruyter-Spira, C., Kohlen, W., Charnikhova, T., van Zeijl, A., van Bezouwen, L., De Ruijter, N., Bouwmeester, H., Physiological effects of the synthetic strigolactone analog GR24 on root system architecture in Arabidopsis: another belowground role for strigolactones? *Plant Physiol.*, *155*, 2, 721–734, 2011.
13. Xie, X., Yoneyama, K., Yoneyama, K., The strigolactone story. *Annu. Rev. Phytopathol.*, *48*, 1, 93–117, 2010.
14. Felemban, A., Braguy, J., Zurbriggen, M.D., Al-Babili, S., Apocarotenoids involved in plant development and stress response. *Front. Plant Sci.*, *10*, 1168, 2019.
15. Wu, F., Gao, Y., Yang, W., Sui, N., Zhu, J., Biological functions of strigolactones and their crosstalk with other phytohormones. *Front. Plant Sci.*, *13*, 821563, 2022.
16. Wang, J.Y., Chen, G.T.E., Braguy, J., Al-Babili, S., Distinguishing the functions of canonical strigolactones as rhizospheric signals. *Trends Plant Sci.*, 29, 925–936, 2024.
17. Koltai, H., Strigolactones are regulators of root development. *New Phytol.*, *190*, 3, 545–549, 2011.
18. Sun, H., Tao, J., Gu, P., Xu, G., Zhang, Y., The role of strigolactones in root development. *Plant Signal. Behav.*, *11*, 1, e1110662, 2016.
19. Kohlen, W., Charnikhova, T., Liu, Q., Bours, R., Domagalska, M.A., Beguerie, S., Ruyter-Spira, C., Strigolactones are transported through the xylem and

play a key role in shoot architectural response to phosphate deficiency in nonarbuscular mycorrhizal host Arabidopsis. *Plant Physiol.*, *155*, 2, 974–987, 2011.

20. Kretzschmar, T., Kohlen, W., Sasse, J., Borghi, L., Schlegel, M., Bachelier, J.B., Martinoia, E., A petunia ABC protein controls strigolactone-dependent symbiotic signalling and branching. *Nat.*, *483*, 7389, 341–344, 2012.
21. Ma, C., Bian, C., Liu, W., Sun, Z., Xi, X., Guo, D., Zheng, X., Strigolactone alleviates the salinity-alkalinity stress of *Malus hupehensis* seedlings. *Front. Plant Sci.*, *13*, 901782, 2022.
22. Liang, G., Liu, J., Zhang, J., Guo, J., Effects of drought stress on photosynthetic and physiological parameters of tomato. *J. Amer. Soc. Hort. Sci.*, *145*, 1, 12–17, 2020.
23. Zhang, J., Huang, Y., Pu, R., Gonzalez-Moreno, P., Yuan, L., Wu, K., Huang, W., Monitoring plant diseases and pests through remote sensing technology: A review. *Comput. Electron. Agric.*, *165*, 104943, 2019.
24. Ha, C.V., Leyva-González, M.A., Osakabe, Y., Tran, U.T., Nishiyama, R., Watanabe, Y., Tran, L.S.P., Positive regulatory role of strigolactone in plant responses to drought and salt stress. *PNAS*, *111*, 2, 851–856, 2014.
25. Liu, J., He, H., Vitali, M., Visentin, I., Charnikhova, T., Haider, I., Cardinale, F., Osmotic stress represses strigolactone biosynthesis in Lotus japonicus roots: exploring the interaction between strigolactones and ABA under abiotic stress. *Planta*, *241*, 1435–1451, 2015.
26. Ruiz-Lozano, J.M., Aroca, R., Zamarreño, Á.M., Molina, S., Andreo-Jiménez, B., Porcel, R. *et al.*, Arbuscular mycorrhizal symbiosis induces strigolactone biosynthesis under drought and improves drought tolerance in lettuce and tomato. *Plant Cell Environ.*, 39, 441–452, 2016.
27. Pérez-Torres, C.A., Lopez-Bucio, J., Cruz-Ramírez, A., Ibarra-Laclette, E., Dharmasiri, S., Estelle, M., Herrera-Estrella, L., Phosphate availability alters lateral root development in Arabidopsis by modulating auxin sensitivity via a mechanism involving the TIR1 auxin receptor. *Plant Cell*, *20*, 12, 3258–3272, 2008.
28. Sedaghat, M., Tahmasebi-Sarvestani, Z., Emam, Y., Mokhtassi-Bidgoli, A., Physiological and antioxidant responses of winter wheat cultivars to strigolactone and salicylic acid in drought. *Plant Physiol. Biochem.*, *119*, 59–69, 2017.
29. Min, Z., Li, R., Chen, L., Zhang, Y., Li, Z., Liu, M., Fang, Y., Alleviation of drought stress in grapevine by foliar-applied strigolactones. *Plant Physiol. Biochem.*, *135*, 99–110, 2019.
30. Li, X. *et al.*, Strigolactone-induced salt stress tolerance in plants is associated with proton pump activation and reactive oxygen species scavenging. *Plant Cell Environ.*, 45, 3, 867–884, 2022.
31. Luqman, M., Shahbaz, M., Maqsood, M.F., Farhat, F., Zulfiqar, U., Siddiqui, M.H., Haider, F.U., Effect of strigolactone on growth, photosynthetic efficiency, antioxidant activity, and osmolytes accumulation in different maize

(*Zea mays* L.) hybrids grown under drought stress. *Plant Signal. Behav.*, *18*, 1, 2262795, 2023.
32. Hu, Q., Zhang, S., Huang, B., Strigolactones promote leaf elongation in tall fescue through upregulation of cell cycle genes and downregulation of auxin transport genes in tall fescue under different temperature regimes. *Int. J. Mol. Sci.*, *20*, 8, 1836, 2019.
33. Omoarelojie, L.O., Kulkarni, M.G., Finnie, J.F., Pospíšil, T., Strnad, M., Van Staden, J., Synthetic strigolactone (rac-GR24) alleviates the adverse effects of heat stress on seed germination and photosystem II function in lupine seedlings. *Plant Physiol. Biochem.*, *155*, 965–979, 2020.
34. Gonai, T. and others, Abscisic acid and gibberellins synergistically induce proteinase inhibitor II gene expression. *Plant Cell Physiol.*, 45, 6, 639–648, 2004.
35. Lechat, M.M., Brun, G., Montiel, G., Véronési, C., Simier, P., Thoiron, S., Delavault, P., Seed response to strigolactone is controlled by abscisic acid-independent DNA methylation in the obligate root parasitic plant, Phelipanche ramosa L. Pomel. *J. Exp. Bot.*, *66*, 11, 3129–3140, 2015.
36. Marzec, M., Roles of strigolactones in response to phosphate starvation. *Phytol.*, 42, 23–29, 2016.
37. Matthews, C., Characterizing the role of the miR156-SPL network in heat stress response in Medicago sativa (Master's thesis), The University of Western Ontario (Canada), 2018.
38. Kapulnik, Y., Resnick, N., Mayzlish-Gati, E., Kaplan, Y., Wininger, S., Hershenhorn, J., Koltai, H., Strigolactones interact with ethylene and auxin in regulating root-hair elongation in Arabidopsis. *J. Exp. Bot.*, *62*, 8, 2915–2924, 2011.
39. Tai, Z., Yin, X., Fang, Z., Shi, G., Lou, L., Cai, Q., Exogenous GR24 alleviates cadmium toxicity by reducing cadmium uptake in switchgrass (*Panicum virgatum*) seedlings. *Int. J. Environ. Res. Public Health*, *14*, 8, 852, 2017.
40. Lv, S., Zhang, Y., Li, C., Liu, Z., Yang, N., Pan, L., Wang, G., Strigolactone-triggered stomatal closure requires hydrogen peroxide synthesis and nitric oxide production in an abscisic acid-independent manner. *New Phytol.*, *217*, 1, 290–304, 2018.
41. Visentin, I., Pagliarani, C., Deva, E., Caracci, A., Turečková, V., Novák, O., Cardinale, F., A novel strigolactone-miR156 module controls stomatal behaviour during drought recovery. *Plant Cell Environ.*, *43*, 7, 1613–1624, 2020.
42. Rasmussen, A., Mason, M.G., De Cuyper, C., Brewer, P.B., Herold, S., Agusti, J., Beveridge, C.A., Strigolactones suppress adventitious rooting in Arabidopsis and pea. *Plant Physiol.*, *158*, 4, 1976–1987, 2012.
43. Sage, R.F. and Kubien, D.S., The temperature response of C3 and C4 photosynthesis. *Plant Cell Environ.*, *30*, 9, 1086–1106, 2007.

44. Breshears, D.D., Fontaine, J.B., Ruthrof, K.X., Field, J.P., Feng, X., Burger, J.R., Hardy, G.E.S.J., Underappreciated plant vulnerabilities to heat waves. *New Phytol.*, *231*, 1, 32–39, 2021.
45. Ruehr, N.K., Grote, R., Mayr, S., Arneth, A., Beyond the extreme: recovery of carbon and water relations in woody plants following heat and drought stress. *Tree Physiol.*, *39*, 8, 1285–1299, 2019.
46. Hoover, D.L., Knapp, A.K., Smith, M.D., Resistance and resilience of a grassland ecosystem to climate extremes. *Ecol.*, *95*, 9, 2646–2656, 2014.
47. Thompson, V., Mitchell, D., Hegerl, G.C., Collins, M., Leach, N.J., Slingo, J.M., The most at-risk regions in the world for high-impact heatwaves. *Nat. Commun.*, *14*, 1, 2152, 2023.
48. Wahid, A., Physiological implications of metabolite biosynthesis for net assimilation and heat-stress tolerance of sugarcane (Saccharum officinarum) sprouts. *J. Plant Res.*, *120*, 219–228, 2007.
49. Bita, C.E. and Gerats, T., Plant tolerance to high temperature in a changing environment: scientific fundamentals and production of heat stress-tolerant crops. *Front. Plant Sci.*, *4*, 273, 2013.
50. Aparecido, L.M., Woo, S., Suazo, C., Hultine, K.R., Blonder, B., High water use in desert plants exposed to extreme heat. *Ecol. Lett.*, *23*, 8, 1189–1200, 2020.
51. Marchin, R.M., Backes, D., Ossola, A., Leishman, M.R., Tjoelker, M.G., Ellsworth, D.S., Extreme heat increases stomatal conductance and drought-induced mortality risk in vulnerable plant species. *Glob. Change Biol.*, *28*, 3, 1133–1146, 2022.
52. Ostria-Gallardo, E., Zúñiga-Contreras, E., Carvajal, D.E., de La Peña, T.C., Gianoli, E., Bascuñán-Godoy, L., Two congeneric shrubs from the atacama desert show different physiological strategies that improve water use efficiency under a simulated heat wave. *Plants*, *12*, 13, 2464, 2023.
53. Allakhverdiev, S.I., Kreslavski, V.D., Klimov, V.V., Los, D.A., Carpentier, R., Mohanty, P., Heat stress: an overview of molecular responses in photosynthesis. *Photosynth. Res.*, *98*, 541–550, 2008.
54. Flexas, J., Bota, J., Cifre, J., Mariano Escalona, J., Galmés, J., Gulías, J., Medrano, H., Understanding down-regulation of photosynthesis under water stress: future prospects and searching for physiological tools for irrigation management. *Ann. Appl. Biol.*, *144*, 3, 273–283, 2004.
55. Swann, A.L., Plants and drought in a changing climate. *Curr. Clim. Change Rep.*, *4*, 192–201, 2018.
56. Noia Junior, R.D.S., do Amaral, G.C., Pezzopane, J.E.M., Toledo, J.V., Xavier, T.M.T., Ecophysiology of C3 and C4 plants in terms of responses to extreme soil temperatures. *Theor. Exp. Plant Physiol.*, *30*, 261–274, 2018.
57. Lobell, D.B., Hammer, G.L., Chenu, K., Zheng, B., McLean, G., Chapman, S.C., The shifting influence of drought and heat stress for crops in northeast Australia. *Glob. Change Biol.*, *21*, 11, 4115–4127, 2015.

58. Mittler, R. and Blumwald, E., Genetic engineering for modern agriculture: challenges and perspectives. *Annu. Rev. Plant Biol.*, *61*, 1, 443–462, 2010.
59. Kole, C., Muthamilarasan, M., Henry, R., Edwards, D., Sharma, R., Abberton, M., Prasad, M., Application of genomics-assisted breeding for generation of climate resilient crops: progress and prospects. *Front. Plant Sci.*, *6*, 563, 2015.
60. Dhankher, O.P. and Foyer, C.H., Climate resilient crops for improving global food security and safety. *Plant Cell Environ.*, *41*, 5, 877–884, 2018.
61. Camejo, D., Rodríguez, P., Morales, M.A., Dell'Amico, J.M., Torrecillas, A., Alarcón, J.J., High-temperature effects on photosynthetic activity of two tomato cultivars with different heat susceptibility. *J. Plant Physiol.*, *162*, 3, 281–289, 2005.
62. Hatfield, J.L. and Prueger, J.H., Temperature extremes: Effect on plant growth and development. *Weather Clim. Extrem.*, *10*, 4–10, 2015.
63. Kaplan, F., Kopka, J., Haskell, D.W., Zhao, W., Schiller, K.C., Gatzke, N., Guy, C.L., Exploring the temperature-stress metabolome of Arabidopsis. *Plant Physiol.*, *136*, 4, 4159–4168, 2004.
64. Marquez-Garcia, B., Shaw, D., Cooper, J.W., Karpinska, B., Quain, M.D., Makgopa, E.M., Foyer, C.H., Redox markers for drought-induced nodule senescence, a process occurring after drought-induced senescence of the lowest leaves in soybean (Glycine max). *Ann. Bot.*, *116*, 4, 497–510, 2015.
65. Jia, K.P., Luo, Q., He, S.B., Lu, X.D., Yang, H.Q., Strigolactone-regulated hypocotyl elongation is dependent on cryptochrome and phytochrome signaling pathways in Arabidopsis. *Mol. Plant*, *7*, 3, 528–540, 2014.
66. Kelly, J.H., Tucker, M.R., Brewer, P.B., The strigolactone pathway is a target for modifying crop shoot architecture and yield. *Biol.*, *12*, 1, 95, 2023.
67. Soliman, S., Wang, Y., Han, Z., Pervaiz, T., El-Kereamy, A., Strigolactones in plants and their interaction with the ecological microbiome in response to abiotic stress. *Plants*, *11*, 24, 3499, 2022.
68. López-Ráez, J.A., Kohlen, W., Charnikhova, T., Mulder, P., Undas, A.K., Sergeant, M.J., Bouwmeester, H., Does abscisic acid affect strigolactone biosynthesis? *New Phytol.*, *187*, 2, 343–354, 2010.
69. Hasanuzzaman, M. and Fujita, M., Plant responses and tolerance to salt stress: Physiological and molecular interventions. *Int. J. Mol. Sci.*, *23*, 9, 4810, 2022.
70. Zhao, S., Zhang, Q., Liu, M., Zhou, H., Ma, C., Wang, P., Regulation of plant responses to salt stress. *Int. J. Mol. Sci.*, *22*, 9, 4609, 2021.
71. Debez, A., Ben Hamed, K., Grignon, C., Abdelly, C., Salinity effects on germination, growth, and seed production of the halophyte Cakile maritima. *Plant Soil*, *262*, 1, 179–189, 2004.
72. Delatorre-Herrera, J. and Pinto, M., Importance of ionic and osmotic components of salt stress on the germination of four quinua (Chenopodium quinoa Willd.) selections. *Chil. J. Agric. Res.*, *69*, 4, 477–485, 2009.

73. Parihar, P., Singh, S., Singh, R., Singh, V.P., Prasad, S.M., Effect of salinity stress on plants and its tolerance strategies: a review. *ESPR*, *22*, 4056–4075, 2015.
74. Xiang, L.X., Hu, L.P., Hu, X.H., Pan, X.B., Ren, W.Q., Response of reactive oxygen metabolism in melon chloroplasts to short-term salinity-alkalinity stress regulated by exogenous γ-aminobutyric acid. *J. Appl. Ecol.*, *26*, 12, 3746–3752, 2015.
75. Guo, R., Shi, L., Yan, C., Zhong, X., Gu, F., Liu, Q., Li, H., Ionomic and metabolic responses to neutral salt or alkaline salt stresses in maize (Zea mays L.) seedlings. *BMC Plant Biol.*, *17*, 1–13, 2017.
76. Jin, X., Liu, T., Xu, J., Gao, Z., Hu, X., Exogenous GABA enhances muskmelon tolerance to salinity-alkalinity stress by regulating redox balance and chlorophyll biosynthesis. *BMC Plant Biol.*, *19*, 1–15, 2019.
77. Cao, Y., Song, H., Zhang, L., New insight into plant saline-alkali tolerance mechanisms and application to breeding. *Int. J. Mol. Sci.*, *23*, 24, 16048, 2022.
78. Shen, B., Jensen, R.G., Bohnert, H.J., Mannitol protects against oxidation by hydroxyl radicals. *Plant Physiol.*, *115*, 2, 527–532, 1997.
79. Tsugane, K., Kobayashi, K., Niwa, Y., Ohba, Y., Wada, K., Kobayashi, H., A recessive Arabidopsis mutant that grows photoautotrophically under salt stress shows enhanced active oxygen detoxification. *Plant Cell*, *11*, 7, 1195–1206, 1999.
80. Malakar, P. and Chattopadhyay, D., Adaptation of plants to salt stress: the role of the ion transporters. *J. Plant Biochem. Biotech.*, *30*, 4, 668–683, 2021.
81. Miller, G.A.D., Suzuki, N., Ciftci-Yilmaz, S.U.L.T.A.N., Mittler, R.O.N., Reactive oxygen species homeostasis and signalling during drought and salinity stresses. *Plant Cell Environ.*, *33*, 4, 453–467, 2010.
82. Zhang, J., Luo, W., Zhao, Y., Xu, Y., Song, S., Chong, K., Comparative metabolomic analysis reveals a reactive oxygen species-dominated dynamic model underlying chilling environment adaptation and tolerance in rice. *New Phytol.*, *211*, 4, 1295–1310, 2016.
83. Xu, J., Kang, Z., Zhu, K., Zhao, D., Yuan, Y., Yang, S., Hu, X., RBOH1-dependent H2O2 mediates spermine-induced antioxidant enzyme system to enhance tomato seedling tolerance to salinity–alkalinity stress. *Plant Physiol. Biochem.*, *164*, 237–246, 2021.
84. Yang, C.W., Xu, H.H., Wang, L.L., Liu, J., Shi, D.C., Wang, D.L., Comparative effects of salt-stress and alkali-stress on the growth, photosynthesis, solute accumulation, and ion balance of barley plants. *Photosynthetica*, *47*, 79–86, 2009.
85. Jia, X., Zhu, Y., Zhang, R., Zhu, Z., Zhao, T., Cheng, L., Wang, Y., Ionomic and metabolomic analyses reveal the resistance response mechanism to saline-alkali stress in Malus halliana seedlings. *Plant Physiol. Biochem.*, *147*, 77–90, 2020.
86. Yang, Y. and Guo, Y., Elucidating the molecular mechanisms mediating plant salt-stress responses. *New Phytol.*, *217*, 2, 523–539, 2018.

87. Guo, R., Yang, Z., Li, F., Yan, C., Zhong, X., Liu, Q., Zhao, L., Comparative metabolic responses and adaptive strategies of wheat (Triticum aestivum) to salt and alkali stress. *BMC Plant Biol.*, *15*, 1–13, 2015.
88. Parida, A.K., Das, A.B., Mittra, B., Effects of salt on growth, ion accumulation, photosynthesis and leaf anatomy of the mangrove, Bruguiera parviflora. *Trees*, *18*, 167–174, 2004.
89. Mahajan, S., Pandey, G.K., Tuteja, N., Calcium-and salt-stress signaling in plants: shedding light on SOS pathway. *Arch. Biochem. Biophys.*, *471*, 2, 146–158, 2008.
90. Li, P., Wang, A., Du, W., Mao, L., Wei, Z., Wang, S., Zhao, L., Insight into the interaction between Fe-based nanomaterials and maize (Zea mays) plants at metabolic level. *Sci. Total Environ.*, *738*, 139795, 2020.
91. Sakamoto, A. and Murata, N., Genetic engineering of glycine betaine synthesis in plants: current status and implications for enhancement of stress tolerance. *J. Exp. Bot.*, *51*, 342, 81–88, 2000.
92. Shabala, S. and Cuin, T.A., Osmoregulation versus osmoprotection: re-evaluating the role of compatible solutes, 2006.
93. Di Martino, C., Delfine, S., Pizzuto, R., Loreto, F., Fuggi, A., Free amino acids and glycine betaine in leaf osmoregulation of spinach responding to increasing salt stress. *New Phytol.*, *158*, 3, 455–463, 2003.
94. Chen, Z., Cuin, T.A., Zhou, M., Twomey, A., Naidu, B.P., Shabala, S., Compatible solute accumulation and stress-mitigating effects in barley genotypes contrasting in their salt tolerance. *J. Exp. Bot.*, *58*, 15-16, 4245–4255, 2007.
95. Marzec, M. and Melzer, M., Regulation of root development and architecture by strigolactones under optimal and nutrient deficiency conditions. *Int. J. Mol. Sci.*, *19*, 7, 1887, 2018.
96. Garnett, T., Conn, V., Kaiser, B.N., Root based approaches to improving nitrogen use efficiency in plants. *Plant Cell Environ.*, *32*, 9, 1272–1283, 2009.
97. Ravazzolo, L., Boutet-Mercey, S., Perreau, F., Forestan, C., Varotto, S., Ruperti, B., Quaggiotti, S., Strigolactones and auxin cooperate to regulate maize root development and response to nitrate. *PCP*, *62*, 4, 610–623, 2021.
98. Matusova, R., Rani, K., Verstappen, F.W., Franssen, M.C., Beale, M.H., Bouwmeester, H.J., The strigolactone germination stimulants of the plant-parasitic Striga and Orobanche spp. are derived from the carotenoid pathway. *Plant Physiol.*, *139*, 2, 920–934, 2005.
99. Abe, S., Sado, A., Tanaka, K., Kisugi, T., Asami, K., Ota, S., Nomura, T., Carlactone is converted to carlactonoic acid by MAX1 in Arabidopsis and its methyl ester can directly interact with AtD14 *in vitro*. *PNAS*, *111*, 50, 18084–18089, 2014.
100. Charnikhova, T.V., Gaus, K., Lumbroso, A., Sanders, M., Vincken, J.P., De Mesmaeker, A., Bouwmeester, H.J., Zealactones. Novel natural strigolactones from maize. *Phytochem.*, *137*, 123–131, 2017.

101. Ma, N., Hu, C., Wan, L., Hu, Q., Xiong, J., Zhang, C., Strigolactones improve plant growth, photosynthesis, and alleviate oxidative stress under salinity in rapeseed (Brassica napus L.) by regulating gene expression. *Front. Plant Sci.*, *8*, 1671, 2017.
102. Danish, S., Hareem, M., Dawar, K., Naz, T., Iqbal, M.M., Ansari, M.J., Datta, R., The role of strigolactone in alleviating salinity stress in chili pepper. *BMC Plant Biol.*, *24*, 1, 209, 2024.

7

# Role of Strigolactones in Heavy Metal Tolerance: A Case Study on Cadmium

**Ishrat Mehmood[1]*, Kaiser Iqbal Wani[2] and Tariq Aftab[2]**

*[1]Department of Biophysics, Panjab University, Chandigarh, India*
*[2]Department of Botany, Aligarh Muslim University, Aligarh, India*

## *Abstract*

Phytohormones are crucial for controlling plant growth and development as well as their response to biotic and abiotic stress, particularly that caused by heavy metals. Heavy metal contamination is a serious environmental issue that endangers all life forms. Of all the heavy metals, cadmium (Cd) is one of the most harmful environmental pollutants in the world, with detrimental impacts on agriculture and significant health issues. Plants in contaminated soils absorb Cd through their roots. The heavy metals that build up in various organs ultimately inhibit plant development and production. High quantities of Cd that bioaccumulate in the food chain are hazardous to human health. The harmful effects of Cd on several metabolic systems are seen in plants. Strigolactones (SLs), a type of endogenous phytohormones produced from carotenoids, govern root growth, shoot branching, and nutrient absorption in response to heavy metal stress and help plants adapt to nutrient-deficient environments. In controlling the growth and development of plants, phytohormones play an endogenous and exogenous role. Since SLs interact with other phytohormones to aid plant stress adaptation, they actively engage in the regulatory networks of plant stress adaptation. These hormones can strengthen the defense mechanisms of plants by enhancing the activity of antioxidant enzymes and reducing $H_2O_2$ and lipoperoxidation. According to studies, SLs may be a good option for Cd detoxification because they lower Cd uptake, increase the expression of genes that code for antioxidant defense, shield plants from oxidative stress by lowering oxidative stress biomarkers, and ultimately lower Cd toxicity. Consequently, our goal in this chapter is to give a thorough description of how SLs function as important regulators in reducing the toxicity of heavy metals in plants.

**Corresponding author*: ishratmehmood943@gmail.com

Tariq Aftab and Kaiser Iqbal Wani (eds.) Strigolactones: Emerging Plant Hormones, (137–160) © 2025 Scrivener Publishing LLC

***Keywords***: Strigolactones, heavy metals, cadmium, toxicity, antioxidant, agriculture

## 7.1 Introduction

Heavy metal pollution is swiftly becoming a great environmental concern to the biosphere because of its potential to endanger both plants and animals [1]. One of the most harmful and pervasive heavy metals in most ecosystems is cadmium (Cd), which may easily spread through polluted sediments, harming plants and other living things as well as posing a threat to food security. Apart from being entirely unnecessary, environmental concentrations of Cd are rising due to both geological and anthropogenic factors, such as mineral mining for phosphate fertilizers and other farming practices [2–4]. Globally, between 5.6 and 38 thousand metric tons of Cd are introduced into soils each year, and many agricultural regions in Europe, North America, and Asia have varied degrees of Cd contamination [5]. Therefore, soil Cd contamination is a major global environmental issue. Around 11 provinces and 25 districts in China are now dealing with problems related to farmlands polluted by Cd, and each year over 12 million metric tons of grains are contaminated with Cd [6]. With a threshold of 100 mg $kg^{-1}$ for agriculture, soil Cd concentrations above 1 mg $kg^{-1}$ are typically considered to indicate anthropogenically polluted soil [7]. It is crucial to create efficient methods for lowering Cd concentrations and easing Cd stress in plants.

Without serving any useful purpose, Cd is incredibly poisonous to plants. Plant roots easily absorb and translocate Cd, endangering ecosystem stability and human health (via the food chain) [8]. Plants might exhibit visible morphological signs such as browning roots, necrosis on leaves, and stunted growth [9]. At physiological levels, too much Cd can interfere with the absorption of nutrients and water and prevent transpiration and photosynthesis [10]. Owing to Cd toxicity, plants have developed many defense mechanisms against excess Cd, such as an active antioxidant system, vacuolar sequestration, and cell wall immobilization [11]. Previous studies have shown that plants associated with Cd detoxification have a specific subcellular distribution of the metal.

Strigolactones (SLs) are a group of carotenoid-derived lactones that were first identified as germination stimulants of some parasitic plants like *Striga lutea*. However, the underlying mechanisms by which these phytohormones aid the detoxification responses have not been fully elucidated [12]. Exogenous and endogenous phytohormones regulate various adaptive mechanisms in plants for the alleviation of heavy metal-induced stress in plants. Their ability to suppress axillary bud growth and exhibit shoot-branching inhibitory action led to their eventual identification as plant hormones [13]. These unique phytohormones not only restrict shoot branching but also regulate root architecture, leaf senescence, reproductive maturity, nutrient uptake, and host plant interactions with soil microorganisms. Moreover, it has been documented that these hormones have a significant role in controlling how plants react to various environmental disturbances [14]. A synthetic SL analogue called GR24 ($C_{17}H_{14}O_5$) has been widely employed recently as a plant hormone that controls plant growth and development and promotes the elongation of root hairs and primary root growth. Additionally, abiotic stressors including salt and drought are suggested to be regulated by this hormone [15, 16].

## 7.2 Cadmium Toxicity on Plants

Plants that are exposed to high levels of Cd have substantial negative effects on a variety of physiological and molecular functions, as will be discussed below.

### 7.2.1 Cd Toxicity on Growth and Development

In many different plant species, Cd, a non-essential metal, can lead to growth irregularities and suppression [17]. The morphological and physiological processes of decreased root cell viability, necrotic and mucilaginous roots, reduced root elongation, seed germination inhibition, nutrient imbalance, leaf rolling, chlorosis, stunted growth, elevated level of oxidative stress, mutagenesis, mortality, and decreased plant yield can all be caused by long-term exposure to Cd [18, 19]. These phytotoxic effects vary from species to species, soil type, and plant growth stage. Even at a relatively low dose of 0.05 μM, it results in chlorosis, stunted growth, and reduced yield [20, 21]. Figure 7.1 provides a mechanistic description of how Cd affects different plant functions.

Phenotypic alterations in leaves lead to reduced chlorophyll concentrations, which minimizes photosynthetic activity and causes chlorosis. At high concentrations, Cd hinders the development of lateral roots in the soil rhizosphere, resulting in a discolored, twisted, and stiff main root [10, 22]. The apical portion of the epidermis and cortical cell layers become abnormally enlarged. Additionally, meristematic cell mitotic division is inhibited by Cd toxicity, which reduces root length and dry biomass and enhances

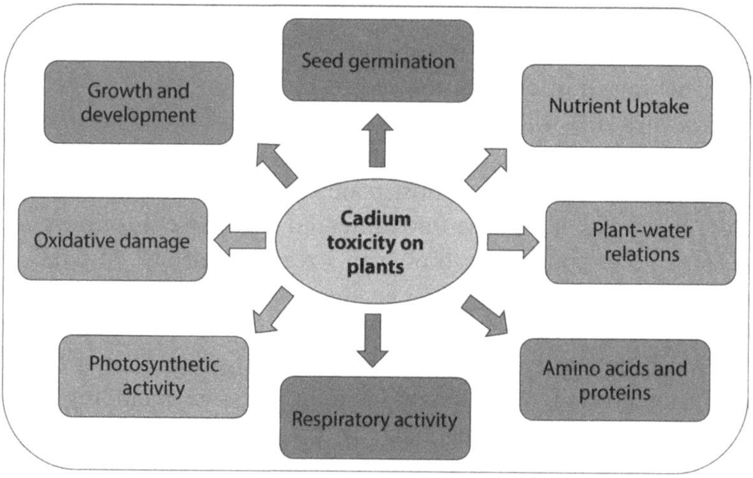


**Figure 7.1** Overview of how Cd toxicity affects different plant functions.

the root diameter [23]. When rice seedlings are treated with Cd, their root development is inhibited and their morphogenesis is altered [24]. Moreover, chromosomal aberrations of root tips and abnormalities of root elongation are brought on by Cd stress in plants. Because of its toxicity, Cd reduces the absorption of minerals and photosynthesis in plants [23, 25].

### 7.2.2 Cd Toxicity Induces Oxidative Damage

Usually, heavy metals induce oxidative stress in plants either directly or indirectly, through the production of Reactive oxygen species (ROS) [26]. The important function that ROS play in mediating biotic and abiotic stressors during plant growth and development has been acknowledged. However, their excess accumulation in plants has been shown to induce phytotoxicity including protein oxidation, DNA deterioration, membrane lipid bilayer damage, and chlorophyll degradation [27, 28]. Under Cd stress, some plants show a significant increase in oxidative stress biomarkers, including *Nicotiana tabacum*, *Phaseolus vulgaris*, *Vigna unguiculata*, *Solanum lycopersicum*, and *Oryza sativa* [29–31]. It is commonly known that oxidative stress is often caused in plant tissues by elevated Cd levels in the soil.

The excess generation and accumulation of ROS can lead to growth abnormalities and physiological disturbances in plants. These disturbances can affect photosynthesis, the assimilation of carbon and nitrogen, and the absorption of water and nutrients, all of which are significant challenges to the production of sustainable crops [32, 33]. As a barrier against Cd stress, Dobrikova *et al.* [34], for example, showed an enhanced accumulation of non-enzymatic antioxidants in the leaves of the plant under Cd stress compared to the control, including phenolic, anthocyanins, carotenoids, and polyphenolics [35, 36]. According to Sattar *et al.*, superoxide $O_2^-$, hydrogen peroxide ($H_2O_2$), malonaldehyde (MDA), and ionic leakage (EL) were all clearly elevated in maize that had been exposed to Cd with a subsequent increase in the activity of superoxide dismutase (SOD), catalase (CAT), ascorbate peroxidase (APX), and glutathione reductase (GR) [37]. Yu *et al.* found that the leaves of Cd-stressed *Brassica campestris* and *Ipomoea aquatics* had higher concentrations of ROS as well as enzymatic and non-enzymatic antioxidants [38]. Similar to this, it has been observed that Cd toxicity reduces photosynthesis and nutrient absorption, which causes a drop in crop growth and quality. There has been a noticeable drop in the dry weight of several different plants [23]. When plants are exposed to Cd, they produce more NADPH oxidase and generate more ROS, which causes plants to necrotize. Altogether, enhanced membrane lipid peroxidation, nucleic acid degradation, and protein degradation are the outcomes of the Cd-induced buildup of ROS [23, 39].

### 7.2.3 Impaired Photosynthetic and Respiratory Activity

Since Cd acts as a powerful inhibitor of photosynthesis, stomatal conductance, and respiratory functions, plants often respond to Cd exposure by impeding these activities. In oilseed, legume, and cereal crops, a linear association was found between reduced transpiration and suppression of photosynthesis, suggesting that Cd buildup in leaves prevents stomata from opening [23]. The photochemical efficiency of PSII, the rate of $CO_2$ absorption, and the activity and efficiency of RuBisCO in *Lactuca sativa* leaves were all negatively impacted by the elevated amounts of Cd. Furthermore, it was discovered that the concentrations of chlorophyll a/b were lower than those of the control [40].

The primary reasons for photosynthetic impairment and growth loss are thought to be the degradative effects of Cd on chlorophyll and its biosynthetic suppression [34]. Studies suggest that the loss of Fe in leaves and the negative impact of Cd on chlorophyll metabolism may be the cause of chlorosis in leaves [41]. Moreover, Cd toxicity interferes with cell division and chloroplast formation in plants by reducing cell size, stomatal conductance, and stomatal density in the epidermis of leaves [42]. Many crops, including *Hordeum vulgare, S. lycopersicum, Z. mays, B. juncea, Lepidium sativum,* and *Triticum aestivum,* have shown decreased chlorophyll concentration under Cd stress, which restricts photosynthesis. The enzymes PEPCase and RUBPCase are essential for plant photosynthesis. By substituting magnesium ions and changing their activity toward oxygenation processes, Cd ions reduce RuBPCase activity. Enzyme activity is hampered as a result of the irreversible dissociation of subunits. Additionally, Cd changes the gas exchange characteristics in the water-oxidizing complex (OEC) of PSII, affecting the photosynthetic pigments and chloroplast structure by substituting $Ca^{2+}$ in oxygen-evolving centres [23].

### 7.2.4 Cd Toxicity on Nutrient Uptake and Plant–Water Relations

Plant root characteristics that are important in water and nutrient uptake are adversely affected by Cd stress. In addition to serving as a point of entry for the uptake of Cd, which causes various physiological changes, plant roots primarily mediate the uptake of water and nutrients from the soil [42]. In addition to competing with other vital plant nutrients for absorption, Cd also has a detrimental effect on plant water status, which in turn disrupts plant metabolism [43]. Under heavy metal stress, proline levels dramatically rise in the leaves, but the relative water content is found to be much lower along with a decreased rate of transpiration [44]. When Cd enters plant leaves through Ca channels, the plant–water relationship is upset, resulting in stomatal closure, which lowers the transpiration rate and inhibits photosynthesis [45]. Research is needed to identify plant species that are Cd tolerant and assess the strategies these plants use to withstand the harmful effects of Cd. Under Cd stress, a notable decrease in P, N, and S concentrations was noted in the aboveground sections of several plant species [46]. When *Solanum lycopersicum* was exposed to 100 μM Cd, for example, the concentration of Ca decreased in the shoot and rose in the root, whereas the concentration of K and Mg was significantly lower throughout the entire plant [47]. It could imply that translocation is more impacted by high Cd concentrations in the portions of plants that are above ground. A thorough grasp of how plants react to Cd stressors and how they acquire nutrients at the genetic level might prove to be a crucial and useful strategy for improving agricultural management [48].

### 7.2.5 Cd Toxicity on Seed Germination

Gibberellic acid, auxin, and abscisic acid are examples of phytohormones that control germination in seeds, an essential step in plant life [49]. On the other hand, exposure to hazardous amounts of Cd can hinder development, impede germination, interfere with the physiological processes of seedlings, and lower agricultural yield [23]. Water absorption in seeds and seedlings is inhibited by Cd, which limits the amount of water available for seed embryo development. As a result, the endosperm's ability to mobilize starch is inhibited, and the transmission of soluble carbohydrates to the seed embryonic axis is compromised [50]. Because Cd inhibits hydrolyzing enzymes, seeds release less starch and have less mobilization from storage. Ca–calmodulin binding sites between Cd and Ca ions are a source of competitiveness for seeds. When it comes to metabolic activation during seed germination, the interaction between calmodulin and Cd is important. Because Cd causes membrane lipid peroxidation, malondialdehyde levels have been found to rise in pea embryos under stress [49]. According to a study, 87.5% less spinach seedlings germinated after being exposed to Cd. Citrus seedlings treated with $CdCl_2$ became chlorotic and wilted. Under Cd stress, parsley seedlings have elevated Cd concentrations without exhibiting any visible indications of stress [51].

### 7.2.6 Cd Toxicity on Amino Acids and Proteins

Variations in the environment can lead to alterations in gene expression, which in turn can alter the protein diversity inside cells [52]. Protein content variations in extremely stressful conditions may serve as molecular markers for the stress responses that plants exhibit [23]. Stress tolerance in plants may be effectively studied by the application of proteomic techniques. Recent years have seen an increasing amount of research demonstrating that plants under stress have a very quick response in terms of proteomics. Generally speaking, heat-shock proteins are proteins that help eukaryotes tolerate stress. When maize is treated with Cd, heat-shock proteins (70 kDa phosphoprotein) are produced. The *Lycopersicon peruvianum* exhibited protective effects when pre-treated with minute heat stress before exposure to Cd since this prevented membrane damage [53–55].

It has been discovered that the pathogen-related protein PrP4A and the heat-shock protein HSP71 shield cells against harm brought on by Cd toxicity as reported in Pea [23]. After 48 h of treatment with $CdCl_2$ (50 μM), a soluble protein of 51 kDa was found in the root tissues underneath the outer periphery and tonoplast plasma membrane. This protein was named Cd toxicity-associated protein [56]. In 2014, Sergeant *et al.* noted that heat-shock and pathogenesis-related proteins are elevated in European aspen (*Populus tremula*) leaves after Cd treatment [57]. Furthermore, in rice, the production of 36 proteins is impacted by Cd. According to Rodríguez-Celma *et al.*, when tomatoes were treated with a low dosage of Cd (10 μM),

41 polypeptides changed, but a greater dose of Cd (100 μM) induced variations in 36 polypeptides [58, 59].

## 7.3 Strigolactone-Mediated Cadmium Tolerance

Because of their recent emergence and function in plant control and signal transduction, SLs have drawn the interest of plant biologists. Strigolactones were initially identified as seed germination stimulators in the roots of parasitic plants like *Orobanche* and *Striga* [60]. Later, it was proposed that because SLs might cause arbuscular mycorrhiza (AM) fungi to branch more, they function as signals originating from the roots of plants, strengthening the symbiotic link between plant roots and AM fungi [61]. These signaling molecules are produced from carotenoid precursors and serve a variety of purposes in enhancing plant resilience to various environmental stresses. Thus, SLs have a major impact on plant growth and development [62].

Strigolactones have been shown to improve a plant's ability to withstand a variety of environmental stressors, such as the toxicity of heavy metals like Cd stress as shown in Table 7.1 [63–66]. It has been demonstrated that barley accumulates less Cd when the synthetic SL analog GR24 is applied [67]. This illustrates how SLs can reduce heavy metal stress and increase a plant's resistance to harmful substances. Furthermore, via activating important stress response mechanisms such as abscisic acid (ABA) accumulation, heat shock proteins, C-REPEAT BINDING FACTOR 1 (CBF1) transcription, and antioxidant enzyme activity, SLs have been demonstrated to favorably modulate plant tolerance to heat and cold shocks [68]. Furthermore, through interactions with other phytohormones like as ABA, cytokinins, and gibberellic acid, SLs regulate how plants respond to and adapt to abiotic stimuli. According to recent research, SLs may be used to increase plant resistance to Cd poisoning, which would help control Cd pollution in agricultural environments [63, 69].

### 7.3.1 Strigolactones Enhance Cd Tolerance by Boosting Antioxidant Defense Mechanisms

Studies have demonstrated that SLs are crucial in enhancing plant antioxidant defense systems to help them withstand the effects of Cd stress. Under Cd stress, exogenous GR24 dramatically lowers $H_2O_2$ and MDA, suggesting that GR24 attenuates oxidative stress and lipid peroxidation caused by Cd [74]. Plants have both enzymatic and non-enzymatic antioxidants to

**Table 7.1** Impact of exogenous GR24 (SL analog) treatment on plants under Cd stress.

| Plant | Cd concentration | GR24 concentration | SL impact | Reference |
|---|---|---|---|---|
| *Galinsoga parviflora* | 0.1 mg $L^{-1}$ | 1 and 2 μmol $L^{-1}$ | Decreased peroxidase activity and increased superoxide dismutase activity<br>Improved Cd phytoremediation | [70] |
| Wheat | 20 mg $kg^{-1}$ | 4 μM | Balance between ROS and antioxidants<br>Reduced Cd accumulation<br>Enhanced NO signaling<br>Upregulation of antioxidant genes | [71] |
| *Artemisia annua* | 20 and 40 mg $kg^{-1}$ soil | 4 μM | Improved growth and biomass<br>ROS homeostasis<br>Improved membrane permeability<br>Reduced Cd uptake | [63] |
| Melon | 300 μmol $L^{-1}$ | 0.1, 0.5, 1, 2, and 3 μmol $L^{-1}$ | Improved root vigor<br>Increased antioxidant enzyme activity<br>Decreased MDA content | [72] |
| Barley | 10 μM | 1 μM | Improved growth<br>Reduced Cd uptake<br>Mediated Cd tolerance via NO and AsA–GSH cycle<br>Improved photosynthesis | [67] |
| Switchgrass | 10 μmol $L^{-1}$ | 1 μmol $L^{-1}$ | Improved shoot biomass and relative water content<br>Improved nutrient uptake<br>Improved chlorophyll content | [73] |

reduce oxidative stress and scavenge excess ROS while preserving cellular homeostasis [75]. Studies have demonstrated that under Cd stress, the addition of GR24 increased the activity of antioxidant enzymes such as CAT, POX, and SOD [66]. Sedaghat *et al.* [76] reported that exogenous GR24 administration on wheat increased the activity of antioxidants, shielding it from drought stress. Exogenous GR24 has also been observed to have a comparable impact on rapeseed when it comes to salt stress. According to some reports, GR24 can increase the antioxidant enzyme capacity to detoxify ROS and provide Cd tolerance [66, 76].

Furthermore, it has been demonstrated that the enzymes of the AsA–GSH cycle and non-enzymatic antioxidants like AsA and GSH work together to coordinate resistance to heavy metal stress. The turnover rates of the AsA–GsH cycle are shown by the redox ratios AsA/DHA and GSH/GSSG. Recent studies have demonstrated a considerable drop in the AsA/DHA and GSH/GSSG ratios in barley during Cd stress. Exogenous GR24 supplementation results in higher AsA/DHA ratios than when Cd stress was applied alone because AsA levels rose but DHA contents fell [67]. Similar alterations were discovered in the contents of GSSG and GSH. Furthermore, under Cd stress, GR24 administration boosted the activities of APX, GPX, GR, DHAR, and MDHAR and promoted AsA and GSH regeneration. According to the research findings, exogenous SLs may help maintain barley's redox state and reduce Cd toxicity by promoting the manufacture of AsA and GSH and the degradation of DHA and GSSG through the AsA–GSH cycle enzymes [67].

### 7.3.2 Regulatory Role of SLs in Photosynthesis for Improving Cadmium Tolerance

The photosynthetic machinery of plants is impacted by Cd stress, which lowers the amount of chlorophyll and carotenoid, results in distorted chloroplasts and decreases the activity of important photosynthetic enzymes like RuBisCO and carbonic anhydrase [77]. Exogenous GR24 application upon stressed plants has shown an ameliorative impact in mitigating these toxicity symptoms. Similarly, it has been noted that rapeseed treated with GR24 under saline stress had higher chlorophyll concentrations. Moreover, it has been discovered that exogenous GR24 enhanced chlorophyll synthesis in addition to upregulating genes related to light harvesting. As a result, SLs may have regulatory roles in reducing the toxicity of Cd by acting as positive regulators of chlorophyll production [64–66].

As a strong inhibitor, Cd suppresses the activity of Calvin cycle enzymes and changes electron transport via photosystems, reducing the production of chlorophyll and stopping the light reaction [78]. Furthermore, because Cd and calcium ($Ca^{2+}$) have the same charge, comparable ionic radius, and behavior, Cd competes with $Ca^{2+}$ and impacts stomatal closure and $CO_2$ conductance in guard cells, which inhibits photosynthesis [79]. These detrimental effects of Cd stress on transpiration rate, stomatal conductance, and net photosynthesis rate have recently been reported in barley [80]. The rate of transpiration, stomatal conductance, and net photosynthesis are all greatly increased by the application of exogenous GR24 [81]. These results show that exogenous SLs may sustain photosynthetic pigment levels and restore photosynthesis in the face of Cd stress, allowing plants to accumulate biomass normally and tolerate Cd.

### 7.3.3 Potential Role of SLs in Improving NO Signaling Under Cadmium Stress

One crucial gaseous signaling molecule for plants is nitric oxide (NO). As both positive and negative regulators of abiotic stress in plants, the interactions between NO and its targets control the expression of different genes [82]. Land plants do not yet have a gene known to encode NOS, yet several species have demonstrated NOS-like activity [83]. It is interesting to note that NO is thought of as a negative regulator that exacerbates disruptions during early exposure to Cd, yet it has protective effects over prolonged Cd stress. Recent research by Qiu *et al.* [67] has demonstrated that long-term exposure to Cd significantly decreased NO concentrations and hindered NOS activity in the roots and shoots of barley.

Additionally, it was demonstrated that adding GR24 to two barley genotypes—W6nk2 (Cd-sensitive) and Zhenong8 (Cd-tolerant)—effectively raised NO levels and improved NOS activity, particularly in W6nk2 [84]. It has been demonstrated that increased NOS causes the production of NO, which functions as a long-distance signal and positive regulator, strengthening the antioxidant system in stressed cells and inducing a defense response in distant organs, therefore reducing Cd toxicity [85]. Furthermore, it has been shown that SLs and NO interact in plants. NO can interact with and modify proteins involved in SL production and signal transduction, indicating that SL functions downstream of NO [86, 87]. Exogenous SL increases NO production in maize roots, but it also suppresses itself, suggesting that NO signaling may be responsible for the SL level negative feedback loop [88]. Thus, strigolactones and NO may work together to lessen the effects of Cd stress.

### 7.3.4 Potential Role of SLs in Regulating Root and Shoot Architecture Under Cadmium Stress

Research studies have indicated that Cd inhibits the growth of plant roots., as reported in switchgrass seedlings under Cd stress. It has been discovered that SLs stimulated the elongation of root hairs via a shared regulatory mechanism, which is generated in tandem with ethylene in Arabidopsis. The root architecture of switchgrass was controlled in the presence of SLs, as evidenced by recent research that demonstrates increased root length, root volume, root area or surface area, and number of root tips displayed upon GR24 supplementation [73]. According to another study, SLs aided in the growth of primary roots [89]. Recent research has highlighted the importance of fine roots (diameter $<0.4$ mm) in enhancing water and nutrient uptake in switchgrass, especially under Cd stress, where the administration of GR24 plays a significant role in promoting the growth of these fine roots [90]. Studies have shown that applying GR24 under Cd stress reduces Cd uptake while increasing Zn and Fe concentrations, indicating a potential strategy to improve mineral element absorption in plants [91].

The lack of vital nutrients needed for plant growth as a result of Cd and other divalent ions competing for the same absorption systems is another explanation of Cd toxicity. According to recent research, applying GR24 to plants resulted in lower concentrations of Cd and higher concentrations of Zn and Fe when compared to Cd stress alone. This supports the idea that adding GR24 lowers Cd uptake and facilitates the uptake of other mineral elements. Nutrient elements are translocated from roots to shoots by transport systems like the ZIP (zinc-regulated transporter) family and the natural resistance-associated macrophage protein (Nramp) family, which are capable of transporting $Fe^{2+}$, $Zn^{2+}$, and $Mn^{2+}$ in addition to non-essential $Cd^{2+}$. Further research is necessary to understand the mechanism of engagement in Cd transport systems in the presence of GR24 [92, 93].

Strigolactones play a crucial role in plant tolerance to Cd stress by regulating various mechanisms. Studies on tomato and *Withania somnifera* have shown that Cd exposure affects shoot architecture by interfering with nutrient uptake, enzymatic activities, and metabolic processes [94, 95]. Additionally, research on Moso bamboo and *Dianthus carthusianorum* calamine ecotype indicates that Cd stress influences root growth, photosynthetic processes, and gene expression related to Cd uptake and detoxification, ultimately impacting shoot development [96–98]. SLs contribute to Cd tolerance by modulating oxidative stress, antioxidant enzyme activities, and heavy metal transport-related gene expression, highlighting their potential role in maintaining shoot architecture under Cd stress conditions [99–101].

These findings collectively emphasize the intricate involvement of SLs in orchestrating plant responses to Cd toxicity and their impact on shoot morphology and growth.

## 7.4 Future Perspectives and Challenges of SLs for Agricultural Practices and Environmental Sustainability

To address global challenges like food security and climate change, future perspectives on sustainable land practices for agricultural practices and environmental sustainability are essential. This is because modern agriculture requires a growing number of specialized interventions during the growing season to handle a wide range of biotic and abiotic issues, necessitating the constant creation of new crop protection solutions. However, the same strategies are insufficient to deal with abiotic stresses such as a shortage of water or nutrients. Plants often utilize signal molecules to adapt to challenging surroundings. These molecules regulate several genetic and metabolic processes. Agrochemical manufacturers currently sell many of these signal molecules as phytoregulators or biofertilizers, with phytohormones (like auxins, which promote flower and fruit growth, and gibberellins, which stimulate fruit ripening and seed germination) playing a major role. Because of their previously reported functions as endogenous hormones and as rhizosphere signal molecules, SLs have caught the interest of the agro-technical industry. To address some important challenges in crop breeding and management, SL research has an optimistic outlook. In SL interactions, genetic variation may present an opportunity to develop plants with improved nutrient uptake efficiency and maybe efficient symbiotic relationships. In particular, SL controls the architecture of the roots and shoots and may thus be a potential target in plant breeding.

At the same time, we should look for strategies to lessen weed parasitism, particularly in Africa where it poses a serious threat to subsistence farmers. SL analogs that can promote suicidal germination of witchweed seeds in the soil before crop planting or plant breeding could be used to achieve this. One of the most effective methods for managing parasitic root weeds is suicidal germination, which is the encouragement of root parasite germination in the absence of host plants. In 2018, Uraguchi *et al.* identified the femtomolar range suicidal germination stimulant for *Striga hermonthica*, Sphynolactone-7 (SPL7) [102]. It is inactive for other biological processes connected to SL, such as preventing shoot branching or hyphal branching of AM fungus. Applying SPL7 to the soil at a concentration higher than 100 PPM before maize planting may help prevent the emergence of *Striga* and shield the crop from parasitic infestation [103]. With no detrimental effects on the shoot and root architecture or AM fungal colonization, this may serve as a perfect foundation for the development of suicidal germination inducers that efficiently reduce the seed bank of *Striga* and other weeds harboring root parasites.

Conversely, SL agonists that just regulate shoot morphology may be used to control the quantity of flowers and fruits. Numerous layers of cross-talk within hormonal networks regulate the manufacture and activity of SL under unfavorable environmental circumstances. As the plant's point of contact with the soil, roots are more susceptible to adverse soil conditions than aerial parts of the plant. The knowledge of how several phytohormones assist the plasticity of root growth has progressed. Further experimental and theoretical research is required to fully understand the diverse functions of SLs and these hormones in the overall level of organization of the plant. Unfortunately, a significant obstacle to the application of these promising biofertilizers in plant protection is the chemical instability of natural SLs in aqueous solution, especially at alkaline pH, which rather quickly hydrolyzes by producing an ABC-formyl lactone and 5-hydroxy butenolide. In addition to this restriction, the large-scale production of natural SLs is currently challenging from a technological and financial standpoint. In actuality, only about 20 different natural SLs have been found and described thus far. Complete chemical synthesis has been achieved but with a poor yield and a labor- and time-intensive process. It is challenging and/or economically impractical to create significant volumes of natural SLs from plants or through organic synthesis for the agrochemical sector since markups on commodity crops are frequently minimal. These variables led to the synthesis of synthetic molecules with a simpler chemical structure than natural SLs, yet bioactivity comparable to that of natural substances. One way to sum up SLs' future prospects in sustainable agriculture is as follows:

1. To get a deeper comprehension of plants' biosynthesis and metabolism, new analytical instruments for determining SL levels in plants must be developed.
2. Determining the genes in charge of the biosynthesis and signaling of SL production, will offer fresh opportunities for genetically engineering plant development and growth.

3. The creation of artificial SL analogs that may be used as herbicides to help suppress weeds without endangering crops by causing "suicidal germination."
4. Recognizing how SLs regulate plant-microbe interactions, particularly AM fungus, and how this knowledge may be applied to increase crop yields and nutrient uptake while reducing the need for chemical fertilizers.
5. The use of SL to increase output in a variety of environmental situations and the ways that future breeding initiatives will provide some positive agricultural outcomes.
6. Low-cost "synthetic SLs" production for broad application in farming environments.

## 7.5 Conclusion

In conclusion, by controlling a variety of developmental and adaption processes, SLs significantly improve plant tolerance to heavy metals like Cd. When plants are exposed to a variety of environmental stressors, such as exposure to heavy metals, SLs accumulate in their tissues and interact with phytohormones to assist in plant stress adaption. This indicates that SL is advantageous in creating a sustainable agricultural production system for places contaminated with Cd. Strigolactone strengthens antioxidant systems, raises the concentrations of nutritional elements, and assists in the upkeep of photosynthesis. Furthermore, by enhancing physiological and biochemical processes, helpful soil microorganisms including plant growth-promoting rhizobacteria (PGPR) and arbuscular mycorrhizal fungus (AMF) help to reduce Cd stress in agricultural plants. These microbes produce stress-tolerant substances like phytohormones and compatible solutes, which lessen the detrimental effects of Cd toxicity on plant growth and development. Consequently, SLs and helpful soil microbes can increase plant tolerance to Cd stress, which will ultimately increase agricultural output.

## References


1. Arora, V., Bithel, N., Singh, R., A Study on Heavy Metal Sources and Pollution: Challenge to Biological and Ecosystem. *BPASc-Botany*, 1, 1–10, 2023.
2. Kubier, A., Wilkin, R.T., Pichler, T., Cadmium in soils and groundwater: A review. *Appl.Geochem.*, *108*, 104388, 2019.

3. Naveed, S., Oladoye, P.O., Alli, Y.A., Toxic heavy metals: A bibliographic review of risk assessment, toxicity, and phytoremediation technology. *SCENV*, 873, 100018, 2023.
4. Priya, A.K., Muruganandam, M., Ali, S.S., Kornaros, M., Clean-up of heavy metals from contaminated soil by phytoremediation: A multidisciplinary and eco-friendly approach. *Toxics*, *11*, 5, 422, 2023.
5. Rashid, A., Schutte, B.J., Ulery, A., Deyholos, M.K., Sanogo, S., Lehnhoff, E.A., Beck, L., Heavy metal contamination in agricultural soil: environmental pollutants affecting crop health. *Agron.*, *13*, 6, 1521, 2023.
6. Wang, J., Wei, H., Pan, B., Accumulation characteristics and probabilistic risk assessment of Cd in agricultural soils across China. *Huan Jing Ke Xue*, *44*, 7, 4006–4016, 2023.
7. Yin, F., Meng, W., Liu, L., Feng, K., Yin, C., Spatial Distribution and Associated Risk Assessment of Heavy Metal Pollution in Farmland Soil Surrounding the Ganhe Industrial Park in Qinghai Province, China. *Land*, *12*, 6, 1172, 2023.
8. Shaari, N.E.M., Tajudin, M.T.F.M., Khandaker, M.M., Majrashi, A., Alenazi, M.M., Abdullahi, U.A., Mohd, K.S., Cadmium toxicity symptoms and uptake mechanism in plants: a review. *Braz. J. Biol.*, *84*, e252143, 2022.
9. Hussain, A., Ali, S., Rizwan, M., Zia-ur-Rehman, M., Yasmeen, T., Hayat, M.T., Hussain, S.M., Morphological and physiological responses of plants to cadmium toxicity, in: *Cadmium toxicity and tolerance in plants*, pp. 47–72, Academic Press, San Diego, CA, 2019.
10. Farid, M., Shakoor, M.B., Ehsan, S., Ali, S., Zubair, M., Hanif, M.S., Morphological, physiological and biochemical responses of different plant species to Cd stress. *Int. J. Chem. Biochem. Sci.*, *3*, 53–60, 2013.
11. Parrotta, L., Guerriero, G., Sergeant, K., Cai, G., Hausman, J.F., Target or barrier? The cell wall of early-and later-diverging plants vs cadmium toxicity: differences in the response mechanisms. *Front. Plant Sci.*, *6*, 123937, 2015.
12. Dell'Oste, V., Spyrakis, F., Prandi, C., Strigolactones, from plants to human health: achievements and challenges. *Mol.*, *26*, 15, 4579, 2021.
13. Schneider, A., Godin, C., Boudon, F., Demotes-Mainard, S., Sakr, S., Bertheloot, J., Light regulation of axillary bud outgrowth along plant axes: an overview of the roles of sugars and hormones. *Front. Plant Sci.*, *10*, 478249, 2019.
14. EL Sabagh, A., Islam, M.S., Hossain, A., Iqbal, M.A., Mubeen, M., Waleed, M., Abdelhamid, M.T., Phytohormones as growth regulators during abiotic stress tolerance in plants. *Front. Agron.*, *4*, 765068, 2022.
15. Yamada, Y., Furusawa, S., Nagasaka, S., Shimomura, K., Yamaguchi, S., Umehara, M., Strigolactone signalling regulates rice leaf senescence in response to a phosphate deficiency. *Planta*, *240*, 399–408, 2014.
16. Ha, C.V., Leyva-González, M.A., Osakabe, Y., Tran, U.T., Nishiyama, R., Watanabe, Y., Tran, L.S.P., Positive regulatory role of strigolactone in plant responses to drought and salt stress. *PNAS*, *111*, 2, 851–856, 2014.

17. Asati, A., Pichhode, M., Nikhil, K., Effect of heavy metals on plants: an overview. *IJAIEM*, *5*, 3, 56–66, 2016.
18. Seth, C.S., Misra, V., Chauhan, L.K.S., Singh, R.R., Genotoxicity of cadmium on root meristem cells of *Allium cepa*: cytogenetic and Comet assay approach. *Ecotoxicol. Environ. Saf.*, *71*, 3, 711–716, 2008.
19. Gill, S.S., Khan, N.A., Anjum, N.A., Tuteja, N., Amelioration of cadmium stress in crop plants by nutrients management: morphological, physiological and biochemical aspects. *Plant Stress*, *5*, 1, 1–23, 2011.
20. Zulfiqar, U., Jiang, W., Xiukang, W., Hussain, S., Ahmad, M., Maqsood, M.F., Mustafa, A., Cadmium phytotoxicity, tolerance, and advanced remediation approaches in agricultural soils; a comprehensive review. *Front. Plant Sci.*, *13*, 773815, 2022.
21. Moravčíková, D. and Žiarovská, J., The effect of cadmium on plants in terms of the response of gene expression level and activity. *Plants*, *12*, 9, 1848, 2023.
22. Arshad, M., Khalid, R., Hina, K., Ullah, S., Ali, M.A., Cadmium removal by aquatic macrophyte (*Nasturtium officinale*) and potential for oil production from the biomass. *PJAAEVS*, *31*, 1, 1–13, 2015.
23. Haider, F.U., Liqun, C., Coulter, J.A., Cheema, S.A., Wu, J., Zhang, R., Farooq, M., Cadmium toxicity in plants: Impacts and remediation strategies. *Ecotoxicol. Environ. Saf.*, *211*, 111887, 2021.
24. Rascio, N., Dalla Vecchia, F., La Rocca, N., Barbato, R., Pagliano, C., Raviolo, M., Gabbrielli, R., Metal accumulation and damage in rice (cv. *Vialone nano*) seedlings exposed to cadmium. *Environ. Exp. Bot.*, *62*, 3, 267–278, 2008.
25. Rizwan, M., Ali, S., Abbas, T., Zia-ur-Rehman, M., Hannan, F., Keller, C., Ok, Y.S., Cadmium minimization in wheat: a critical review. *Ecotoxicol. Environ. Saf.*, *130*, 43–53, 2016.
26. Berni, R., Luyckx, M., Xu, X., Legay, S., Sergeant, K., Hausman, J.F., Guerriero, G., Reactive oxygen species and heavy metal stress in plants: Impact on the cell wall and secondary metabolism. *Environ. Exp. Bot.*, *161*, 98–106, 2019.
27. Abbas, T., Rizwan, M., Ali, S., Adrees, M., Zia-ur-Rehman, M., Qayyum, M.F., Murtaza, G., Effect of biochar on alleviation of cadmium toxicity in wheat (Triticum aestivum L.) grown on Cd-contaminated saline soil. *Environ. Sci. Pollut. Res. Int.*, *25*, 25668–25680, 2018.
28. Younis, U., Malik, S.A., Rizwan, M., Qayyum, M.F., Ok, Y.S., Shah, M.H.R., Ahmad, N., Biochar enhances the cadmium tolerance in spinach (*Spinacia oleracea*) through modification of Cd uptake and physiological and biochemical attributes. *Environ. Sci. Pollut. Res. Int.*, *23*, 21385–21394, 2016.
29. Noreen, S., Akhter, M.S., Noman, A., Ummara, U., Mahmood, S., Hashim, M., Future Perspectives of Oxidative Stress and Antioxidant Defence System in Plants, in: *Developing Climate-Resilient Crops*, pp. 53–84, CRC Press, Boca Raton, FL., 2021.
30. Faizan, M., Bhat, J.A., Hessini, K., Yu, F., Ahmad, P., Zinc oxide nanoparticles alleviates the adverse effects of cadmium stress on *Oryza sativa* via

modulation of the photosynthesis and antioxidant defense system. *Ecotoxicol. Environ. Saf.*, *220*, 112401, 2021.
31. Kolahi, M., Kazemi, E.M., Yazdi, M., Goldson-Barnaby, A., Oxidative stress induced by cadmium in lettuce (*Lactuca sativa* Linn.): Oxidative stress indicators and prediction of their genes. *PPB*, *146*, 71–89, 2020.
32. Sharma, M., Gupta, S.K., Deeba, F., Pandey, V., Effects of reactive oxygen species on crop productivity: an overview, in: *Reactive Oxygen Species in Plants: Boon Or Bane-Revisiting the Role of ROS*, pp. 117–136, 2017.
33. Mishra, S., Spaccarotella, K., Gido, J., Samanta, I., Chowdhary, G., Effects of heat stress on plant-nutrient relations: An update on nutrient uptake, transport, and assimilation. *Int. J. Mol. Sci.*, *24*, 21, 15670, 2023.
34. Dobrikova, A.G. and Apostolova, E.L., Damage and protection of the photosynthetic apparatus under cadmium stress, in: *Cadmium toxicity and tolerance in plants*, pp. 275–298, Academic Press, San Diego, CA, 2019.
35. Dobrikova, A.G., Apostolova, E.L., Hanć, A., Yotsova, E., Borisova, P., Sperdouli, I., Moustakas, M., Cadmium toxicity in *Salvia sclarea* L.: An integrative response of element uptake, oxidative stress markers, leaf structure and photosynthesis. *Ecotoxicol. Environ. Saf.*, *209*, 111851, 2021.
36. Dobrikova, A., Apostolova, E., Adamakis, I.D.S., Hanć, A., Sperdouli, I., Moustakas, M., The combined impact of excess zinc and cadmium on elemental uptake, leaf anatomy and pigments, antioxidant capacity, and function of photosynthetic apparatus in clary sage (*Salvia sclarea* L.). *Plants*, *11*, 18, 2407, 2022.
37. Sattar, A., Sher, A., Ijaz, M., Ul-Allah, S., Abbas, T., Hussain, S., Ibrahim, M.F., Silicon and Strigolecton Application Alleviates the Adversities of Cadmium Toxicity in Maize by Modulating Morpho-Physiological and Antioxidants Defense Mechanisms. *Agron.*, *13*, 9, 2352, 2023.
38. Yu, F., Yao, Y., Feng, J., Wang, X., Ma, J., Liu, K., Li, Y., Enterobacter sp. FM-1 inoculation influenced heavy metal-induced oxidative stress in pakchoi (*Brassica campestris* L. ssp. Chinensis Makino) and water spinach (*Ipomoea aquatic* F.) cultivated in cadmium and lead co-contaminated soils. *PLSOA2*, *459*, 155–171, 2021.
39. Hu, C.H., Wang, P.Q., Zhang, P.P., Nie, X.M., Li, B.B., Tai, L., Chen, K.M., NADPH oxidases: The vital performers and center hubs during plant growth and signaling. *Cells*, *9*, 2, 437, 2020.
40. Dias, M.C., Monteiro, C., Moutinho-Pereira, J., Correia, C., Gonçalves, B., Santos, C., Cadmium toxicity affects photosynthesis and plant growth at different levels. *Acta Physiol. Plant.*, *35*, 1281–1289, 2013.
41. Muradoglu, F., Gundogdu, M., Ercisli, S., Encu, T., Balta, F., Jaafar, H.Z., Zia-Ul-Haq, M., Cadmium toxicity affects chlorophyll a and b content, antioxidant enzyme activities and mineral nutrient accumulation in strawberry. *Biol. Res.*, *48*, 1–7, 2015.
42. Tran, T.A. and Popova, L.P., Functions and toxicity of cadmium in plants: recent advances and future prospects. *Turk. J. Bot.*, *37*, 1, 1–13, 2013.

43. El Rasafi, T., Oukarroum, A., Haddioui, A., Song, H., Kwon, E.E., Bolan, N., Rinklebe, J., Cadmium stress in plants: A critical review of the effects, mechanisms, and tolerance strategies. *Crit. Rev. Env. Sci. Tec.*, *52*, 5, 675–726, 2022.
44. Islam, M.M., Hoque, M.A., Okuma, E., Banu, M.N.A., Shimoishi, Y., Nakamura, Y., Murata, Y., Exogenous proline and glycinebetaine increase antioxidant enzyme activities and confer tolerance to cadmium stress in cultured tobacco cells. *J. Plant Physiol.*, *166*, 15, 1587–1597, 2009.
45. Rasheed, A., Jie, H., He, P., Lv, X., Ali, B., Ma, Y., Jie, Y., A Systematic Review on the Improvement of Cd Stress Tolerance in Ramie Crop, Limitations and Future Prospective. *Agron.*, *13*, 7, 1793, 2023.
46. Zhu, Y.X., Zhuang, Y., Sun, X.H., Du, S.T., Interactions between cadmium and nutrients and their implications for safe crop production in Cd-contaminated soils. *Crit. Rev. Env. Sci. Tec.*, *53*, 24, 2071–2091, 2023.
47. Awad-Allah, E.F., Elsokkary, I.H., Lindberg, S.M., Phosphorus protects against cadmium-induced phytotoxicity by stimulating plasma membrane H+-ATPase activity. *J. Plant Nutr.*, *46*, 16, 3918–3932, 2023.
48. Li, L., Guo, B., Feng, C., Liu, H., Lin, D., Growth, physiological, and temperature characteristics in Chinese cabbage pakchoi as affected by Cd-stressed conditions and identifying its main controlling factors using the PLS model. *BMC Plant Biol.*, *22*, 1, 571, 2022.
49. Huybrechts, M., Cuypers, A., Deckers, J., Iven, V., Vandionant, S., Jozefczak, M., Hendrix, S., Cadmium and plant development: An agony from seed to seed. *Int. J. Mol. Sci.*, *20*, 16, 3971, 2019.
50. Vijayaragavan, M., Prabhahar, C., Sureshkumar, J., Natarajan, A., Vijayarengan, P., Sharavanan, S., Toxic effect of cadmium on seed germination, growth and biochemical contents of cowpea (*Vigna unguiculata* L.) plants. *Int. Multidiscip. Res. J.*, *1*, 5, 1–7, 2011.
51. Bautista, O.V., Fischer, G., Cárdenas, J.F., Cadmium and chromium effects on seed germination and root elongation in lettuce, spinach and Swiss chard. *Agron. Colomb.*, *31*, 1, 48–57, 2013.
52. Kieffer, P., Schröder, P., Dommes, J., Hoffmann, L., Renaut, J., Hausman, J.F., Proteomic and enzymatic response of poplar to cadmium stress. *J. Proteomics*, *72*, 3, 379–396, 2009.
53. Jali, P., Pradhan, C., Das, A.B., Effects of cadmium toxicity in plants: a review article. *Sch. Acad. J. Biosci.*, *4*, 12, 1074–1081, 2016.
54. Popova, L.P., Maslenkova, L.T., Ivanova, A., Stoinova, Z., Role of salicylic acid in alleviating heavy metal stress, in: *Environmental adaptations and stress tolerance of plants in the era of climate change*, pp. 447–466, 2012.
55. Yabuta, Y., Functions of heat shock transcription factors involved in response to photooxidative stresses in Arabidopsis. *Biosci. Biotechnol., Biochem.*, *80*, 7, 1254–1263, 2016.
56. Mittra, B., Sharma, S., Das, A.B., Henry, S.L., Das, T.K., Ghosh, P., Mohanty, P., A novel cadmium-induced protein in wheat: characterization and localization in root tissue. *Biol. Plant.*, *52*, 343–346, 2008.

57. Sergeant, K., Kieffer, P., Dommes, J., Hausman, J.F., Renaut, J., Proteomic changes in leaves of poplar exposed to both cadmium and low-temperature. *EEB*, *106*, 112–123, 2014.
58. Rodríguez-Celma, J., Rellán-Álvarez, R., Abadía, A., Abadía, J., López-Millán, A.F., Changes induced by two levels of cadmium toxicity in the 2-DE protein profile of tomato roots. *J. Proteomics*, *73*, 9, 1694–1706, 2010.
59. Semane, B., Dupae, J., Cuypers, A., Noben, J.P., Tuomainen, M., Tervahauta, A., Vangronsveld, J., Leaf proteome responses of Arabidopsis thaliana exposed to mild cadmium stress. *J. Plant Physiol.*, *167*, 4, 247–254, 2010.
60. Sato, D., Awad, A.A., Takeuchi, Y., Yoneyama, K., Confirmation and quantification of strigolactones, germination stimulants for root parasitic plants Striga and Orobanche, produced by cotton. *Biosci. Biotechnol., Biochem.*, *69*, 1, 98–102, 2005.
61. Mitra, D., BE, G.S., Khoshru, B., De Los Santos Villalobos, S., Belz, C., Chaudhary, P., Mohapatra, P.K.D., Impacts of arbuscular mycorrhizal fungi on rice growth, development, and stress management with a particular emphasis on strigolactone effects on root development. *Commun. Soil Sci. Plant Anal.*, *52*, 14, 1591–1621, 2021.
62. Selwal, N., Wani, A.K., Akhtar, N., Kaur, M., Jassal, P.S., Molecular insights of Strigolactone biosynthesis, signalling pathways, regulatory roles, and hormonal crosstalks in plant systems. *S. Afr. J. Bot.*, *160*, 9–22, 2023.
63. Wani, K.I., Naeem, M., Khan, M.M.A., Aftab, T., Insights into strigolactone (GR24) mediated regulation of cadmium-induced changes and ROS metabolism in *Artemisia annua*. *J. Hazard. Mater.*, 448, 130899, 2023.
64. Wani, K.I., Zehra, A., Choudhary, S., Naeem, M., Khan, M.M.A., Castroverde, C.D.M., Aftab, T., Mechanistic insights into strigolactone biosynthesis, signaling, and regulation during plant growth and development. *J. Plant Growth Regul.*, 39, 1, 1–17, 2020.
65. Wani, K.I., Zehra, A., Choudhary, S., Naeem, M., Khan, M.M.A., Khan, R., Aftab, T., Exogenous strigolactone (GR24) positively regulates growth, photosynthesis, and improves glandular trichome attributes for enhanced artemisinin production in *Artemisia annua*. *J. Plant Growth Regul.*, *42*, 8, 4606–4615, 2023.
66. Wani, K.I., Abdullah, Naeem, M., Aftab, T., Understanding the Interplay Between Strigolactone and Nitric Oxide in Alleviating Cadmium-Induced Toxicity in *Artemisia annua*. *EEB*, 205, 105856, 2024.
67. Qiu, C.W., Zhang, C., Wang, N.H., Mao, W., Wu, F., Strigolactone GR24 improves cadmium tolerance by regulating cadmium uptake, nitric oxide signalling and antioxidant metabolism in barley (*Hordeum vulgare* L.). *Environ. Pollut.*, 273, 116486, 2021.
68. Chi, C., Xu, X., Wang, M., Zhang, H., Fang, P., Zhou, J., Yu, J., Strigolactones positively regulate abscisic acid-dependent heat and cold tolerance in tomato. *Hortic. Res.*, *8*, 226, 2021.

69. Bhoi, A., Yadu, B., Chandra, J., Keshavkant, S., Contribution of strigolactone in plant physiology, hormonal interaction and abiotic stresses. *Planta*, *254*, 2, 28, 2021.
70. Yang, Y., Zheng, Q., Zhou, K., Xiao, Y., Huang, C., Hu, R., Wang, J., Effects of exogenous strigolactone on the cadmium accumulation in *Galinsoga parviflora* Cav. *Chem. Ecol.*, 40, 3, 1–13, 2024.
71. Shah, T., Khan, Z., Asad, M., Imran, A., Niazi, M.B.K., Alsahli, A.A., Alleviation of cadmium toxicity in wheat by strigolactone: Regulating cadmium uptake, nitric oxide signalling, and genes encoding antioxidant defence system. *PPB*, 202, 107916, 2023.
72. Chen, X., Shi, X., Ai, Q., Han, J., Wang, H., Fu, Q., Transcriptomic and metabolomic analyses reveal that exogenous strigolactones alleviate the response of melon root to cadmium stress. *Hortic. Plant J.*, 8, 5, 637–649, 2022.
73. Tai, Z., Yin, X., Fang, Z., Shi, G., Lou, L., Cai, Q., Exogenous GR24 alleviates cadmium toxicity by reducing cadmium uptake in switchgrass (Panicum virgatum) seedlings. *Int. J. Environ. Res. Public Health*, 14, 8, 852, 2017.
74. Zhang, X.H., Ma, C., Zhang, L., Su, M., Wang, J., Zheng, S., Zhang, T.G., GR24-mediated enhancement of salt tolerance and roles of H2O2 and $Ca^{2+}$ in regulating this enhancement in cucumber. *J. Plant Physiol.*, *270*, 153640, 2022.
75. Garcia-Caparros, P., De Filippis, L., Gul, A., Hasanuzzaman, M., Ozturk, M., Altay, V., Lao, M.T., Oxidative stress and antioxidant metabolism under adverse environmental conditions: a review. *Bot. Rev.*, *87*, 421–466, 2021.
76. Sedaghat, M., Sarvestani, Z.T., Emam, Y., Bidgoli, A.M., Sorooshzadeh, A., Foliar-applied GR24 and salicylic acid enhanced wheat drought tolerance. *Russ. J. Plant Physiol.*, *67*, 733–739, 2020.
77. Wahid, A., Arshad, M., Farooq, M., Cadmium phytotoxicity: responses, mechanisms and mitigation strategies: a review, in: *Organic Farming, Pest Control and Remediation of Soil Pollutants: Organic farming, pest control and remediation of soil pollutants*, pp. 371–403, 2010.
78. Ivanov, A.A. and Kosobryukhov, A.A., Ecophysiology of plants under cadmium toxicity: photosynthetic and physiological responses, in: *Plant Ecophysiology and Adaptation under Climate Change: Mechanisms and Perspectives I: General Consequences and Plant Responses*, pp. 429–484, 2020.
79. Perfus-Barbeoch, L., Leonhardt, N., Vavasseur, A., Forestier, C., Heavy metal toxicity: cadmium permeates through calcium channels and disturbs the plant water status. *Plant J.*, *32*, 4, 539–548, 2002.
80. Chen, F., Wang, F., Sun, H., Cai, Y., Mao, W., Zhang, G., Wu, F., Genotype-dependent effect of exogenous nitric oxide on Cd-induced changes in antioxidative metabolism, ultrastructure, and photosynthetic performance in barley seedlings (Hordeum vulgare). *J. Plant Growth Regul.*, *29*, 394–408, 2010.
81. Luqman, M., Shahbaz, M., Maqsood, M.F., Farhat, F., Zulfiqar, U., Siddiqui, M.H., Haider, F.U., Effect of strigolactone on growth, photosynthetic

efficiency, antioxidant activity, and osmolytes accumulation in different maize (*Zea mays* L.) hybrids grown under drought stress. *Plant Signal. Behav.*, *18*, 1, 2262795, 2023.
82. Asgher, M., Per, T.S., Masood, A., Fatma, M., Freschi, L., Corpas, F.J., Khan, N.A., Nitric oxide signaling and its crosstalk with other plant growth regulators in plant responses to abiotic stress. *Environ. Sci. Pollut. Res. Int.*, *24*, 2273–2285, 2017.
83. Jeandroz, S., Wipf, D., Stuehr, D.J., Lamattina, L., Melkonian, M., Tian, Z., Wendehenne, D., Occurrence, structure, and evolution of nitric oxide synthase–like proteins in the plant kingdom. *Sci. Signal.*, *9*, 417, re2–re2, 2016.
84. Chen, F., Dong, J., Wang, F., Wu, F., Zhang, G., Li, G., Wei, K., Identification of barley genotypes with low grain Cd accumulation and its interaction with four microelements. *Chemosphere*, *67*, 10, 2082–2088, 2007.
85. Arasimowicz-Jelonek, M., Floryszak-Wieczorek, J., Izbiańska, K., Costs and benefits of nitric oxide generation in plants exposed to cadmium. *Adv. Bot. Res.*, *77*, 97–121, 2016.
86. Korek, M. and Marzec, M., Strigolactones and abscisic acid interactions affect plant development and response to abiotic stresses. *BMC Plant Biol.*, *23*, 1, 314, 2023.
87. Lu, X., Liu, X., Xu, J., Liu, Y., Chi, Y., Yu, W., Li, C., Strigolactone-mediated trehalose enhances salt resistance in tomato seedlings. *Horticulturae*, *9*, 7, 770, 2023.
88. Kee, Y.J., Ogawa, S., Ichihashi, Y., Shirasu, K., Yoshida, S., Strigolactones in rhizosphere communication: multiple molecules with diverse functions. *PCP*, *64*, 9, 955–966, 2023.
89. Alvi, A.F., Sehar, Z., Fatma, M., Masood, A., Khan, N.A., Strigolactone: an emerging growth regulator for developing resilience in plants. *Plants*, *11*, 19, 2604, 2022.
90. Song, G., Zhang, J., Wang, Y., Ji, Y., Fang, Z., Cai, Q., Xu, B., Overexpression of PvBiP2 improved biomass yield and cadmium tolerance in switchgrass (*Panicum virgatum* L.). *J. Hazard. Mater.*, *446*, 130648, 2023.
91. Gomes, L.A., Costa, J., Santos, F., Fernando, A.L., Environmental and socio-economic impact assessment of the switchgrass production in heavy metals contaminated soils, in: *International Conference Innovation in Engineering*, 2021, June, Springer International Publishing, Cham, pp. 410–419.
92. Wang, S., Liu, J.B., Kang, J., Xu, B.C., Chen, Y.L., Root growth and morphology of switchgrass (*Panicum virgatum* L.) and bush clover (*Lespedeza davurica* S.) in mixed plantation under varying soil water and phosphorus supply conditions. *Pak. J. Bot.*, *53*, 4, 1227–1237, 2021.
93. Zang, H., He, J., Zhang, Q., Li, X., Wang, T., Bi, X., Zhang, Y., Ectopic expression of PvHMA2. 1 enhances cadmium tolerance in *Arabidopsis thaliana*. *Int. J. Mol. Sci.*, *24*, 4, 3544, 2023.

94. Liu, C.X., Yang, T., Zhou, H., Ahammed, G.J., Qi, Z.Y., Zhou, J., The E3 Ubiquitin Ligase Gene Sl1 Is Critical for Cadmium Tolerance in *Solanum lycopersicum* L. *Antige.*, *11*, 3, 456, 2022.
95. Mishra, B., Sangwan, R.S., Mishra, S., Jadaun, J.S., Sabir, F., Sangwan, N.S., Effect of cadmium stress on inductive enzymatic and nonenzymatic responses of ROS and sugar metabolism in multiple shoot cultures of Ashwagandha (*Withania somnifera* Dunal). *Protoplasma*, *251*, 1031–1045, 2014.
96. Yang, F., Chang, Y.Z., Zheng, Y.T., Pan, X., Ji, H., Shao, J.F., Physiological and transcriptomic characterization of cadmium toxicity in Moso bamboo (*Phyllostachys edulis*), a non-timber forest species. *Tree Physiol.*, *43*, 7, 1250–1264, 2023.
97. Marques, D.N., Nogueira, M.L., Gaziola, S.A., Batagin-Piotto, K.D., Freitas, N.C., Alcantara, B.K., Azevedo, R.A., New insights into cadmium tolerance and accumulation in tomato: dissecting root and shoot responses using cross-genotype grafting. *Environ. Res.*, *216*, 114577, 2023.
98. Muszyńska, E., Hanus-Fajerska, E., Ciarkowska, K., Studies on lead and cadmium toxicity in *Dianthus carthusianorum* calamine ecotype cultivated *in vitro*. *Plant Biol.*, *20*, 3, 474–482, 2018.
99. Wei, K. and Guo, T., The Changes of Tolerance, Accumulation and Oxidative Stress Response to Cadmium in Tobacco Caused by Introducing Datura stramonium L. Genes. *Agron.*, *13*, 3, 882, 2023.
100. Narayanan, M. and Ma, Y., Metal tolerance mechanisms in plants and microbe-mediated bioremediation. *Environ. Res.*, *222*, 115413, 2023.
101. Al-Khayri, J.M., Banadka, A., Rashmi, R., Nagella, P., Alessa, F.M., Almaghasla, M.I., Cadmium toxicity in medicinal plants: An overview of the tolerance strategies, biotechnological and omics approaches to alleviate metal stress. *Front. Plant Sci.*, *13*, 1047410, 2023.
102. Uraguchi, D., Kuwata, K., Hijikata, Y., Yamaguchi, R., Imaizumi, H., Am, S., Tsuchiya, Y., A femtomolar-range suicide germination stimulant for the parasitic plant *Striga hermonthica*. *Sci.*, *362*, 6420, 1301–1305, 2018.
103. Hajam, A.H., Ali, M.S., Singh, S.K., Bashri, G., Dubey, G., Implications of strigolactones in plant biology: Achievements, future perspectives, and challenges, in: *Strigolactones*, pp. 221–236, 2024.

# 8

# Strigolactone Interplay with Other Phytohormones Under Stressed and Normal Conditions

**Ishrat Mehmood[1]*, Kaiser Iqbal Wani[2] and Tariq Aftab[2]**

[1]*Department of Biophysics, Panjab University, Chandigarh, India*
[2]*Department of Botany, Aligarh Muslim University, Aligarh, India*

### *Abstract*
Strigolactones (SLs) are small carotenoid-derived signaling compounds, and a class of butenolide-bearing phytohormones, which control various developmental and adaptation processes in plants. Different growth aspects are attributed to these, including the patterning of root and shoot architecture, germination of seeds, senescence of leaves, nutrition acquisition, symbiotic and parasitic relationships, and the mediation of plant responses to biotic and abiotic stimuli. Nutrient-deficient plants produce more SLs, which helps with root architecture and arbuscular mycorrhizal fungal symbiosis. This leads to an increase in nutrient absorption. The interaction of SLs with other phytohormones, including auxin, cytokinin, abscisic acid, and gibberellins during abiotic stress responses, suggests that SLs play a proactive role in the regulatory mechanisms of plant stress tolerance. Crosstalk between hormone signaling in plants during development and environmental response can happen either sequentially or simultaneously. The precise underlying mechanism is not clear because of the diverse effects of phytohormones and the differing locations of their activities. In response to a variety of environmental conditions such as heat, cold, drought, salinity, heavy metals, and nutrient deprivation, these SLs accumulate in plant tissues. Many hormone response pathways are regulated by SLs, which help plants overcome demanding environmental conditions and mitigate adverse effects on crop yield as a whole. In this chapter, we will present the latest insights about SL-associated hormonal networks in plant development and their responses to various environmental stressors.

**Corresponding author*: ishratmehmood943@gmail.com

Tariq Aftab and Kaiser Iqbal Wani (eds.) Strigolactones: Emerging Plant Hormones, (161–192) © 2025 Scrivener Publishing LLC

***Keywords:*** Strigolactones, abiotic stress, phosphate, nitrogen, salinity, light, chilling

## 8.1 Introduction

Plants are frequently subjected to unfavorable environmental circumstances, resulting in abiotic stressors and lower yield. Phytohormones, often known as plant hormones, are important mediators of plant physiological and developmental processes [1]. Strigolactones (SLs) are a class of novel phytohormones initially found in 1996 in the root exudates of *Gossypium hirsutum* as a crystalline germination stimulant of the parasitic plant *Striga lutea*. Later on, two groups of researchers characterized them as plant hormones, independently in 2008. Based on their chemical structure, SLs were classed as canonical and non-canonical SLs [2, 3]. These hormones are tiny, diffusible molecules that may readily pass between cells and act as signaling molecules in normal and stressful situations [4]. Under normal conditions, SLs work with other phytohormones to coordinate plant development and growth, particularly in nitrogen and phosphorus-inadequate environments [5, 6]. The interaction between SLs and other phytohormones is complex. It varies depending on the physiological and environmental context, as they regulate both below and above-ground plant morphogenesis, such as leaf senescence, shoot branching, reproductive development, root hair density, and adventitious root formation [7, 8]. Strigolactones mediate the plant's response and adaptability to salt stress, drought, and pathogen attack. Coordinated biochemical alterations in a complex plant development network are necessary for these activities, necessitating two or more hormone signals to communicate and cooperate [9]. Under normal conditions, the interplay between SLs and other phytohormones helps to regulate plant growth and development. This connection is mediated by a network of phytohormones, including gibberellins, auxin cytokinins, jasmonates, abscisic acid, salicylic acid, and ethylene. Under normal circumstances, SLs function in tandem with other phytohormones to control several aspects of growth and development in plants. This interaction can result in crosstalk, synergistic effects, or antagonistic effects between SLs and other phytohormones [10–12].

## 8.2 Role of SLs in Abiotic Stress Tolerance

When the plant encounters particular environmental challenges, the SL levels rise to optimize and boost the growth methods to suit the circumstances.

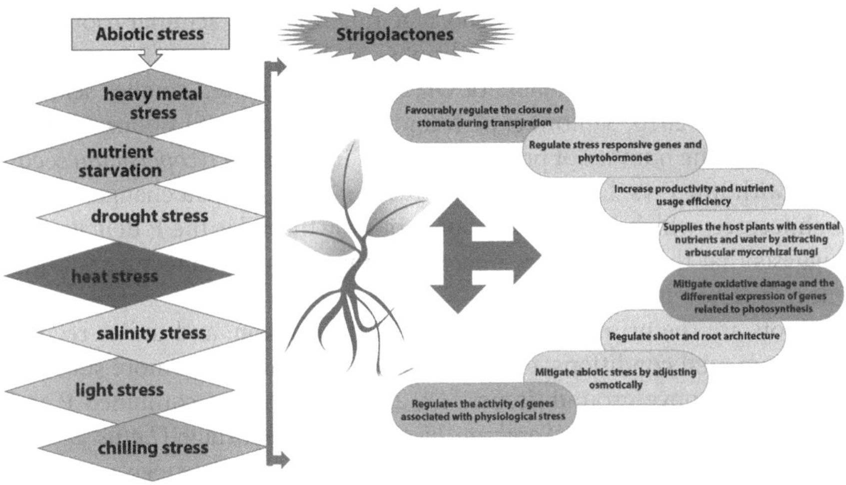


**Figure 8.1** The proposed scheme illustrates how SLs help plants adapt to abiotic stressors.

Abiotic stressors such as salt, drought, extreme temperatures, excess heavy metals, nutrient deficiency, etc., pose a major threat to agricultural plants and thereby, world food security [13, 14]. Strigolactones can confer tolerance to numerous abiotic stressors in plants (Figure 8.1).

### 8.2.1 SLs During Nutrient Starvation

Being a structural element of cell macromolecules and a key player in fundamental metabolic processes, phosphorus (P) is a macronutrient that is vital to plants. To combat P-starvation, plants have developed a variety of ways to obtain localized P-sources. Phosphorous shortage causes structural alterations in root hair density and root architecture in plants [9, 15]. Strigolactone control over lateral root (LR) production is controlled by inorganic phosphate (Pi) content. Strigolactones suppress LR production when Pi is available in adequate quantities, but it is triggered when Pi levels are low [16]. Thus, SLs have therefore been proposed as plant architecture regulators, particularly in response to Pi [11]. Inorganic phosphate starvation-induced (PSI) genes are strongly associated with plants in response to inadequate Pi environments and are commonly employed as markers in such situations [17]. These genes get transcribed under Pi stress, however in the SL mutant *Arabidopsis thaliana*, Pi was unable to influence *PSI* expression, possibly because these genes are regulated by SL. Under P-deficient conditions, *A. thaliana* exhibited decreased primary root development and increased LR density [16, 18, 19].

Another component that Pi shortage might alter is seminal root elongation. Under P-inadequate conditions, SL-biosynthesis (*d27* and *d10*) or SL-insensitive (*d3*) mutants of *Oryza sativa* exhibited smaller seminal roots [20]. In 2020, Santoro *et al.* found that exogenous SLs boosted fresh weight and the number of primary and LRs in *Solanum lycopersicum* seedlings with low P vs. high P [21]. Furthermore, SLs cause a change in root architecture, which is important for root growth under the P deficit. This suggests that the effect of *rac*-GR24 therapy on root architecture is influenced by P levels [18, 22].

Roots are the primary locations for SL production, and under regular developmental conditions, their production is significantly more substantial than in above-ground tissues [18, 23]. Unlike roots, combined P and nitrogen (N) deficiency (*Sorghum bicolour*) or P shortage (*Lotus japonicus*) had little or no effect on SL content in shoots. This suggests that SL levels rise largely in the roots during nutritional deficiency rather than in above-ground organs, at least in the case of *S. bicolour* and *L. japonicus*. In *S. bicolour*, concurrent N and P deprivation had a very low influence on root SL exudation than individual P or N shortage [18, 24]. A parallel phenomenon was observed in the roots of *O. sativa*, where the limitation of both P and N resulted in reduced SL exudation than P shortage alone. SLs can operate as signaling molecules under nutritional stress because N and Pi deficits cause their accumulation in root tissues. As a result, SLs encourage Pi uptake and movement in cases of severe deficiency [18, 25, 26]. In wild-type (WT) *O. sativa*, P or N shortage increased SL synthesis by increasing the expression of *D27*, *D17*, and *D10* while reducing *D53, D14*, and *D3*, expression [27, 28]. In *A. thaliana* roots, N deprivation enhanced *MAX4* and *MAX3* transcription, which might imply that SLs build in root tissues as a result of N starvation. The *PDR1*, SL transporter-encoding gene is expressed more in the roots under P-deficient conditions, which are required for SL translocation from root to shoot in *L. japonicus* and *Petunia hybrida* [18, 23, 29].

The emergence of arbuscular mycorrhizal fungus (AMF) symbiosis necessitates a higher degree of collaboration between two partners because of a precisely tuned molecular relationship [30]. In nutrient-deficient conditions, the host plants create signaling molecules and exude them into the rhizosphere to establish this symbiotic association, that promotes AMF and hyphae development [31, 32]. Phosphorous is considered as the soil's least accessible important nutrient. Based on the low mobility of P in soil solutions, plants rapidly develop a P-depletion zone around their roots [33]. In *A. thaliana*, the low-P environment promotes SL secretion and biosynthesis by roots, increasing plant-AMF contact and activating AMF branching.

Phosphate is absorbed through the fine exterior hyphae of AMF, which extend outside the depletion zone. *Trifolium pratense* was shown to generate large quantities of SL in low P circumstances [18, 34–36]. Additionally, Pan *et al.* discovered that in *Zea mays*, lower P activates the *ZmCCD7*, the SL-biosynthetic gene [37]. Rhizobia and leguminous plants build a symbiotic relationship and acquire N from root nodules; in this environment, only P shortage enhances SL discharge, attracting AM symbionts due to the availability of P [38]. However, in non-leguminous plants such as *S. bicolour*, both P and N shortages promote SL discharge, showing that these plants rely on AM symbionts to meet their N and P requirements [18].

### 8.2.2 SLs During Drought Stress

One of the most prevalent abiotic stressors that plants face is water shortage, which disrupts growth and hurts their survival and output. Drought stress disrupts the electron transport chain (ETC) in plants, causing oxidative stress and the accumulation of reactive oxygen species (ROS), which damages certain cell organelles such as mitochondria, peroxisomes, and chloroplasts [39]. To maximize water intake, plants have developed several defense strategies including stomatal closure, osmotic modification, morphological adaptation, and antioxidant system enhancement [40, 41]. In 2016, Ruiz-Lozano *et al.* observed during drought stress, a decrease in SL levels was found to be positively connected to the reduced expression of *SlCCD7,* the SL-biosynthetic gene in *S. lycopersicum* [42]. Drought stress induced by Polyethylene glycol enhanced the production of $H_2O_2$ and malondialdehyde (MDA) in *Vitis vinifera* while decreasing chlorophyll concentration and altering the antioxidant defense system. However, the GR24 treatment showed increased tolerance to drought stress via regulating photosynthetic rate and chlorophyll components [43, 44]. In addition, it was found that using GR24 reduced MDA and $H_2O_2$ levels, suggesting that SL may serve as a ROS scavenger and reduce lipid peroxidation (LPO) under drought conditions in *Triticum aestivum* [44, 45]. Max mutants (*max4*, *max3*, and *max2*) of *A. thaliana* lose more water than WT, possibly because of the differing transpiration rates [46]. Exogenous SL treatment did restore the drought-sensitive traits of SL-deficient *max4* and *max3* mutants to the same extent while WT had no meaningful obvious on the SL-response *max2* mutants [47]. This literature further reinforces the hypothesis that SLs operate as mutual regulators of stomatal closure. SLs also have an indirect mechanism for stomatal closure, namely control of abscisic acid production [48]. A comparison of *O. sativa* SL mutants (*d27, d17, d10,* and *d3*) found that *d27* generated much less abscisic acid than other *d* mutants, specifying that *D27* may be involved in the control of abscisic acid biosynthesis. In contrast to the high abscisic acid level in SL biosynthetic/signaling mutants, the lower abscisic acid content of *d27* revealed the probable function of *OsD27* in abscisic acid production [18, 49, 50]. In these mutant lines, this demonstrates that abscisic acid concentration is strongly associated with the expression of the *D27* gene. The control of SL and abscisic acid homeostasis is linked to *D27*, the *β-carotene isomerase*-producing gene, which improves tolerance to drought in plants [51, 52]. The *hvd14.d* gene in one of the *Hordeum vulgare* mutants was found to be hypersensitive to drought stress, impairing photosynthesis, disorganizing chloroplast structure, lowering relative water content, altering stomatal density, and delaying stomatal closure when compared to the WT ones. More studies revealed that the SL mutant's drought sensitivity might be attributed to a disturbance in abscisic acid metabolism and/or signaling pathways. This indicated an abscisic acid-dependent link between the drought stress response and SL signaling [53]. Some studies found that the triple mutant *smx7,8, l6* of *A. thaliana* is more drought tolerant than WT. The triple mutant during the drought stress produced more anthocyanin, had better detoxifying abilities, and was more sensitive to abscisic acid. They also discovered that the mutant had greater transcript levels of *LEA76*, *LEA18*, and *LEA/ABR* but fewer expressions of *QQS*, *WRKY46*, *PDH2*, *PDH1*, and genes than the WT [18, 54].

Plants can develop symbiotic relationships with a range of rhizosphere-dwelling microbes in addition to their fundamental defense mechanisms against environmental stresses. These interactions can help reduce the symptoms of stress [55]. The AM symbiosis diminishes stress-induced negative consequences in many of the instances tested, making host plants more drought-tolerant. Due to the absence of AM symbiosis, *Lactuca sativa* and *S. lycopersicum* plants cultivated under drought conditions showed a reduction in the SL level and an increase in abscisic acid concentration [42].

### 8.2.3 SLs During Salinity Stress

Salinization has emerged as one of the most pressing environmental and agricultural issues in recent years. More than 7% of land is currently covered with salt-containing soil, and by the middle of the twenty-first century, it is expected that this figure will rise to 50% [56, 57]. However, exogenous administration of GR24 can reduce the limiting effects of salt stress, as shown by the rise in the photosynthetic capacity of *Brassica napus* [58]. Ha *et al.* discovered that *A. thaliana* SL-signaling (*max2*) and SL-deficient (*max4* and *max3*) mutants were more susceptible to salt stress than their WT counterparts throughout the vegetative/germination phases of growth [47]. In some studies, *S-24* and *PARI-73*, two *T. aestivum* cultivars, were chosen to investigate the impact of GR24 against salt stress. It was that found the growth parameters were drastically decreased; however, the seed population treated with GR24 exhibited a distinct growth rate [18]. It was found that salt-stressed *O. sativa* had increased levels of ROS and MDA while causing cellular damage, but seedlings treated with GR24 had much lower levels of both these stress biomarkers, allowing the plant to grow normally [59–61]. Similarly, high MDA and $H_2O_2$ levels have been observed in *Salvia nemorosa* in saline conditions. Furthermore, the content of essential oil was found to be less. The exogenous administration of GR24 reduced MDA buildup and $H_2O_2$ while improving essential oil output [62]. Pre-sowing treatment with GR24 in *Helianthus annuus* under salt stress significantly reversed various growth parameters, such as net $CO_2$ assimilation, photosynthetic pigments, and stomatal conductance. The concentrations of $Na^+$, $K^+$, and $Ca^{2+}$ in the roots and shoots were likewise increased by this treatment. Similarly, during salt stress, *H. annuus* displayed significant increases in MDA, $Na^+$, and $H_2O_2$, as well as a significant decrease in soluble proteins, callus biomass, water potential, $K^+$, and $Ca^{2+}$. These results were additionally reversed by the injection of GR24 [63, 64].

Strigolactone and abscisic acid are both very crucial to developing AMF symbiosis and regulation of salt stress responses [65]. In 2018, Ren *et al.* discovered that the external administration of abscisic acid, $H_2O_2$, and NaCl increased the SL-biosynthetic gene expression in *Sesbania cannabina*, including *MAX2* (in shoot) and *CCD8* and *CCD7* (in root) [66]. Tungstate and dimethyl thiourea (DMTU; an $H_2O_2$ scavenger) have similarly been shown to suppress the *CCD8* and *CCD7* expressions in AMF-associated *S. cannabina* during salinity stress. In addition, the pre-treatment of DMTU and TIS108 (SL biosynthesis inhibitor) inhibited abscisic acid-induced SL production. As a result, abscisic acid may be considered important for SL biosynthesis induced by AMF under salt exposure. Similarly, a positive correlation was found between SL biosynthesis and salt exposure in *L. sativa* roots injected with AMF. Furthermore, the abscisic acid biosynthetic gene exhibited a favorable connection with SL biosynthesis under stress. Mycorrhizal *L. sativa* grows better under salt stress than non-mycorrhizal ones. This suggests that the association of these two phytohormones improves plant salt tolerance under symbiotic conditions [18, 52, 66, 67].

### 8.2.4 SLs During Light Stress

Light is an omnipresent environmental component that cannot be evaded, unlike other sources of stress. This component also affects the SL level in plants, which obliquely supports the requirement for SLs in light signal cascades [68]. The MAX2/ORE9 F-box proteins, which are encoded by pleiotropic photosignaling (PPS), promote photomorphogenesis and inhibit hypocotyl elongation. Blue, red, and far-red light all cause hypersensitivity in the *A. thaliana pps* mutant. In a light-dependent way, SL slows the development of hypocotyls, and this inhibition necessitates the presence of functional *Elongated Hypocotyl5* (*HY5*). Hypocotyl development is negatively regulated by *HY5*, while the quantity of *HY5* is favorably regulated by SLs. GR24 regains the elongated hypocotyl phenotypes of the *hy2* and *hy1* mutants, but not of the *hy5* mutant; this may be because the SL *hy5* mutant is insensitive. However, another element that encourages hypocotyl extension in dark environments is *Constitutive Photomorphogenic1* (*COP1*). The nuclear exclusion of COP1 mediated by GR24 exogenously treated the cell, stabilizing *HY5* and reducing hypocotyl elongation in the process. SL functions as a complement to light in dark environments because both SL and light inhibit *COP1* activity [18, 69]. It was observed that in *A. thaliana*, the expressions of *SMXL7* and *SMXL6* (SL repressors) genes are directly upregulated to enhance shoot branching by phytochrome A mediated light signaling transcription factors, *FAR1* and *FHY3*. Additionally, these two transcription factors interact with *Squamosa Promoter Binding Protein-Like9* (*SPL*) and *SPL15*, which encourage shoot branching and repress *BRC1* [18, 70, 71].

Phytochromes often hinder mesocotyl development and are important in the light absorption process by plants. Any disruption to SL biosynthesis or signaling causes mesocotyl to elongate when in a dark environment [72, 73]. The interaction between cytokinin and SL to control mesocotyl development in SL-related d mutants (*d3*, *d14*, *d27-1*, *d17-1*, and *d10-1*) of *O. sativa* cultivated in dark circumstances is discussed later in this review. The *d27-1*, *d17-1*, and *d10-1* mutants showed inhibition of mesocotyl development upon administration of GR24, but the SL-insensitive *d14* and *d3* mutants did not exhibit this effect [18, 74]. In one study, the mesocosms of *O. sativa* d14-1 and d10-1 were found to have largely expressed CK-responsive type-A response regulator (RR) genes (*OsRR11*, *OsRR9*/10, *OsRR7*, *OsRR5*, and *OsRR4*), while the mesocotyl of these mutants showed a repressed expression of *OsTCP5* (*Teosinte Branched1/Cycloidea/PCFs5*), which inhibits proliferation and growth. GR24 caused the *OsTCP5* expression to be repressed in the SL biosynthesis mutant (*d10-1*), while

GR24 treatment did not impact *OsTCP5* expression in the SL-signaling mutant (*d14-1*) [75]. In the *d14* and *d3* mutants of *O. sativa*, Kameoka and Kyozuka likewise observed a similar outcome, with both strains being GR24-insensitive [76]. By suppressing photosynthetic traits and disrupting assimilative metabolisms, low light stress caused many metabolic alterations in *S. lycopersicum*, which were remedied by using GR24 [77].

### 8.2.5 SLs During Heat Stress

Heat stress is the primary factor limiting the development of roots of cold-season plants since the development of roots needs relatively low temperatures. As a result, these plants are very sensitive to rising external temperatures. By observing changes in the expression patterns of genes such as *Proliferating Cell Nuclear Antigen* (*PCNA*), *Cyclin-D2* (*CycD2*), and *Cyclin-Dependent Kinase B* (*CDKB*) linked to the cell cycle and the cell division in root tips in heat-stressed *Festuca arundinacea*, the positive effects of GR24 on the elongation of crown roots and several root cells have been noted. Heat stress in *A. thaliana* mutants with SL defects has shown the SL-mediated control of germination [18, 78]. According to some studies, the GR24 treatment reduced the negative impact of heat exposure on seed germination in *Lupinus polyphyllus* [79]. The GR24 scavenges ROS by enhancing antioxidant activities during the germination period. It has

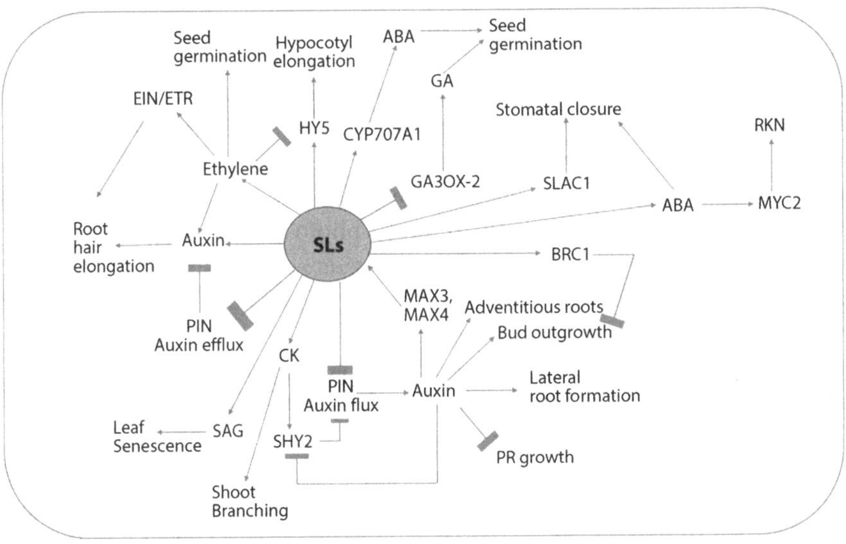


**Figure 8.2** Strigolactone interactions with different phytohormones.

been found that during warm stratification the administration of SLs in *Pasteuria ramosa* seeds reverses the temperature-induced dormancy. When GR24 is applied exogenously, the expression of *CYP707A1*, the catabolic gene of abscisic acid, gets enhanced, which leads to the breakdown of abscisic acid and the germination of seeds even in the presence of heat exposure [5, 80, 81]. According to Toh *et al.*, under heat stress, the max mutant's seed germination response in *A. thaliana* was much more sensitive than WT's [82]. The *max1-1* mutant phenotypes were restored by exogenous GR24 therapy, but not *max2-1. L. sativa* showed a reduction at high temperatures in the rate of seed germination which was moreover boosted by the application of SLs by lowering the abscisic acid/gibberellic acid ratio [83, 84]. The aforementioned studies imply that the SLs and gibberellic acid may be informally communicating to track seed germination (Figure 8.2).

### 8.2.6 SLs During Chilling Stress

In various parts of the world, low temperatures have a detrimental impact on agricultural output and result in financial loss. It has been shown that, in comparison to their WT counterparts, *Pisum sativum rms4* mutants and *A. thaliana max* (*max4-1* and *max2-1*) significantly decreased their leaf area and photosynthetic rate with cooling temperature. Since the genes *max* and *rms4* are linked to SLs, as previously noted, it is reasonable to believe that the SL effectively contributes to these plants' ability to withstand low temperatures [85]. In *B. rapa*, after chilling stress, changes were revealed in growth parameters; in contrast, GR24 treatment significantly enhanced cell viability, photosynthesis, antioxidant enzymes, proline content, and soluble protein and decreased relative conductivity. They also observed that GR24 dramatically increased the expression of genes linked to cold, including MAPK, NADPH oxidase, and antioxidant enzymes. Chilling treatment significantly reduces photosystem II quantum efficiency, promotes dehydration, and upsets water status as observed in *Vigna radiata*, according to the findings of another research. Lipid peroxidation was significantly decreased and soluble sugar, proline concentration and the activities of phenylpropanoid and lipoxygenase pathway enzymes were all raised under the GR24 therapy, which helped to alleviate chilling stress [78, 86].

While the expressions of *Os01g0701500* and *D10* genes were shown to be downregulated by low temperatures, during both lower and normal temperatures, In *O. sativa* the gene expression study indicated identical levels of *D17* and *D27* in the roots of non-mycorrhizal plants.

However, it has been observed that at low temperatures, mycorrhizal *O. sativa* roots exhibit considerable up-regulation of all the aforementioned genes. It has been noted in the past that the AMF promotes the expression of the genes responsible for SL biosynthesis in host plants, hence increasing the accumulation of SLs [18, 87, 88]. Therefore, it is reasonable to believe that the SL may have regulated how plants responded to cold stress. The role of SLs during chilling stress is still poorly understood. In the future, further investigation will undoubtedly be required to determine the precise mechanism by which SLs help plants reduce chilling stress.

### 8.2.7 SLs During Heavy Metal Stress

Based on the extent of their harm to human health, heavy metals are regarded as the second most dangerous environmental issue worldwide. Application of different phytohormones may be able to mitigate the heavy metals toxicity in plants, which manifests as retardation of growth and development as a result of heavy metal accumulation in their organs [89]. Depending on the kind of metal element, heavy metals can have different toxicities. In 2017, in *Panicum virgatum*, the effects of Cd and its absorption were examined. Growth characteristics, photosynthetic rate, chlorophyll content, and antioxidant activities were all suppressed, whereas ROS levels were elevated. In contrast, the GR24-added Cd-grown plants showed better growth, chlorophyll content, and antioxidant activities with reduced Cd accumulation [18]. Recently in 2021, Mostofa *et al.* conducted a comparative analysis of the function of SL in *O. sativa* WT, *d17* and *d10* mutants under arsenic (As) stress. *OsCATA*, *OsAPX2*, *OsAPX1*, *OsCuZnSOD2*, and *OsCuZnSOD1* are among the antioxidant enzymes whose mRNA levels are comparatively greater in WT roots as compared to SLs mutant roots. On the other hand, it has been shown that mutant plants expressed higher levels of the transporter genes (*OsPT8*, *OsPT4*, *OsPT2*, and *OsPT1*) than WT plants [90].

The bioavailability of metal ions from the soil is mostly dependent on membrane transporters, and the co-transportation of several metal (oid) cations is facilitated by the plasma membrane of roots. These transporters play a vital role in both metal-induced stress responses and preserving the physiological concentration of metal ions within cells [91, 92]. Similar in structure to Pi, arsenic (V) is readily released via plant roots via Pi transporters. The Pi transporter1 family (Pht1; TC 2.A.1.9, the PHS family) is home to several high-affinity Pi transporters, some of which are also involved in the transport of As (V) in plants. Consequently, the Pi shortage in plants is brought on by As toxicity. As previously mentioned in this chapter, SLs are essential to plant defense mechanisms against nutrient deficits, which, in turn, trigger the expression of SL biosynthesis genes (e.g., *D27*, *D17*, *D10*, *OsMAX1*, etc.) in roots [93, 94]. Therefore, it stands to reason that the SLs might help in reducing the stress caused by heavy metals in plants. To ascertain the molecular mechanisms by which SLs work on plants stressed by heavy metals to mitigate their detrimental effects, more investigation is necessary.

## 8.3 Interplay of SLs with Other Phytohormones

### 8.3.1 Crosstalk Between SLs and Abscisic Acid

Plant responses and resistance to both biotic and abiotic stresses are mediated by SLs, according to a growing body of experimental research. To control physiological processes and help plants adapt to their environment, abscisic acid and SLs have an essential link. In the well-watered and dehydrated SL signaling-deficient mutant *max2* of *Arabidopsis*, the abscisic acid importer genes *ABCG40/AT1G15520* and *ABCG22/AT5G06530* were downregulated. Strigolactones trigger tolerance to salt stress and drought via triggering abscisic acid signaling [54, 95]. The delayed stomatal closure tolerance to drought associated with abscisic acid insensitivity was demonstrated to be impaired in the *L. japonicus* SL biosynthetic mutant [96]. Pre-treating grapevine and rice seedlings with GR24 reduced the negative effects of salt stress and improved stomatal closure, demonstrating a similar abscisic acid-SL interaction. SLs impact stomatal closure by abscisic acid synthesis, transport, and sensitivity. A recent research has shown that rice mutants *D10*, *D17*, and *D13*, which are weak in SL biosynthesis and SL signaling, accumulated more abscisic acid than WT plants and were thus able to promote drought tolerance. However, on the other hand, drought might affect the low-abscisic acid-producing line *D27*, suggesting that *D27* might be involved in the abscisic acid signaling system [49, 59, 97, 98]. In addition, the exact method by which *D27* connects abscisic acid to SL is still unclear. The beneficial function of SLs in the response to heat and cold stress is linked to abscisic acid biosynthesis, in addition to their significance in drought resistance. Heat and cold tolerance in tomatoes were increased by *GR24* application, but the abscisic acid-deficient mutant lessened the benefits of GR24. This suggests that SL, at least in part through abscisic acid-dependent mechanisms, enables plants to adapt and overcome these stressors.

SLs and abscisic acid have a role in controlling branching or tillering, with abscisic acid functioning downstream of SLs and *BRC1* in *Arabidopsis* [99, 100].

By releasing *BRC1* transcriptional repression and breaking down *SMXL6*, SLs that mediate axillary bud outgrowth contribute to the activation of *AtNCED3/OsNCED1* expression, the induction of *HB40/OsHOX12* expression, and the promotion of abscisic acid accumulation in the shoot bases of rice or the lateral buds of *Arabidopsis* [99, 101].

Although it does not have as much of an impact as SLs, abscisic acid supply does prevent the development of unproductive upper tillers and the growth of tiller buds in rice. Abiotic stress causes abscisic acid levels to rise quickly; however, various species may have differing amounts of SL. To enhance *S. cannabina* resistance to and acclimatization to salt stress, abscisic acid positively regulates the SL levels and the expression of signaling genes. In a similar vein, rice root extracts stimulate plant resistance to water scarcity by upregulating the transcript levels of SL biosynthetic genes [49, 66], but in *L. japonicus* and tomato, during osmotic stress, both the SL gene expression and the SL level dropped [88]. More research is needed to determine whether SL signal upregulation may be used to breed crop types that are thought to be resistant to drought. The hypothesis that SL signals may disrupt and suppress the activity of abscisic by affecting SNF1-related protein kinases and type 2C protein phosphatases, important elements of abscisic acid signaling, merits additional investigation to ascertain if SLs do, in fact, function as feedback modulators of abscisic acid activities [95, 102].

### 8.3.2 Interplay of SLs with Jasmonic Acid

During vegetative development, plant–insect and plant–pathogen interactions, secondary metabolism, wounding, and other situations, an array of physiological activities are mediated by jasmonates, a class of plant oxylipins [103]. There is currently insufficient experimental evidence to characterize the nature of SL-jasmonic acid crosstalk adequately. Nonetheless, the findings that an SL-deficient tomato mutant (*Sl-ccd8*) had lower levels of jasmonates and the expression of *PINII*, the jasmonate-dependent gene, confers tomato resistance to *Botrytis cinerea* suggest a relationship between SLs and jasmonates in disease tolerance [104, 105]. An orthologue of *Ph-PDR1*, *Nicotiana tabacum PDR6* (*Nt-PDR6*), the SL transporter gene, is subject to some effect from methyl jasmonates (plant defense signaling chemical), suggesting a further connection to this relationship. Despite being substantially stimulated by P deficiency and implicated in the shoot branching control, *Nt-PDR6* has not been linked to SL transport [106]. According to a rather contradictory finding, neither the GR24 supply nor the *Mucor* sp. inoculation had any effect on the accumulation of jasmonic acid in WT plants, *At-max2* and *At-max1* [107]. It is known that both hormones gather responses throughout developmental processes that are similar to one another, such as senescence, mesocotyl growth, and plant-microbe interactions, which suggests that crosstalk between SL-jasmonate is likely to play a role in these events and cannot be completely ruled out, even though the combined observations are insufficient to draw firm conclusions about the nature of crosstalk between SL-jasmonate.

### 8.3.3 SLs and Salicylic Acid Crosstalk

A key component of defense against pathogens, plant sensing, and tolerance to abiotic stresses is salicylic acid. Its impact on plant ROS state accounts for a substantial portion of its management in this regard. Senescence, stomatal closure/conductance, and ability to tolerate drought are all under the control of SA and are all regulated by SL [107, 108]. Only interactions between plants and endophytic fungi have shown salicylic acid–SL interactions thus far. In these interactions, GR24 increased the buildup of SA, while *max2*, an SL signaling mutant had lower concentrations of salicylic acid [109]. This suggests that by triggering salicylic acid signaling, SLs may affect how plants defend against infection. It is unknown how exactly SLs cause salicylic acid accumulation—whether it is by direct transcriptional regulation of the genes that produce SA or by triggering ROS or oxidative outbursts, which are known to cause SA formation. In light of the latest findings on the significance of SLs in abscisic acid-induced stomatal closure during droughts and senescence, investigating SL–SA crosstalk in these domains might yield essential insights for breeding initiatives aimed at creating drought-tolerant crops with longer fruit shelf lives [47, 107, 110, 111].

### 8.3.4 Integration of SLs with Ethylene Signaling

It has been shown that SLs and ethylene signals combine to govern some growth and developmental processes. These include the extension of root hair, germination of seeds, hypocotyl development, and leaf senescence (Figure 8.2). The COP1–HY5 complex integrates light and hormone signals to control hypocotyl growth during hypocotyl development. By upregulating *HY5* expression in an *MAX2*-dependent manner, SLs prevent hypocotyl elongation in the presence of light, whereas ethylene increases *COP1*-mediated *HY5* breakdown to encourage hypocotyl elongation [112, 113]. All of these studies together point to the possibility that SLs and ethylene signaling work against one another to control hypocotyl development. How SLs counteract the growth-promoting effects of ethylene on the development of hypocotyl in light is one key concern as hypocotyl growth is quickly stopped in light. Additionally, SL is reliant on ethylene signaling to regulate the lengthening of root hair. This is demonstrated by the fact that *At-etr1* and *At-ein2*, the ethylene signaling mutants, showed decreased sensitivity to GR24 treatment and that the effects of SL on elongation of root hair were reduced when ethylene synthesis was blocked. Despite the presence of GR24, treatment with the ethylene biosynthesis inhibitor aminoethoxyvinylglycine in *Arabidopsis* did not increase root hair elongation [114, 115]. Instead, GR24 treatment alone did this, suggesting that ethylene is required to encourage root hair elongation mediated by SL. Moreover, *AtACS2* transcription, a gene implicated in the production of ethylene, was improved by GR24 therapy [67].

Strigolactones employ ethylene, which may then attract auxin signaling suggesting that there is a three-way hormonal interplay that regulates the root hair growth. However, it has been reported that ethylene does not always require auxin signaling to influence the development of root hair and can act independently. This is demonstrated by the fact that a transcriptional complex, consisting of a positive regulator of hair cells and an ethylene-activated factor, co-activates (*ROOT HAIR DEFECTIVE 6-LIKE*), the hair length-determining gene and the other root hair genes [116]. An early discovery revealed that SLs promoted ethylene production in *Striga* seeds before germination, further indicating SLs and ethylene crosstalk.

Although ethylene-independent routes may potentially be used, SLs also promoted leaf senescence via activating ethylene-mediated senescence signals. When considered collectively, the information indicates that ethylene functions independently of SLs and also demonstrates that changes in tissue ethylene concentration trigger SL-mediated reactions that necessitate ethylene signaling. Ethylene may work downstream of SLs, as evidenced by its independent influence on adventitious root initiation sites in the *Arabidopsis* hypocotyl [117, 118]. Despite this, it is still unknown which specific molecular channels SLs employ to trigger the transcriptional regulation of ethylene synthesis.

The theory that SLs may affect ethylene signaling in a phytochrome-dependent manner is worth exploring as light is a powerful regulator of ethylene biosynthesis through processes mediated by phytochrome/phytochrome interaction factors [119]. Additionally, phytochrome/cryptochrome is used to trigger certain SL responses. Similar to gibberellins, available experimental evidence is currently insufficient to describe a clear and comprehensive picture of SL–ethylene crosstalk's role in controlling plant developmental processes. To comprehend the role of SL-ethylene hormonal and signaling crosstalk(s) in processes like root development, fruit ripening, and senescence will require answers on how *D53/SMXL* activity and *D14/MAX2* signaling affect essential ethylene signaling factors and ethylene-responsive TFs, and vice versa [107]. Combining these hormones likely results in a purposefully designed network that controls how a plant grows and reacts to unfavorable growth circumstances.

### 8.3.5 Strigolactones and Gibberellins

Many methods for lowering the height of several crops suggest that gibberellic acid is one of the main variables influencing plant height. SLs employ the F-box protein (RMS4/D3/MAX2) and an α/β-fold hydrolase (D14/DAD2), similar to the Gibberellic acid signaling pathway. Different species, including *O. sativa*, *A. thaliana*, *S. lycopersicum*, *P. sativum*, *P. hybrida*, and *L. japonicus*, have been shown to exhibit shorter stature traits in their SL-deficient and SL-response mutants compared to their WT counterparts [18, 52]. Some studies suggest SL–gibberellin crosstalk, although this is still mainly unknown. Numerous studies have been conducted to determine if SLs and gibberellic acids may be correlated—for example, it was found that gibberellin and GR24 caused an additive transcriptional change in a collection of overlapping genes [120]. Additionally, their research indicated that gibberellin had an antagonistic influence on the increased expression of several ABC-type transporters produced by SL. Additionally, an *in silico* study revealed that gibberellin regulates SL production, which was later subsequently shown to be mediated via a GID1-DELLA signal pathway in an experimental setting. Although this has not yet been proven experimentally, the finding that *OSH1* (*ORYZA SATIVA HOMEOBOX 1*) expression is increased in largely branched gibberellin-deficient mutants and SL-signaling mutants that implies gibberellin and SLs share a similar molecular mechanism in the control of branching and tillering [67, 107]. There have been attempts made to establish a direct molecular relationship between SLs and gibberellin signaling downstream components. In this regard, D14 interacts with SLENDER RICE1 (SLR1), a DELLA protein that functions as a downstream signal component and gibberellin signaling regulator in an SL-dependent manner. However, the relationship's functional importance and/or downstream effect(s) are still mostly unknown, and there is currently no experimental evidence of SL-induced SLR1 degradation [121, 122].

The studies from the recent past investigations on the molecular relationship between the SL and gibberellin signaling support the functional independence of both hormones—for instance, SL signaling did not affect the stability and concentration of DELLA proteins. In *O. sativa* and *L. japonicus*, it was shown that the usage of bioactive forms of *GA4*, *GA3*,

*GA1*, and *GA* adversely affected SL production. In *O. sativa* SL-insensitive mutants, *d14-1* and *d3-1*, *GA3* treatment decreased the number of endogenous SLs, providing evidence that gibberellic acid regulates SL production independently of SL signaling [52]. The observation that SLs increase the cell number rather than the length to drive internode elongation independently of gibberellin provides more evidence for functional diversification for both hormones. In light of the aforementioned, it seems likely that the crosstalk between gibberellin and SL is restricted, if it exists at all, and that gibberellin's regulation of SL synthesis merely modifies the content of SL. Regardless of the situation, more research is necessary to fully understand the functions of the individual DELLA protein and the SL receptor as well as the real nature of the SL–gibberellin crosstalk [18, 67].

### 8.3.6 Cross-Talk Between SLs and Auxins

Auxins, important growth promoters have been shown to increase the expression of SL-associated genes in *A. thaliana*, *P. sativum*, *O. sativa*, and *Chrysanthemum morifolium*, including *CCD7* and *CCD8* [18, 123]. SLs and auxin work together to control shoot branching and root growth. The control of shoot branching by SLs is closely associated with PIN-dependent auxin transport, more precisely with its canalization. This is corroborated by the observations that auxin efflux carrier PIN1 buildup is enhanced on the basal plasma membrane of *D27-1* and *MAX4-5*, the SL biosynthesis-deficient mutants and that auxin transport and PIN1 endocytosis during SL biosynthesis in *Arabidopsis* can be reduced by GR24 treatment, but not in response mutants. Through *MAX2*- and *D14*-mediated signaling pathways, SLs block clathrin-mediated endocytosis of PIN proteins and auxin feedback on PIN polarity [95, 124]. Exogenously SLs can, however, still reduce bud outgrowth in shoots treated with the auxin transport inhibitor *1-N57* naphthylphthalamic acid, indicating the possibility of an additional SL-mediated mechanism for bud growth suppression [95]. This implies that SL influences bud expansion directly, without relying on polar auxin transit. Examining a few potential possibilities as downstream mediators of SL signaling may help to clarify several uncertainties regarding the precise mechanism behind SL action and perception. By altering the distribution of auxin, SLs control the formation of lateral roots and root hairs throughout root development. PIN, the auxin efflux protein, is primarily responsible for polar auxin transport [125]. This protein forms the foundation for root initiation and elongation by generating local auxin maxima. In rice, GR24 downregulated the expression of *PIN* family genes, which decreased the distribution of

IAA and altered the synthesis of adventitious root [126]. However, in the presence of exogenous auxin, GR24 treatment did not affect the *PIN* gene expression level in the primary root tip of *Arabidopsis*. Elevated expression of several auxin transporters and increased auxin stem conductivity belonging to the *PIN* family (*PIN3* and *PIN1*) are seen in the *A. thaliana max* mutant. The *max* mutant's increased auxin concentration upregulates the expression of the *MAX3* and *MAX4* genes and promotes positive feedback regulation of the *MAX* pathway [67]. Consequently, the low SL would raise the auxin levels and start the SL biosynthesis all over again. This suggests that the *MAX* pathway's primary role is to change *PIN* expression in the stems. Furthermore, the expression levels of *OsPIN*—particularly *OsPIN9*—also dramatically dropped in the *O. sativa* nodes, in response to GR24 treatment, indicating that GR24 lowers the ability for auxin transportation [52, 127]. Zhang *et al.* found that in *P. sativum* and *A. thaliana*, auxin feedback on clathrin-mediated endocytosis of *PIN* proteins and PIN polarity are two canalization-dependent developmental processes that both exogenous and endogenous SLs interact with [124]. These findings reveal a non-transcriptional mechanism by which SL uncoupled the auxin feedback on trafficking and *PIN* polarity and controls the development and regeneration of vascular tissue.

In the control of the development of root hair similar interactions between auxin and SLs are seen, where high root hair length and density are the outcomes of the SL-mediated decrease of auxin buildup within root cells. P-deficient circumstances are typified by these root responses. ROOT HAIR DEFECTIVE SIX-LIKE 4, the auxin-responsive element of the bHLH transcription factor positively regulates genes implicated in cell processes essential to the growth of root hair under the P-deficient state. These contextual details elucidate the possibility that RSL4 may serve as a shared integrator for the SL-auxin crosstalk that modulates root hair elongation [67, 128]. Interacting mechanisms governed by intricate hormone signaling propel root hair development. It is still not clear how these signaling elements at the molecular level cause SL biosynthesis and signaling based on the P state. To learn more about signaling networks, future research should concentrate on cloning the genes present in root hair mutants and doing reverse genetics and mutant complementation investigations.

### 8.3.7 Emerging Perspectives on SLs and Cytokinins

The relationships between SLs and CKs are as diverse as physiological activities. The way that cytokinins and SLs interact appears to change depending on physiological events. In adventitious rooting, they each regulate distinct processes and work separately while antagonistically regulating the expansion of axillary buds and working in concert to govern lateral development [20]. Strigolactones and cytokinins have a direct interaction in buds, and they work together to support *BRC1* transcriptional regulation in *Arabidopsis* and peas as well as FCI transcription in rice. It is known that *BRC1* functions as an important core of regulatory signal controlling bud outgrowth, which, in turn, influences the potential of bud activation in many species [95]. Auxin inhibits bud expansion by a mechanism mediated by the antagonistic actions of cytokinin and SL. By triggering cytokinin catabolism through *cytokinin oxidase/dehydrogenase 9* (*OsCKX9*) activity, SLs in rice modify the shoot architecture. Thus, through interaction with auxin, SLs may impact the cytokinin content in addition to inducing activation of *OsCKX9* [129–131]. Moreover, it is shown that elevated sugar levels blocked SL perception, most notably by SL signaling directly. Additionally, it was shown that sugars increased the levels of cytokinin, which has an antagonistic effect on SLs. The precise function of cytokinin in the sugar response is yet unknown though. *HEXOKINASE1* during branching modulates the sugar signaling pathway and enables plants to modify the shoot architecture and interaction with cytokinin and SLs [132]. GR24 inhibits primary root extension by altering the transcription of the PIN gene, which is controlled by Short Hypocotyl2 (SHY2) through cytokinin signaling factors. While the ARR1 (cytokinin response transcription factor) directly binds to certain promoter areas of the SHY2 protein and stimulates its synthesis, inhibiting the PIN genes, auxin inhibits PIN activity through SHY2-mediated suppression of lateral root growth. SLs may affect the endogenous distribution and concentrations of each hormone, hence coordinating the control of root (meristem) development [11, 133, 134]. In terms of how plants respond to drought, cytokinins and SLs have opposing regulatory functions. Cytokinins and cytokinin-signaling factors were discovered to adversely influence plant drought adaptation in *Arabidopsis* CK-signaling and CK-depleted mutants [135]. Additionally, it was discovered that, in the dark, they had an antagonistic connection in the *d10-1* SL-deficient mutant, while this negative phenomenon was not seen in the *d14-1* SL-insensitive mutant. Conversely, SLs positively regulate physiological phenotypes linked to drought tolerance by altering stomatal conductance and stomatal density [18]. Furthermore, the SL signaling-deficient mutant *MAX2* was shown to downregulate the cytokinin catabolism genes (*CKX5*, *CKX3*, *CKX2*, and *CKX1*) during dehydration in comparison to WT plants. Because *MAX2* appears to be shared by both the SL signaling and karrikin signaling pathways, it is suggested that the SL signal might had an antagonistic impact on the cytokinin content [107, 136, 137]. Extensive research on SL biosynthesis and signaling mutants under drought stress may validate this.

## 8.4 Future Directions and Implications for Agriculture

Strigolactone-mediated control of plant growth has been thoroughly investigated since the recent revelation of SLs' hormonal role. To address various biotic and abiotic challenges, modern agriculture requires an increasing number of specialized interventions during the growing season; thus, new crop protection solutions need to be developed regularly. Conventional breeding has lately been connected to using a new class of pesticide chemicals. For plants to adapt to rapidly changing or harsh abiotic conditions, phenotypic plasticity is essential. A crucial physiological strategy for improving agricultural productivity, resistance, and nutrient absorption and utilization is the alteration of SL signaling pathways to create the optimal crop design. The possibility of genetically modifying crops has grown, thanks to new molecular technology. Transgenic rice, for instance, has enhanced grain output by controlling tillering through activated SL response pathways. Moreover, achieving a prospective breeding stock may benefit from ontogenetic manipulation of the SL transport signal. It has great potential to shape plant architecture, increase AM colonization, and improve performance and resistance by using SL analogues. Large-scale agricultural uses demand more affordable supplies of SL analogues. Like other phytohormones, under unfavorable environmental circumstances, numerous layers of interaction in hormonal pathways control SL activity and biosynthesis. Roots are more susceptible to unfavorable soil conditions compared to the aerial components of plants since they serve as the contact between the soil and the plant. The way that roots perceive their surroundings affects the morphology of plants. The knowledge of how many phytohormones support the root development flexibility has advanced. These processes require more components, and it is necessary to determine the spatial and temporal interactions between them. Further experimental and theoretical study is required to fully understand the diverse roles that these hormones and SLs play in the overall structure of the plant. Optimizing endogenous phytohormone levels is essential to maximize the stress-responsive crosstalk among different hormones in dynamic conditions.

## 8.5 Conclusion

Hormones are a major factor in influencing the structure and responses of plants to various environmental stimuli. Plant secondary metabolites called SLs are produced from carotenoid compounds. To control a plant's growth mechanisms, they affect several parts of the plant, such as root development, shoot branching, senescence of the leaves, and seed germination. Through experimental observations, the relationships between chemically varied phytohormones in plant growth have been identified. Plant developmental patterns are shaped by an entire circuit of molecular signals consisting of these interactions in response to both internal and exterior stimuli. Stress may affect the signaling, cross-talk, and production of SL with other plant hormones. In recent years, SLs have gained a lot of interest due to their important functions in controlling certain physiological and molecular processes as well as helping plants adapt to many abiotic conditions including cold, heat, drought, nutrient deficiencies, salt, and heavy metals. The hormonal effects of jasmonates, abscisic acid, and salicylic acid play a crucial role in allowing plants to adapt to harsh and restricting environmental situations. All of these sectors together show how much SLs have impacted contemporary agriculture. Therefore, using SLs may hold great potential for developing novel strategies and technological advancements that align with the changing tenets of sustainable agriculture.

## References


1. Bhatt, D., Nath, M., Sharma, M., Bhatt, M.D., Bisht, D.S., Butani, N.V., Role of growth regulators and phytohormones in overcoming environmental stress, in: *Protective chemical agents in the amelioration of plant abiotic stress: biochemical and molecular perspectives*, pp. 254–279, 2020.
2. Zwanenburg, B., Pospíšil, T., Ćavar Zeljković, S., Strigolactones: new plant hormones in action. *Planta*, *243*, 1311–1326, 2016.
3. Yoneyama, K., Xie, X., Yoneyama, K., Kisugi, T., Nomura, T., Nakatani, Y., McErlean, C.S., Which are the major players, canonical or non-canonical strigolactones? *J. Exp. Bot.*, *69*, 9, 2231–2239, 2018.
4. Saed-Moucheshi, A., Pakniyat, H., Pirasteh-Anosheh, H., Azooz, M.M., Role of ROS as signalling molecules in plants, in: *Oxidative damage to plants*, pp. 585–620, Academic Press, London, United Kingdom, 2014.
5. Jiang, W., fei Lu, C., Xu, X., Riaz, M.W., Aimin, L.V., Shao, Q., Strigolactones: Biosynthetic regulation, hormonal interaction, and their involvement in abiotic stress adaption. *Sci. Hortic.*, *325*, 112689, 2024.
6. Janeeshma, E., Habeeb, H., Shackira, A.M., Sinisha, A.K., Mirshad, P.P., Khoshru, B., Mitra, D., Strigolactone and analogues: a new generation of

plant hormones with multifactorial benefits in environmental sustainability. *EEB*, 215, 105775, 2024.
7. Rameau, C., Goormachtig, S., Cardinale, F., Bennett, T., Cubas, P., Strigolactones as plant hormones, in: *Strigolactones-biology and applications*, pp. 47–87, 2019.
8. López-Ráez, J.A., How drought and salinity affect arbuscular mycorrhizal symbiosis and strigolactone biosynthesis? *Planta, 243*, 1375–1385, 2016.
9. Ha, S. and Tran, L.S., Understanding plant responses to phosphorus starvation for improvement of plant tolerance to phosphorus deficiency by biotechnological approaches. *Crit. Rev. Biotechnol., 34*, 1, 16–30, 2014.
10. Faizan, M., Faraz, A., Sami, F., Siddiqui, H., Yusuf, M., Gruszka, D., Hayat, S., Role of strigolactones: Signalling and crosstalk with other phytohormones. *Open Life Sci., 15*, 1, 217–228, 2020.
11. Cheng, X., Ruyter-Spira, C., Bouwmeester, H., The interaction between strigolactones and other plant hormones in the regulation of plant development. *Front. Plant Sci., 4*, 199, 2013.
12. Sharma, P., Jha, A.B., Dubey, R.S., Strigolactones: Coordination with other Phytohormones and Enhancement of Abiotic Stress Responses. *EEB*, 215, 105782, 2024.
13. Alvi, A.F., Sehar, Z., Fatma, M., Masood, A., Khan, N.A., Strigolactone: an emerging growth regulator for developing resilience in plants. *Plants, 11*, 19, 2604, 2022.
14. Soliman, S., Wang, Y., Han, Z., Pervaiz, T., El-Kereamy, A., Strigolactones in plants and their interaction with the ecological microbiome in response to abiotic stress. *Plants, 11*, 24, 3499, 2022.
15. Hawkesford, M.J., Cakmak, I., Coskun, D., De Kok, L.J., Lambers, H., Schjoerring, J.K., White, P.J., Functions of macronutrients, in: *Marschner's mineral nutrition of plants*, pp. 201–281, Academic Press, Cambridge, Massachusetts, USA, 2023.
16. Mayzlish-Gati, E., De-Cuyper, C., Goormachtig, S., Beeckman, T., Vuylsteke, M., Brewer, P.B., Koltai, H., Strigolactones are involved in root response to low phosphate conditions in Arabidopsis. *Plant Physiol., 160*, 3, 1329–1341, 2012.
17. Zhang, Y., van Dijk, A.D.J., Scaffidi, A., Flematti, G.R., Hofmann, M. *et al.*, Rice cytochrome P450 MAX1 homologs catalyze distinct steps in strigolactone biosynthesis. *Nat. Chem. Biol.*, 10, 1028–33, 2014.
18. Bhoi, A., Yadu, B., Chandra, J., Keshavkant, S., Contribution of strigolactone in plant physiology, hormonal interaction and abiotic stresses. *Planta, 254*, 2, 28, 2021.
19. Sun, L., Tian, J., Zhang, H., Liao, H., Phytohormone regulation of root growth triggered by P deficiency or Al toxicity. *J. Exp. Bot., 67*, 12, 3655–3664, 2016.
20. Kaniganti, S., Bhattacharya, J., Petla, B.P., Reddy, P.S., Strigolactone, a neglected plant hormone, with a great potential for crop improvement: Crosstalk with other plant hormones. *EEB, 204*, 105072, 2022.

21. Santoro, V., Schiavon, M., Gresta, F., Ertani, A., Cardinale, F., Sturrock, C.J., Schubert, A., Strigolactones control root system architecture and tip anatomy in Solanum lycopersicum L. plants under P starvation. *Plants*, *9*, 5, 612, 2020.
22. Sun, H., Li, W., Burritt, D.J., Tian, H., Zhang, H., Liang, X., Tran, L.S.P., Strigolactones interact with other phytohormones to modulate plant root growth and development. *Crop J.*, *10*, 6, 1517–1527, 2022.
23. Mostofa, M.G., Li, W., Nguyen, K.H., Fujita, M., Tran, L.S.P., Strigolactones in plant adaptation to abiotic stresses: An emerging avenue of plant research. *Plant Cell Environ.*, *41*, 10, 2227–2243, 2018.
24. Yoneyama, K., Xie, X., Kim, H.I., Kisugi, T., Nomura, T., Sekimoto, H., Yoneyama, K., How do nitrogen and phosphorus deficiencies affect strigolactone production and exudation? *Planta*, *235*, 1197–1207, 2012.
25. Czarnecki, O., Yang, J., Weston, D.J., Tuskan, G.A., Chen, J.G., A dual role of strigolactones in phosphate acquisition and utilization in plants. *Int. J. Mol. Sci.*, *14*, 4, 7681–7701, 2013.
26. Tariq, A., Ullah, I., Sardans, J., Graciano, C., Mussarat, S., Ullah, A., Peñuelas, J., Strigolactones can be a potential tool to fight environmental stresses in arid lands. *Environ. Res.*, 226, 115966, 2023.
27. Zhou, F., Lin, Q., Zhu, L., Ren, Y., Zhou, K. *et al.*, D14–SCFD3-dependent degradation of D53 regulates strigolactone signalling. *Nature*, 504, 406–10, 2013.
28. Sun, H., Tao, J., Liu, S., Huang, S., Chen, S., Xie, X., Xu, G., Strigolactones are involved in phosphate-and nitrate-deficiency-induced root development and auxin transport in rice. *J. Exp. Bot.*, *65*, 22, 6735–6746, 2014.
29. Moturu, T. and Koča, J., Deciphering the Role of Strigolactone Hormone in Plant Developmental Processes, 2020.
30. Wipf, D., Krajinski, F., van Tuinen, D., Recorbet, G., Courty, P.E., Trading on the arbuscular mycorrhiza market: from arbuscules to common mycorrhizal networks. *New Phytol.*, *223*, 3, 1127–1142, 2019.
31. Andreo-Jimenez, B., Ruyter-Spira, C., Bouwmeester, H.J., Lopez-Raez, J.A., Ecological relevance of strigolactones in nutrient uptake and other abiotic stresses, and in plant-microbe interactions below-ground. *PLSOA2*, *394*, 1–19, 2015.
32. Singh, I. and Giri, B., Arbuscular mycorrhiza mediated control of plant pathogens, in: *Mycorrhiza-nutrient uptake, biocontrol, ecorestoration*, pp. 131–160, 2017.
33. Kunwar, V.S., Lamichhane, J., Gauchan, D.P., Strategies to improve phosphorus availability in a sustainable agricultural system. *IJISRT*, *3*, 323–331, 2018.
34. Mbodj, D., Effa-Effa, B., Kane, A., Manneh, B., Gantet, P., Laplaze, L., Grondin, A., Arbuscular mycorrhizal symbiosis in rice: establishment, environmental control and impact on plant growth and resistance to abiotic stresses. *Rhizosphere*, 8, 12–26, 2018.

35. Ge, S., He, L., Jin, L., Xia, X., Li, L., Ahammed, G.J., Zhou, Y., Light-dependent activation of HY5 promotes mycorrhizal symbiosis in tomato by systemically regulating strigolactone biosynthesis. *New Phytol.*, *233*, 4, 1900–1914, 2022.
36. Etesami, H., Jeong, B.R., Glick, B.R., Contribution of arbuscular mycorrhizal fungi, phosphate–solubilizing bacteria, and silicon to P uptake by plant. *Front. Plant Sci.*, *12*, 699618, 2021.
37. Pan, X., Zheng, H., Zhao, J., Xu, Y., Li, X., ZmCCD7/ZpCCD7 encodes a carotenoid cleavage dioxygenase mediating shoot branching. *Planta*, *243*, 1407–1418, 2016.
38. Yoneyama, K., Xie, X., Kusumoto, D., Sekimoto, H., Sugimoto, Y., Takeuchi, Y., Yoneyama, K., Nitrogen deficiency as well as phosphorus deficiency in sorghum promotes the production and exudation of 5-deoxystrigol, the host recognition signal for arbuscular mycorrhizal fungi and root parasites. *Planta*, *227*, 125–132, 2007.
39. Zhang, H., Zhao, Y., Zhu, J.K., Thriving under stress: how plants balance growth and the stress response. *Dev. Cell*, *55*, 5, 529–543, 2020.
40. Lamalakshmi Devi, E., Kumar, S., Basanta Singh, T., Sharma, S.K., Beemrote, A., Devi, C.P., Wani, S.H., Adaptation strategies and defence mechanisms of plants during environmental stress, in: *Medicinal plants and environmental challenges*, pp. 359–413, 2017.
41. Hasanuzzaman, M., Nahar, K., Gill, S.S., Fujita, M., Drought stress responses in plants, oxidative stress, and antioxidant defense, in: *Climate change and plant abiotic stress tolerance*, pp. 209–250, 2013.
42. Ruiz-Lozano, J.M., Aroca, R., Zamarreño, Á.M., Molina, S., Andreo-Jiménez, B., Porcel, R., López-Ráez, J.A., Arbuscular mycorrhizal symbiosis induces strigolactone biosynthesis under drought and improves drought tolerance in lettuce and tomato. *Plant Cell Environ.*, *39*, 2, 441–452, 2016.
43. Meng, J.F., Xu, T.F., Wang, Z.Z., Fang, Y.L., Xi, Z.M., Zhang, Z.W., The ameliorative effects of exogenous melatonin on grape cuttings under water-deficient stress: antioxidant metabolites, leaf anatomy, and chloroplast morphology. *J. Pineal Res.*, *57*, 2, 200–212, 2014.
44. Luqman, M., Shahbaz, M., Maqsood, M.F., Farhat, F., Zulfiqar, U., Siddiqui, M.H., Haider, F.U., Effect of strigolactone on growth, photosynthetic efficiency, antioxidant activity, and osmolytes accumulation in different maize (*Zea mays* L.) hybrids grown under drought stress. *Plant Signal. Behav.*, *18*, 1, 2262795, 2023.
45. Song, M., Hu, N., Zhou, S., Xie, S., Yang, J., Ma, W., He, D., Physiological and RNA-Seq Analyses on Exogenous Strigolactones Alleviating Drought by Improving Antioxidation and Photosynthesis in Wheat (*Triticum aestivum* L.). *Antige.*, *12*, 10, 1884, 2023.
46. Bu, Q., Lv, T., Shen, H., Luong, P., Wang, J., Wang, Z., Huq, E., Regulation of drought tolerance by the F-box protein MAX2 in Arabidopsis. *Plant Physiol.*, *164*, 1, 424–439, 2014.

47. Ha, C.V., Leyva-González, M.A., Osakabe, Y., Tran, U.T., Nishiyama, R., Watanabe, Y., Tran, L.S.P., Positive regulatory role of strigolactone in plant responses to drought and salt stress. *PNAS*, *111*, 2, 851–856, 2014.
48. Lv, S., Zhang, Y., Li, C., Liu, Z., Yang, N., Pan, L., Wang, G., Strigolactone-triggered stomatal closure requires hydrogen peroxide synthesis and nitric oxide production in an abscisic acid-independent manner. *New Phytol.*, *217*, 1, 290–304, 2018.
49. Haider, I., Andreo-Jimenez, B., Bruno, M., Bimbo, A., Floková, K., Abuauf, H., Ruyter-Spira, C., The interaction of strigolactones with abscisic acid during the drought response in rice. *J. Exp. Bot.*, *69*, 9, 2403–2414, 2018.
50. Abuauf, H.W., Enzymatic and biological studies on the Arabidopsis enzyme DWARF27 and its homolog DWARF27-like 1, 2018.
51. Gulyás, Z., Moncsek, B., Hamow, K.Á., Stráner, P., Tolnai, Z., Badics, E., Soós, V., D27-LIKE1 isomerase has a preference towards trans/cis and cis/cis conversions of carotenoids in Arabidopsis. *Plant J.*, *112*, 6, 1377–1395, 2022.
52. Rani, V., Sengar, R.S., Garg, S.K., Mishra, P., Shukla, P.K., RETRACTED ARTICLE: Physiological and Molecular Role of Strigolactones as Plant Growth Regulators: A Review. *Mol. Biotechnol.*, 65, 1, 1–1, 2023.
53. Marzec, M., Daszkowska-Golec, A., Collin, A., Melzer, M., Eggert, K., Szarejko, I., Barley strigolactone signalling mutant hvd14. d reveals the role of strigolactones in abscisic acid-dependent response to drought. *Plant Cell Environ.*, *43*, 9, 2239–2253, 2020.
54. Korek, M. and Marzec, M., Strigolactones and abscisic acid interactions affect plant development and response to abiotic stresses. *BMC Plant Biol.*, *23*, 1, 314, 2023.
55. Khan, N., Ali, S., Shahid, M.A., Mustafa, A., Sayyed, R.Z., Curá, J.A., Insights into the interactions among roots, rhizosphere, and rhizobacteria for improving plant growth and tolerance to abiotic stresses: a review. *Cells*, *10*, 6, 1551, 2021.
56. Butcher, K., Wick, A.F., DeSutter, T., Chatterjee, A., Harmon, J., Soil salinity: A threat to global food security. *AJ*, *108*, 6, 2189–2200, 2016.
57. Ben-Dor, E., Taylor, R.G., Hill, J., Demattê, J.A.M., Whiting, M.L., Chabrillat, S., Sommer, S., Imaging spectrometry for soil applications. *Adv. Agron.*, *97*, 321–392, 2008.
58. Wani, K.I., Zehra, A., Choudhary, S., Naeem, M., Khan, M.M.A., Khan, R., Aftab, T., Exogenous strigolactone (GR24) positively regulates growth, and photosynthesis, and improves glandular trichome attributes for enhanced artemisinin production in Artemisia annua. *J. Plant Growth Regul.*, *42*, 8, 4606–4615, 2023.
59. Ling, F., Su, Q., Jiang, H., Cui, J., He, X., Wu, Z., Zhao, Y., Effects of strigolactone on photosynthetic and physiological characteristics in salt-stressed rice seedlings. *Sci. Rep.*, *10*, 1, 6183, 2020.
60. Hasanuzzaman, M., Bhuyan, M.B., Parvin, K., Bhuiyan, T.F., Anee, T.I., Nahar, K., Fujita, M., Regulation of ROS metabolism in plants under environmental

stress: A review of recent experimental evidence. *Int. J. Mol. Sci.*, *21*, 22, 8695, 2020.

61. Hossain, M.A., Hoque, T.S., Zaid, A., Wani, S.H., Mostofa, M.G., Henry, R., Targeting the Ascorbate-Glutathione Pathway and the Glyoxalase Pathway for Genetic Engineering of Abiotic Stress-Tolerance in Rice, in: *Molecular breeding for rice abiotic stress tolerance and nutritional quality*, pp. 398–427, 2021.
62. Sharifi, P. and Bidabadi, S.S., Strigolactone could enhances gas-exchange through augmented antioxidant defense system in *Salvia nemorosa* L. plants subjected to saline conditions stress. *Ind. Crop Prod.*, *151*, 112460, 2020.
63. Sarwar, Y. and Shahbaz, M., Modulation in growth, photosynthetic pigments, gas exchange attributes and inorganic ions in sunflower (*Helianthus annuus* L.) by strigolactones (GR24) achene priming under saline conditions. *Pak. J. Bot.*, *52*, 1, 23–31, 2020.
64. Zulfiqar, H., Shahbaz, M., Ahsan, M., Nafees, M., Nadeem, H., Akram, M., Fahad, S., Strigolactone (GR24) induced salinity tolerance in sunflower (*Helianthus annuus* L.) by ameliorating morpho-physiological and biochemical attributes under *in vitro* conditions. *J. Plant Growth Regul.*, *40*, 2079–2091, 2021.
65. Aroca, R., Ruiz-Lozano, J.M., Zamarreño, Á.M., Paz, J.A., García-Mina, J.M., Pozo, M.J., López-Ráez, J.A., Arbuscular mycorrhizal symbiosis influences strigolactone production under salinity and alleviates salt stress in lettuce plants. *J. Plant Physiol.*, *170*, 1, 47–55, 2013.
66. Ren, C.G., Kong, C.C., Xie, Z.H., Role of abscisic acid in strigolactone-induced salt stress tolerance in arbuscular mycorrhizal *Sesbania cannabina* seedlings. *BMC Plant Biol.*, 18, 1–10, 2018.
67. Bhoi, A., Yadu, B., Chandra, J., Keshavkant, S., Cross-talk of strigolactones with abscisic acid, gibberellins, ethylene, and other hormones, in: *Strigolactones*, pp. 103–126, Academic Press, Cambridge, Massachusetts, USA, 2024.
68. Nicoletti, M. and di Clima, F.P., *Gemmotherapy, and the Scientific Foundations of a Modern Meristemotherapy*, Cambridge Scholars Publishing, Newcastle upon Tyne, United Kingdom, 2020.
69. Shen, H., Luong, P., Huq, E., The F-box protein MAX2 functions as a positive regulator of photomorphogenesis in Arabidopsis. *Plant Physiol.*, *145*, 4, 1471–1483, 2007.
70. Zhang, H., Wang, L., Gao, Y., Guo, Y., Zheng, N., Xu, X., Yang, W., Genome-wide identification of SMXL Gene family in soybean and expression analysis of GmSMXLs under Shade stress. *Plants*, *11*, 18, 2410, 2022.
71. Xie, Y., Liu, Y., Ma, M., Zhou, Q., Zhao, Y., Zhao, B., Wang, H., Arabidopsis FHY3 and FAR1 integrate light and strigolactone signaling to regulate branching. *Nat. Commun.*, *11*, 1, 1955, 2020.

72. Zhan, J., Lu, X., Liu, H., Zhao, Q., Ye, G., Mesocotyl elongation, an essential trait for dry-seeded rice (*Oryza sativa* L.): a review of physiological and genetic basis. *Planta*, *251*, 1–14, 2020.
73. Feng, F., Mei, H., Fan, P., Li, Y., Xu, X., Wei, H., Luo, L., Dynamic transcriptome and phytohormone profiling along the time of light exposure in the mesocotyl of rice seedling. *Sci. Rep.*, *7*, 1, 11961, 2017.
74. Hu, Z., Yan, H., Yang, J., Yamaguchi, S., Maekawa, M., Takamure, I., Nakazono, M., Strigolactones negatively regulate mesocotyl elongation in rice during germination and growth in darkness. *PCP*, *51*, 7, 1136–1142, 2010.
75. Hu, Z., Yamauchi, T., Yang, J., Jikumaru, Y., Tsuchida-Mayama, T., Ichikawa, H., Nakazono, M., Strigolactone and cytokinin act antagonistically in regulating rice mesocotyl elongation in darkness. *PCP*, *55*, 1, 30–41, 2014.
76. Kameoka, H. and Kyozuka, J., Downregulation of rice DWARF 14 LIKE suppress mesocotyl elongation via a strigolactone independent pathway in the dark. *JGG*, *42*, 3, 119–124, 2015.
77. Lu, T., Yu, H., Li, Q., Chai, L., Jiang, W., Improving plant growth and alleviating photosynthetic inhibition and oxidative stress from low-light stress with exogenous GR24 in tomato (*Solanum lycopersicum* L.) seedlings. *Front. Plant Sci.*, *10*, 431739, 2019.
78. Faizan, M., Cheng, S.H., Tonny, S.H., Robab, M.I., Specific roles of strigolactones in plant physiology and remediation of heavy metals from contaminated soil. *PPB*, *192*, 186–195, 2022.
79. Omoarelojie, L.O., Kulkarni, M.G., Finnie, J.F., Van Staden, J., Strigolactone analog (rac-GR24) enhances chilling tolerance in mung bean seedlings. *S. Afr. J. Bot.*, *140*, 173–181, 2021.
80. Hajam, A.H., Ali, M.S., Singh, S.K., Bashri, G., Dubey, G., Implications of strigolactones in plant biology: Achievements, future perspectives, and challenges, in: *Strigolactones*, pp. 221–236, 2024.
81. Lechat, M.M., Brun, G., Montiel, G., Véronési, C., Simier, P., Thoiron, S., Delavault, P., Seed response to strigolactone is controlled by abscisic acid-independent DNA methylation in the obligate root parasitic plant, *Phelipanche ramosa* L. Pomel. *J. Exp. Bot.*, *66*, 11, 3129–3140, 2015.
82. Toh, S., Kamiya, Y., Kawakami, N., Nambara, E., McCourt, P., Tsuchiya, Y., Thermoinhibition uncovers a role for strigolactones in Arabidopsis seed germination. *PCP*, *53*, 1, 107–117, 2012.
83. Gonai, T., Kawahara, S., Tougou, M., Satoh, S., Hashiba, T., Hirai, N., Yoshioka, T., Abscisic acid in the thermoinhibition of lettuce seed germination and enhancement of its catabolism by gibberellin. *J. Exp. Bot.*, *55*, 394, 111–118, 2004.
84. Sohn, S.I., Pandian, S., Kumar, T.S., Zoclanclounon, Y.A.B., Muthuramalingam, P., Shilpha, J., Ramesh, M., Seed dormancy and pre-harvest sprouting in rice—an updated overview. *Int. J. Mol. Sci.*, *22*, 21, 11804, 2021.
85. Cooper, J.W., Hu, Y., Beyyoudh, L., Yildiz Dasgan, H., Kunert, K., Beveridge, C.A., Foyer, C.H., Strigolactones positively regulate chilling tolerance in

*Pisum sativum* and in *Arabidopsis thaliana*. *Plant Cell Environ.*, 41, 1298–1310, 2018, https://doi.org/10.1111/pce.13147.
86. Zhang, X., Zhang, L., Sun, Y., Zheng, S., Wang, J., Zhang, T., Hydrogen peroxide is involved in strigolactone induced low temperature stress tolerance in rape seedlings (*Brassica rapa* L.). *PPB, 157*, 402–415, 2020.
87. Hepworth, J.A., Comparative analysis of the MAX pathway (Doctoral dissertation), University of York, York, United Kingdom, 2012.
88. Liu, Z., Li, Y., Wang, J., He, X., Tian, C., Different respiration metabolism between mycorrhizal and non-mycorrhizal rice under low-temperature stress: a cry for help from the host. *J. Agric. Sci., 153*, 4, 602–614, 2015.
89. Sytar, O., Kumari, P., Yadav, S., Brestic, M., Rastogi, A., Phytohormone priming: regulator for heavy metal stress in plants. *J. Plant Growth Regul., 38*, 739–752, 2019.
90. Mostofa, M.G., Rahman, M.M., Nguyen, K.H., Li, W., Watanabe, Y., Tran, C.D., Tran, L.S.P., Strigolactones regulate arsenate uptake, vacuolar sequestration and antioxidant defense responses to resist arsenic toxicity in rice roots. *J. Hazard. Mater., 415*, 125589, 2021.
91. Ali, W., Mao, K., Zhang, H., Junaid, M., Xu, N., Rasool, A., Yang, Z., A comprehensive review of the basic chemical behaviours, sources, processes, and endpoints of trace element contamination in paddy soil-rice systems in rice-growing countries. *J. Hazard. Mater., 397*, 122720, 2020.
92. de Saint Germain, A., Ligerot, Y., Dun, E.A., Pillot, J.P., Ross, J.J., Beveridge, C.A., Rameau, C., Strigolactones stimulate internode elongation independently of gibberellins. *Plant Physiol., 163*, 2, 1012–1025, 2013.
93. Li, N., Wang, J., Song, W.Y., Arsenic uptake and translocation in plants. *PCP*, 57, 1, 4–13, 2016.
94. Li, N., Wang, J., Song, W.Y., Arsenic uptake and translocation in plants, *PCP*, 57, 1, 4–13, 2016.
95. Wu, F., Gao, Y., Yang, W., Sui, N., Zhu, J., Biological functions of strigolactones and their crosstalk with other phytohormones. *Front. Plant Sci., 13*, 821563, 2022.
96. Liu, J., He, H., Vitali, M., Visentin, I., Charnikhova, T., Haider, I., Cardinale, F., Osmotic stress represses strigolactone biosynthesis in *Lotus japonicus* roots: exploring the interaction between strigolactones and ABA under abiotic stress. *Planta, 241*, 1435–1451, 2015.
97. Min, Z., Li, R., Chen, L., Zhang, Y., Li, Z., Liu, M., Fang, Y., Alleviation of drought stress in grapevine by foliar-applied strigolactones. *PPB, 135*, 99–110, 2019.
98. Visentin, I., Pagliarani, C., Deva, E., Caracci, A., Turečková, V., Novák, O., Cardinale, F., A novel strigolactone-miR156 module controls stomatal behaviour during drought recovery. *Plant Cell Environ., 43*, 7, 1613–1624, 2020.

99. Wang, L., Wang, B., Yu, H., Guo, H., Lin, T., Kou, L., Li, J., Transcriptional regulation of strigolactone signalling in Arabidopsis. *Nature*, *583*, 7815, 277–281, 2020.
100. González-Grandío, E., Pajoro, A., Franco-Zorrilla, J.M., Tarancón, C., Immink, R.G., Cubas, P., Abscisic acid signaling is controlled by a BRANCHED1/HD-ZIP I cascade in Arabidopsis axillary buds. *PNAS*, *114*, 2, E245–E254, 2017.
101. Liu, X., Hu, Q., Yan, J., Sun, K., Liang, Y., Jia, M., Wang, B., ζ-Carotene isomerase suppresses tillering in rice through the coordinated biosynthesis of strigolactone and abscisic acid. *Mol. Plant*, *13*, 12, 1784–1801, 2020.
102. Vlad, F., Rubio, S., Rodrigues, A., Sirichandra, C., Belin, C., Robert, N., Merlot, S., Protein phosphatases 2C regulate the activation of the Snf1-related kinase OST1 by abscisic acid in Arabidopsis. *Plant Cell*, *21*, 10, 3170–3184, 2009.
103. Santino, A., Taurino, M., De Domenico, S., Bonsegna, S., Poltronieri, P., Pastor, V., Flors, V., Jasmonate signaling in plant development and defense response to multiple (a) biotic stresses. *Plant Cell Rep.*, *32*, 1085–1098, 2013.
104. Raza, A., Javed, R., Zahid, Z., Sharif, R., Hafeez, M.B., Ghouri, M.Z., Siddiqui, M.H., Strigolactones for sustainable plant growth and production under adverse environmental conditions, in: *Plant Performance Under Environmental Stress: Hormones, Biostimulants and Sustainable Plant Growth Management*, pp. 129–166, 2021.
105. Torres-Vera, R., García, J.M., Pozo, M.J., López-Ráez, J.A., Expression of molecular markers associated to defense signaling pathways and strigolactone biosynthesis during the early interaction tomato-Phelipanche ramosa. *Physiol. Mol. Plant Pathol.*, *94*, 100–107, 2016.
106. Xie, X., Wang, G., Yang, L., Cheng, T., Gao, J., Wu, Y., Xia, Q., Cloning and characterization of a novel Nicotiana tabacum ABC transporter involved in shoot branching. *Physiol. Plant.*, *153*, 2, 299–306, 2015.
107. Omoarelojie, L.O., Kulkarni, M.G., Finnie, J.F., Van Staden, J., Strigolactones and their crosstalk with other phytohormones. *Ann. Bot.*, *124*, 5, 749–767, 2019.
108. Santisree, P., Jalli, L.C.L., Bhatnagar-Mathur, P., Sharma, K.K., Emerging roles of salicylic acid and jasmonates in plant abiotic stress responses, in: *Protective chemical agents in the amelioration of plant abiotic stress: biochemical and molecular perspectives*, pp. 342–373, 2020.
109. Rozpądek, P., Domka, A.M., Nosek, M., Ważny, R., Jędrzejczyk, R.J., Wiciarz, M., Turnau, K., The role of strigolactone in the cross-talk between Arabidopsis thaliana and the endophytic fungus Mucor sp. *Front. Microbiol.*, *9*, 307553, 2018.
110. Tian, M.Q., Jiang, K., Takahashi, I., Li, G.D., Strigolactone-induced senescence of a bamboo leaf in the dark is alleviated by exogenous sugar. *J. Pestic. Sci.*, *43*, 3, 173–179, 2018.

111. Visentin, I., Vitali, M., Ferrero, M., Zhang, Y., Ruyter-Spira, C., Novák, O., Cardinale, F., Low levels of strigolactones in roots as a component of the systemic signal of drought stress in tomato. *New Phytol.*, *212*, 4, 954–963, 2016.
112. Jia, K.P., Luo, Q., He, S.B., Lu, X.D., Yang, H.Q., Strigolactone-regulated hypocotyl elongation is dependent on cryptochrome and phytochrome signaling pathways in Arabidopsis. *Mol. Plant*, *7*, 3, 528–540, 2014.
113. Yu, Y., Wang, J., Zhang, Z., Quan, R., Zhang, H., Deng, X.W., Huang, R., Ethylene promotes hypocotyl growth and HY5 degradation by enhancing the movement of COP1 to the nucleus in the light. *PLos Genet.*, *9*, 12, e1004025, 2013.
114. Kapulnik, Y., Resnick, N., Mayzlish-Gati, E., Kaplan, Y., Wininger, S., Hershenhorn, J., Koltai, H., Strigolactones interact with ethylene and auxin in regulating root-hair elongation in Arabidopsis. *J. Exp. Bot.*, *62*, 8, 2915–2924, 2011.
115. Lee, H.Y. and Yoon, G.M., Strigolactone elevates ethylene biosynthesis in etiolated Arabidopsis seedlings. *Plant Signal. Behav.*, *15*, 11, 1805232, 2020.
116. Feng, Y., Xu, P., Li, B., Li, P., Wen, X., An, F., Guo, H., Ethylene promotes root hair growth through coordinated EIN3/EIL1 and RHD6/RSL1 activity in Arabidopsis. *PNAS*, *114*, 52, 13834–13839, 2017.
117. Ueda, H. and Kusaba, M., Strigolactone regulates leaf senescence in concert with ethylene in Arabidopsis. *Plant Physiol.*, *169*, 1, 138–147, 2015.
118. Rasmussen, A., Hu, Y., Depaepe, T., Vandenbussche, F., Boyer, F.D., Van Der Straeten, D., Geelen, D., Ethylene controls adventitious root initiation sites in Arabidopsis hypocotyls independently of strigolactones. *J. Plant Growth Regul.*, *36*, 897–911, 2017.
119. Zdarska, M., Dobisová, T., Gelová, Z., Pernisová, M., Dabravolski, S., Hejátko, J., Illuminating light, cytokinin, and ethylene signalling crosstalk in plant development. *J. Exp. Bot.*, *66*, 16, 4913–4931, 2015.
120. Lantzouni, O., Klermund, C., Schwechheimer, C., Largely additive effects of gibberellin and strigolactone on gene expression in *Arabidopsis thaliana* seedlings. *Plant J.*, *92*, 5, 924–938, 2017.
121. Sun, H., Guo, X., Zhu, X., Gu, P., Zhang, W., Tao, W., Zhang, Y., Strigolactone and gibberellin signaling coordinately regulate metabolic adaptations to changes in nitrogen availability in rice. *Mol. Plant*, *16*, 3, 588–598, 2023.
122. Xiong, G., Wang, Y., Li, J., Action of strigolactones in plants, in: *The Enzymes*, vol. 35, pp. 57–84, Academic Press, Cambridge, Massachusetts, USA, 2014.
123. Wani, A.K., Akhtar, N., Garg, S.S., Rahayu, F., Prakash, A., Singh, R., Strigolactones in plants: web networking with phytohormonal crosstalk, in: *Strigolactones, alkamides and karrikins in plants*, pp. 59–73, CRC Press, Boca Raton, Florida, USA, 2023.
124. Zhang, J., Mazur, E., Balla, J., Gallei, M., Kalousek, P., Medveďová, Z., Friml, J., Strigolactones inhibit auxin feedback on PIN-dependent auxin transport canalization. *Nat. Commun.*, *11*, 1, 3508, 2020.

125. Jan, M., Muhammad, S., Jin, W., Zhong, W., Zhang, S., Lin, Y., Wang, G., Modulating root system architecture: cross-talk between auxin and phytohormones. *Front. Plant Sci.*, *15*, 1343928, 2024.
126. Sun, H., Tao, J., Hou, M., Huang, S., Chen, S., Liang, Z., Zhang, Y., A strigolactone signal is required for adventitious root formation in rice. *Ann. Bot.*, *115*, 7, 1155–1162, 2015.
127. Xu, J., Zha, M., Li, Y., Ding, Y., Chen, L., Ding, C., Wang, S., The interaction between nitrogen availability and auxin, cytokinin, and strigolactone in the control of shoot branching in rice (Oryza sativa L.). *Plant Cell Rep.*, *34*, 1647–1662, 2015.
128. Vijayakumar, P., Datta, S., Dolan, L., ROOT HAIR DEFECTIVE SIX-LIKE 4 (RSL 4) promotes root hair elongation by transcriptionally regulating the expression of genes required for cell growth. *New Phytol.*, *212*, 4, 944–953, 2016.
129. Dun, E.A., de Saint Germain, A., Rameau, C., Beveridge, C.A., Antagonistic action of strigolactone and cytokinin in bud outgrowth control. *Plant Physiol.*, *158*, 1, 487–498, 2012.
130. Duan, J., Yu, H., Yuan, K., Liao, Z., Meng, X., Jing, Y., Li, J., Strigolactone promotes cytokinin degradation through transcriptional activation of CYTOKININ OXIDASE/DEHYDROGENASE 9 in rice. *PNAS*, *116*, 28, 14319–14324, 2019.
131. Zha, M., Zhao, Y., Wang, Y., Chen, B., Tan, Z., Strigolactones and cytokinin interaction in buds in the control of rice tillering. *Front. Plant Sci.*, *13*, 837136, 2022.
132. Wang, M., Molecular regulatory network of BRANCHED1 (BRC1) expression in axillary bud of Rpsa sp. in response to sugar and auxin (Doctoral dissertation), Agrocampus Ouest, Rennes, France, 2019.
133. Salvi, E., Rutten, J.P., Di Mambro, R., Polverari, L., Licursi, V., Negri, R., Ten Tusscher, K., A self-organized PLT/Auxin/ARR-B network controls the dynamics of root zonation development in Arabidopsis thaliana. *Dev. Cell*, *53*, 4, 431–443, 2020.
134. Martínez, C., Espinosa-Ruiz, A., Prat, S., Gibberellins and plant vegetative growth, in: *Annual Plant Reviews, Volume 49: Gibberellins, The*, pp. 285–322, 2016.
135. Hai, N.N., Chuong, N.N., Tu, N.H.C., Kisiala, A., Hoang, X.L.T., Thao, N.P., Role and regulation of cytokinins in plant response to drought stress. *Plants*, *9*, 4, 422, 2020.
136. Yang, T., Lian, Y., Wang, C., Comparing and contrasting the multiple roles of butenolide plant growth regulators: strigolactones and karrikins in plant development and adaptation to abiotic stresses. *Int. J. Mol. Sci.*, *20*, 24, 6270, 2019.
137. Bennett, T. and Leyser, O., Strigolactone signalling: standing on the shoulders of DWARFs. *Curr. Opin. Plant Biol.*, *22*, 7–13, 2014.

9

# Strigolactone Analogs: Synthesis, Structural Features and Biological Activity

**Roheela Ahmad[1], Pirzada Mohammad Haris[2], Zubair Altaf Reshi[3*], Tahir Ahmad Sheikh[4], Ayman Javed[1] and Inayat Mustafa Khan[1]**

[1]*Division of Soil Science and Agricultural Chemistry, Faculty of Agriculture, SKUAST-Kashmir, Sopore, India*
[2]*Department of Geology, Aligarh Muslim University, Aligarh, India*
[3]*Department of Botany, Aligarh Muslim University, Aligarh, India*
[4]*Division of Agronomy, Faculty of Agriculture, SKUAST-Kashmir, Sopore, India*

***Abstract***
Strigolactones (SLs) are a remarkable class of plant hormones that have attracted more attention recently. Their function has broadened to include a variety of physiological activities in plants. They were originally identified as signaling molecules for host detection by parasitic plants and for promoting symbiotic associations between plants and arbuscular mycorrhizal fungi (AMF). Even though plants only produce trace amounts of them, the purification procedure may jeopardize their stability. Thus, it is crucial to synthesize them completely in order to verify the structures that are obtained via spectroscopic and other physical investigations. Plant roots secrete SLs, which are essential for initiating seed germination in root parasitic weeds like *Striga* and *Orobanche* species and for stimulating hyphal branching in AM fungi. Strigolactones have recently been identified as a novel family of plant hormones controlling shoot growth, in addition to their roles in the rhizosphere. The common C–D ring structure that has been identified in natural SLs is thought to be crucial to their biological activity. By hydroxylating at positions C-4, C-5, and C-9, respectively, 5-deoxystrigol, the basic SL, can have its A–B ring structure modified to produce a range of derivatives, including orobanchol, strigol, and sorgomol. These hydroxy-strigolactones may subsequently undergo acetylation and maybe other derivatizations as well. The finding of 2′-epi-strigolactones in the root exudates of tobacco, rice, pea, and other plant species reveals that certain plants synthesize both epimers, even though the C-2′-(R)

**Corresponding author*: zubairreshi26@gmail.com

Tariq Aftab and Kaiser Iqbal Wani (eds.) Strigolactones: Emerging Plant Hormones, (193–210) © 2025 Scrivener Publishing LLC

stereochemistry was first thought to be essential for strong germination stimulation activity.

***Keywords*:** Strigolactones, total synthesis, analogues, parasitic weeds, arbuscular mycorrhizal fungi

## 9.1 Introduction

In recent years, strigolactones (SLs) have drawn a lot of attention as a unique family of plant hormones [1, 2]. A major achievement was the isolation of strigol from cotton root exudate about 50 years ago [3]. However, it took over 20 years to completely understand its chemical makeup [4, 5]. Despite this novel ability—which is to stimulate the germination of parasitic weeds such as *Striga* and *Orobanche* sp.—SLs were first given little attention in the field of plant research. Organic chemists faced an enormous task in synthesizing strigol as its whole, and the development of SL chemistry was contingent upon the isolation of other naturally occurring SLs from other host plants utilized by these parasites.

A structural motif common to all natural SLs is made up of three fused rings that create the ABC scaffold. The D-ring, a butenolide, is joined to the ABC scaffold by an enol ether unit [6, 7]. The stereochemistry of the BC junction distinguishes the two families of SLs that have been found thus far: the strigol and orobanchol families have 8bS stereochemistry, whereas the latter has 8bR stereochemistry (Figure 9.1) [6]. It is important to remember that separating SLs from root exudates is a labor-intensive procedure, and understanding their structures presents substantial difficulties.

SLs are terpenoid lactones that are generated from carotenoid pigments, and it has been verified that they are present in several plant species. In these cases, many SLs are usually recognized together [8]. Plants primarily synthesize SL in their root tissues, while there is some indication that synthesis can also take place in the lower portions of the shoot. However, it is suggested that SLs, their metabolites, or other unknown secondary messengers must go from roots to shoots to cause a notable decrease in shoot branching [9].

This idea is reinforced by the discovery that *Arabidopsis* xylem sap contains the SL orobanchol, which suggests that SLs produced from roots are transported to the shoot [10, 11]. Furthermore, it has been discovered that SLs have several functions in plant physiology, such as modulating root development, promoting adventitious root production, and controlling secondary growth [12]. Moreover, SLs have been implicated in shoot responses to low phosphorus (Pi) environments in several studies [13].

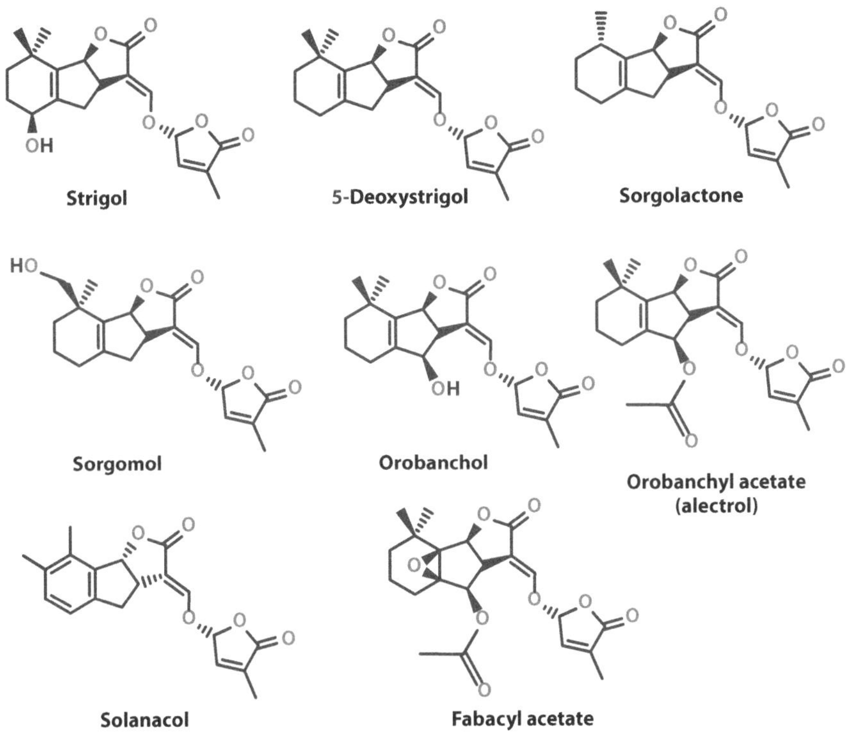


**Figure 9.1** Two classes of SLs.

## 9.2 Synthesis of Natural SLS

### 9.2.1 Introduction to Synthesis

- To produce SL analogs, compounds with alterations to increase desirable qualities are crafted to resemble the structure of natural SLs.
- Strategies for building the target compounds are usually determined by using organic synthesis techniques, such as retrosynthetic analysis.

Three crucial steps make up the method for the comprehensive synthesis of SLs [16]: First is the ABC scaffold's synthesis, the A-and/or B-ring's intended capabilities are incorporated in the second phase, and the D-ring is attached in the third. Alternately, the order can be reversed, with the D-ring attached first and the suitable substituents introduced in the A and B rings after that.

There are a lot of steps involved in building the ABC scaffold (Scheme 9.1). Mesityl oxide is a useful starting material. It is first converted to a derivative of cyclohexene carboxylic acid and then goes through a Nazarov cyclization to produce the AB-ring system. The ABC scaffold possesses one effective stereogenic unit at the BC junction, where only a cis-BC lactone is feasible. Hence, the synthesis results in racemic material as no chiral reagents are utilized.

### 9.2.2 Attachment of the D-Ring

In general, the D-ring attachment always proceeds according to the same protocol. First, the ABC scaffold is formylated with (m)ethyl formate when a suitable base is present. Next, the enolate that is produced is treated with either chloro or bromo butenolide. Potassium tert-butoxide is normally the suitable base, though other bases such as sodium hydride, potassium carbonate, and some tertiary amine bases have also been used. Mechanistically, it is not entirely clear whether this coupling takes place via an intermediate oxonium ion or through an SN1 pathway with SN2 characteristics.

The easiest way to obtain the necessary bromo butenolide is to use N-bromo-succinimide/dibenzoyl peroxide or N-bromo-succinimide plus light for radical bromation of the matching butenolide. There are times when the coupling reaction exhibits unpredictable behavior for unclear reasons.

### 9.2.3 Modification of the D-Ring and Side Chains

- Changes made to the D-ring and side chains can impact the effectiveness, specificity, and endurance of SL analogs.
- Structural adjustments may involve substituting, adding, or removing functional groups as well as altering stereochemistry.
- Systematic design based on structure–activity relationships (SAR) is utilized to refine analog structures for particular biological functions.

### 9.2.4 Examples of Total Synthesis

In 1974, MacAlpine and his colleagues [14] accomplished the first effective total synthesis of strigol. It was the same procedure as described earlier, with a small adjustment. In this example, the hydroxyl (OH) group was

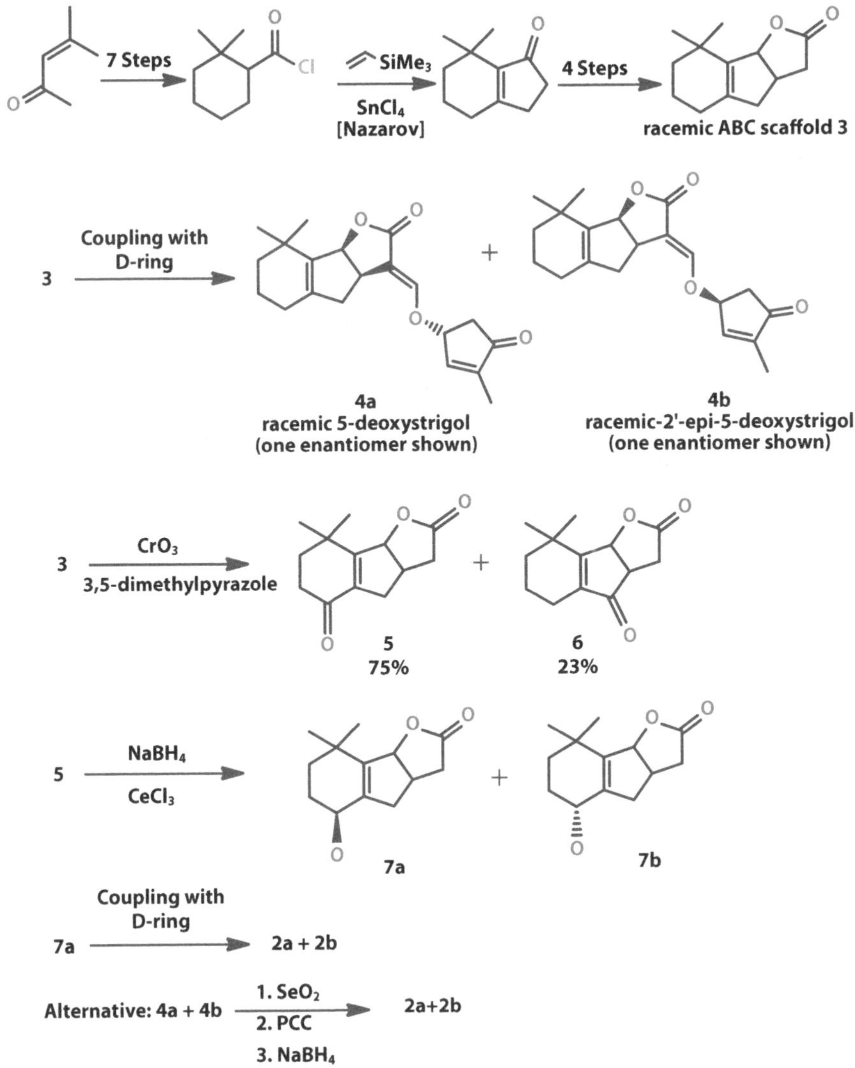


**Scheme 9.1** Total synthesis of strigol.

added to the A-ring in two stages before the C-ring closed. The dimethyl cyclohexanone was successfully annealed via a Nazarov cyclization. A combination of racemic strigol and its 2′-epimer was obtained by coupling the OH group-containing ABC scaffold with the D-ring.

An extended sequence was employed in an alternate method to synthesize the same essential intermediate from a condensation product of methyl vinyl ketone and nitromethane. Another method for obtaining this intermediate comes via a condensation reaction using ethyl acetoacetate and mesityl oxide, two basic starting ingredients [19]. However, all these techniques for producing the ABC scaffold needed complex, multistep processes, none of which were suitable for large-scale synthesis. Indeed, the elegance of these strigol syntheses took precedence over their usefulness. The key features of each route are enumerated in Scheme 9.2.

The prototype compound for naturally occurring SLs is 5-deoxy-strigol, an SL without substituents on the A- and B-rings. Scheme 9.1 included the synthesis of both epimers as racemates already. After the corresponding racemates were separated using a chiral HPLC column (Chiralpack AD), all four stereoisomers were obtained in enantiopure form [15] (Scheme 9.3). [16] shows an elegant and alternative method of manufacturing the four stereoisomers by the use of chiral [2+2] intramolecular cycloaddition of ketene-iminium salts. The enantiopure ABC scaffold is obtained by 11 steps in this process, which starts with commercially available Hageman's ester and is too long to be deemed practical.

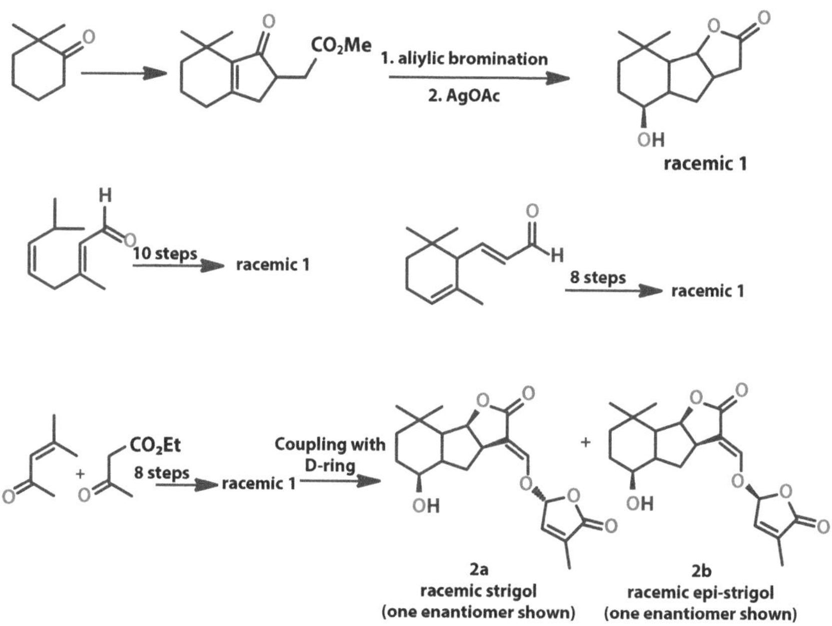


**Scheme 9.2** Essentials of the early strigol synthesis.

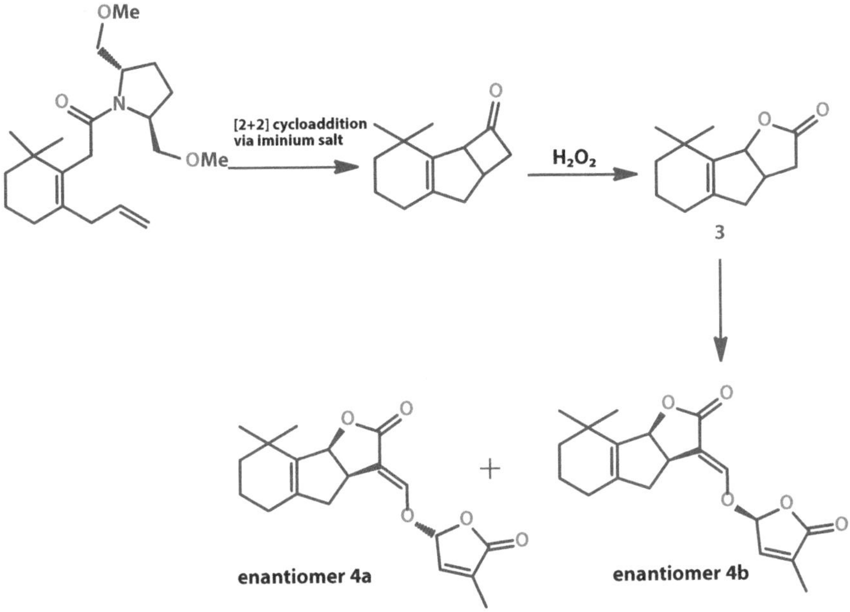


**Scheme 9.3** Asymmetric synthesis of 5-deoxystrigol.

For the preparation of enantiopure SLs, the availability of the homochiral D-ring synthon is strategically important. Its preparation on a 3-g scale demonstrates that this condition is met [17]. Matsui *et al.* [18] synthesized the enantiopure ABC scaffold in 13 stages using methyl (S)-(−)-citronellate as a starting material from the chiral pool (Scheme 9.4). Two enantiopure epimers were produced as a diastereomeric mixture by coupling with the D-ring later on, and this combination was easily separated.

Notably, the sequence used by Matsui *et al.*, [18] did not yield cis-isomers since it was not possible to extract citronellate with the opposite chirality from the chiral pool.

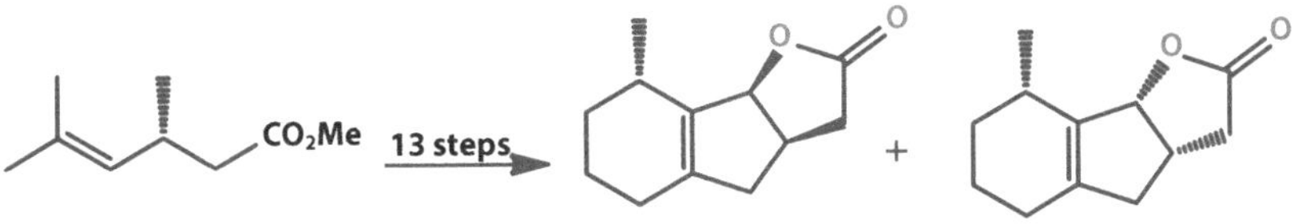


**Scheme 9.4** Synthesis of the ABC scaffold from the chiral pool.

Most likely, the most prevalent SL found in nature is orobanchol. Its structure was first established based on spectroscopic data and the hypothesis that it would share stereochemical characteristics with (+)-strigol after it was isolated from red clover [19]. Ueno *et al.* [20], however, showed that the true structure of orobanchol found in nature is a stereoisomer of the structure that was first suggested. The ABC scaffold is the starting point for the complete synthesis of the SL with the initially allocated structure. It is oxidized with $CrO_3$ to produce a mixture of keto compounds in an unfavorable ratio of 3:1 [21].

### 9.2.5 Lessons Learned from Total Synthesis

It is evident from the overview of total synthesis approaches for natural SLs that the ABC scaffold preparation is frequently a laborious procedure. However, there is a scalable method for producing the basic ABC system. It can be difficult to introduce the right substituents, especially when trying to selectively oxidize the A- or B-ring. Diversifying options in this regard is essential, as ensuring proper stereochemistry is imperative.

The complete synthesis of 7-oxo-5-deoxystrigol serves as an example of how the previously described strategy can be applied in a variety of situations [22]. Once 2,2-dimethylcyclohexane-1,3-dione is used to produce the beginning material with an oxo group next to the gem-dimethyl, the typical procedure for building the ABC scaffold and affixing the D-ring is followed (Scheme 9.5).

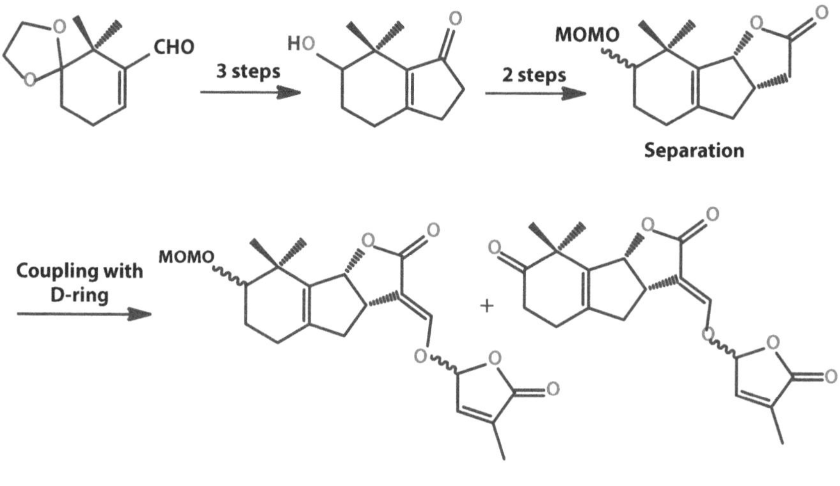


**Scheme 9.5** Synthesis of 7-oxo-5-deoxystrigol.

## 9.3 Structures of Strigolactones

### 9.3.1 Essential Structural Features and Structural Diversity

Figure 9.2 shows the natural SLs that have been isolated thus far. The first SLs to be identified and isolated from cotton root exudates were strigol (1) and strigyl acetate (2) [3, 23]. Later, strigol was discovered in the exudates from the roots of several *Striga* hosts, such as proso millet

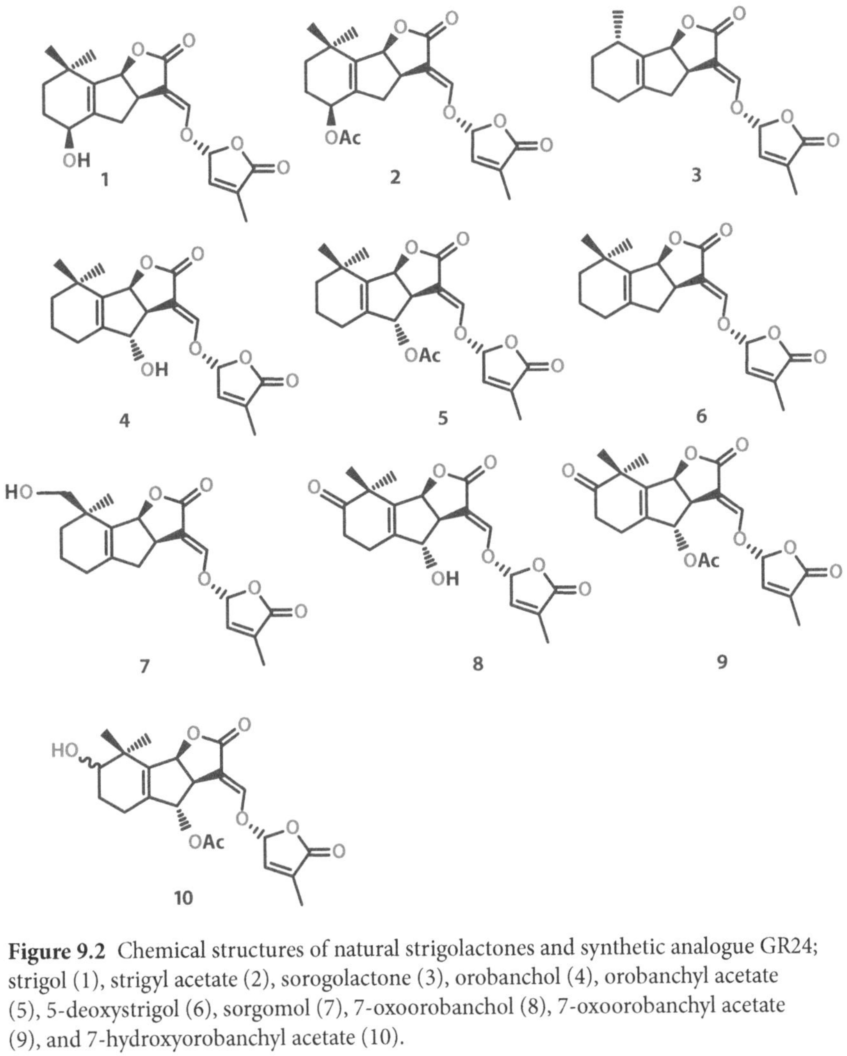

**Figure 9.2** Chemical structures of natural strigolactones and synthetic analogue GR24; strigol (1), strigyl acetate (2), sorogolactone (3), orobanchol (4), orobanchyl acetate (5), 5-deoxystrigol (6), sorgomol (7), 7-oxoorobanchol (8), 7-oxoorobanchyl acetate (9), and 7-hydroxyorobanchyl acetate (10).

(*Pennisetum glaucum* R.Br.), maize (*Zea mays* L.), and sorghum (*Sorghum bicolor* (L.) Monch [24]. The root exudates of cowpea (*Vigna unguiculata* Auct.) [14] and sorghum were used to extract sorgolactone (3) and alectrol, respectively.

The use of SLs by *Striga* and *Orobanche* was confirmed by the separation of orobanchol (4) and alectrol as the first *Orobanche* germination stimulants from red clover (*Trifolium pratense* L.) root exudates [19]. Orobanchyl acetate has been discovered as alectrol (5) [19, 25].

Root exudates from a variety of plant species, including monocots and dicots, have been found to contain 5-deoxystrigol (6), the first branching factor for AM fungi from *Lotus japonicus* (Regel) Larsen root exudates. This finding suggests that other SLs are derived from 5-deoxystrigol. Strigol (1) or orobanchol (4) are created by the allylic hydroxylation of 5-deoxystrigol (6), while sorgomol (7), the third hydroxy-strigolactone, is generated by hydroxylation at the homoallylic position [26]. These hydroxy-strigolactones can conjugate with sugars and amino acids as well as undergo acetylation. Sorgolactone (3) is produced by further oxidizing the hydroxymethyl group of sorgomol (7) and then decarboxylating it [27, 28].

Orobanchol (4) is one of the three hydroxy-strigolactones that seems to be most widely distributed throughout the plant world. Recently, root exudates from flax (*Linum usitatissimum* L.) and cucumber (*Cucumis sativus* L.) were shown to contain 7-oxoorobanchol (8), 7-oxoorobanchyl acetate (9), and 7-hydroxyorobanchol acetate (10) [29].

The C–D ring moiety of SLs has been extensively studied in regard to SAR and has been found to be important for promoting the germination of root parasitic weed seeds [30]. This structure is present in every natural SL that has been isolated thus far, as shown in Figure 9.2. Before 2′-epiorobanchol (12) was found in tobacco (*N. tabacum*) root exudates [31], natural SLs were thought to have 2′-(R) stereochemistry. Furthermore, 5-deoxystrigol, orobanchol, and their 2′-epimers appear to be produced and released by rice plants (*Oryza sativa* L. cv. Nipponbare) (Yoneyama K, unpublished). Legume plants that yield detectable levels of fabacyl acetate (ent-2′-epi-4a, 8a-epoxy-orobanchyl acetate) (13) but not its epimer include *Pisum sativum* L. [29]. These results imply that only one enantiomer may be released into the rhizosphere and that the binding of the D-ring to the ABC component may not be a stereoselective process.

## 9.4 Biological Activity of Strigolactones

### 9.4.1 Germination Stimulation Activity

Strigolactones were discovered almost 50 years ago to be germination boosters for root parasite plants, especially those in the genera *Striga*, *Orobanche*, and *Phelipanche* [32–34] that negatively impact numerous important food crops. In many areas, these parasites cause large agricultural losses by attaching themselves to plant roots and depriving them of nutrients and water [35–37]. These parasites are notable for producing a large number of tiny seeds that can survive in the soil for 15 to 20 years [36, 38, 39]. The radicle of the seed clings to the host root during germination and develops into a haustorium, which ultimately results in the growth of flowers and the production of seeds [40–42].

All SLs generally show germination activity against root parasite plants belonging to the families *Striga*, *Orobanche*, and *Phelipanche*; in fact, this activity frequently outperforms the synthetic counterpart GR24 by a factor of up to 100 [43]. Kim *et al.* [43] found that the lipophilicity of SL molecules determines the germination-stimulating action of naturally occurring SLs on *Orobanche minor* seeds. The highest overall activity is shown by hydroxylated SLs, whereas their acetates show significantly lesser responses. Moreover, Ueno *et al.* [20] discovered that critical structural prerequisites for germination induction include an oxygenated substituent at C-4, the structure of the tricyclic lactone, and the D-ring.

Analysis of the data shows that different parasitic plant species may have different germination stimulation of SLs on their seeds [43–45]. Furthermore, distinct germination tests could produce different results. Authors frequently assess the biological activity of putative germination stimulants using the standardized germination bioassay outlined by Mangnus *et al.* [30]. Notably, Matusova *et al.* [46] and Song *et al.* [47] observed that the temperature during *Orobanche* and *Striga* seeds' preconditioning greatly influenced how they responded to the germination stimulants applied, which may have contributed to the variations in germination activity that have been documented. Furthermore, their reaction could be influenced by the seeds' place of origin.

So far, every plant species that has been studied has been found to release SL combinations. Within these mixes, individual SLs can differ even amongst cultivars of the same plant species (e.g., sorghum). Additionally, it appears that different growth phases and environmental factors affect the amounts and ratios of SLs. As a result, in the field, hyphae and spores of AM fungus, as well as seeds of root parasites, are exposed to continuously varying combinations of SLs. Therefore, it is essential to look into antagonistic and/or synergistic interactions between various SLs to comprehend why these organisms are host-specific.

### 9.4.2 Hyphal Branching

Strigolactones have been identified as essential molecules for communication between plant roots and AM fungi [15, 48]. Plants and AM fungi have a symbiotic connection in which the plant receives water and phosphate in return for carbon. In AM fungi, SLs released by roots promote hyphal branching, which aids in the fungal development in the direction of the associated plant roots. The fungi provide nutrients from the earth in return for the carbohydrates they take from the host plants. Furthermore, it has been documented that AM fungus shields host plants from disease and drought-related damage [49–51].

It is hypothesized that plants produced SLs as a means of enabling AM fungi to colonize their roots and that parasitic plants use these signaling molecules to identify possible hosts [2, 48, 52, 53]. When it comes to the enol ether bridge in the C-D portion of the SL molecules, the structural requirements of SLs as hyphal branching factors in AM fungi are comparable, if not exact, to those noted for the germination of seeds of root parasitic weeds. Based on the substitution pattern in the AB moiety of SLs, research data from the past show a large variance in hyphal branching activity across AM fungi. Moreover, stereochemistry is important, as shown by Akiyama *et al.* [15] who found that natural SLs are more active than their enantiomers. Furthermore, the hyphal branching inducers varied in both the induced hyphal branching pattern and active concentration. Clarifying the structure–activity link of SLs with respect to their role as branching factors for AM fungi is important. The intelligent creation of artificial SL analogs with active branching activity will be made easier with the use of such information.

### 9.4.3 Shoot Branching Inhibition

A new class of plant hormones known to prevent shoot branching are SLs or their derivatives [13, 54]. As yet, bioassays have only shown the synthetic SL GR24 to be effective in preventing shoot branching; the bioactive form or forms of this novel hormone are still unknown. The levels of orobanchyl acetate and fabacyl acetate [54] and 2′-epi-5-deoxystrigol [13], respectively, were much lower in mutants like pea *ccd8* and rice *d10*. Thus, it is possible that additional synthetic and natural SLs that stimulate the germination of parasite seeds also have the ability to prevent shoot branching. It is crucial to create sensitive and quick bioassays to identify the hormone's bioactive form or forms.

## 9.5 Conclusion

Mixtures of SLs that plants release into the rhizosphere are likely to expose AM fungus and root parasite seeds to these substances when they are in close proximity to any live root. Despite this, the hyphal branching of AM fungus and the seed germination of parasites are frequently reduced in the proximity of non-host roots. Therefore, it seems that, in addition to SLs, other signaling molecules that have an antagonistic or synergistic interaction with SLs are also involved in the hyphal branching of AM fungi and the germination of root parasitic weeds.

To understand why plants make so many SLs and how each helps root parasites and AM fungi recognize their hosts, more research is necessary. To create budget-friendly control plans for root parasitic weeds, SLs must be recognized as host recognition signals in both parasitism and symbiosis.

## References


1. Tsuchiya, Y. and McCourt, P., Strigolactones: a new hormone with a past. *Curr. Opin. Plant Biol.*, 12, 556–561, 2009.
2. Bouwmeester, H.J., Roux, C., Lopez-Raez, J.A., Becard, G., Rhizosphere communication of plants, parasitic plants and AM fungi. *Trends Plant Sci.*, *12*, 5, 224–230, 2007.
3. Cook, C.E., Whichard, L.P., Turner, B., Wall, M.E., Egley, G.H., Germination of witchweed (Striga lutea Lour.): isolation and properties of a potent stimulant. *Sci.*, 154, 3753, 1189–1190, 1966.
4. Heather, J.B., Mittal, R.S.D., Sih, C.J., Synthesis of the witchweed seed germination stimulant (+)-strigol. *J. Am. Chem. Soc.*, 98, 12, 3661–3669, 1976.
5. Brooks, D.W., Bevinakatti, H.S., Powell, D.R., The absolute structure of (+)-strigol. *J. Org. Chem.*, 50, 20, 3779–3781, 1985.
6. Zwanenburg, B. and Pospíšil, T., Structure and activity of strigolactones: new plant hormones with a rich future. *Mol. Plant*, 6, 38–62, 2013.

7. Zwanenburg, B., Mwakaboko, A.S., Reizelman, A., Anilkumar, G., Sethumadhavan, D., Structure and function of natural and synthetic signaling molecules in parasitic weed germination. *Pest Manag. Sci.*, 65, 478–491, 2009.
8. Xie, X., Yoneyama, K., Harada, Y., Fusegi, N., Yamada, Y., Ito, S., Yoneyama, K., Fabacyl acetate, a germination stimulant for root parasitic plants from Pisum sativum. *Phytochem.*, *70*, 2, 211–215, 2009.
9. Dailey, O.D. and Jr., A., New synthetic route to (±)-strigol. *J. Org. Chem.*, 52, 1984–1989, 1987.
10. Kohlen, W., Charnikhova, T., Liu, Q., Bours, R., Domagalska, M.A., Beguerie, S., Verstappen, F., Leyser, O., Bouwmeester, H.J., Ruyter-Spira, C., Strigolactones are transported through the xylem and play a key role in shoot architectural response to phosphate deficiency in non-AM host Arabidopsis thaliana. *Plant Physiol.*, 155, 2, 974–987, 2011.
11. Seto, Y., Kameoka, H., Yamaguchi, S., Kyozuka, Y., Recent advances in strigolactone research: chemical and biological aspects. *PCP*, 53, 11, 1843–1853, 2012.
12. Kapulnik, Y., Delaux, P.-M., Resnick, N., Mayzlish-Gati, E., Wininger, S., Bhattacharya, C., Sejalon-Delmas, N., Combier, J.-P., Becard, G., Belausov, E., Beeckman, T., Dor, E., Hershenhorn, J., Koltai, H., Strigolactones affect lateral root formation and root-hair elongation in Arabidopsis. *Planta*, 233, 1, 209–216, 2011a.
13. Umehara, M., Hanada, A., Yoshida, S., Akiyama, K., Arite, T., TakedaKamiya, N. *et al.*, Inhibition of shoot branching by new terpenoid plant hormones. *Nature*, 455, 195–200, 2008.
14. MacAlpine, G.A., Raphael, R.A., Shaw, A., Taylor, A.W., Wild, H.-J., Synthesis of the germination stimulant (±)-strigol. *J. Chem. Soc., Chem. Commun.*, 20, 834–835, 1974.
15. Akiyama, K., Ogasawara, S., Ito, S., Hayashi, H., Structural requirements of strigolactones for hyphal branching in AM fungi. *PCP*, 51, 7, 1104–1117, 2010.
16. Lachia, M., Dakas, P.-Y., De Mesmaeker, A., Asymmetric synthesis of the four stereoisomers of 5-deoxystrigol. *Tetrahedr. Lett.*, 55, 48, 6577–6581, 2014.
17. Cavar, S., Zwanenburg, B., Tarkowski, P., Strigolactones: occurrence, structure, and biological activity in the rhizosphere. *Phytochem. Rev.*, 14, 691–711, 2015.
18. Matsui, J., Bando, M., Kido, M., Takeuchi, Y., Mori, K., Plant regulators 2: Synthesis of (±)- and (+)-sorgolactone, the germination stimulant from Sorghum bicolor. *Eur. J. Org. Chem.*, 1999, 2183–2194, 1999.
19. Yokota, T., Sakai, H., Okuno, K., Yoneyama, K., Takeuchi, Y., Alectrol and orobanchol, germination stimulants for *Orobanche minor*, from its host red clover. *Phytochem.*, 49, 7, 1967–1973, 1998.

20. Ueno, K., Nomura, S., Muranaka, S., Mizutani, M., Takikawa, H., Sugimoto, Y., Ent-2′ -epi-Orobanchol and its acetate, as germination stimulants for *Striga gesnerioides* seeds isolated from cowpea and red clover. *J. Agric. Food. Chem.*, 59, 19, 10 485–10 490, 2011b.
21. Matsui, J., Yokota, T., Bando, M., Takeuchi, J., Mori, K., Synthesis and structure of orobanchol, the germination stimulant for Orobanche minor. *Eur. J. Org. Chem.*, 1999, 9, 2201–2210, 1999.
22. Tanaka, M., Sugimoto, Y., Kuse, M., Takikawa, H., Synthesis of 7-oxo-5-deoxystrigol, a 7-oxygenated strigolactone analog. *Biosci. Biotechnol., Biochem.*, 77, 832–835, 2013.
23. Cook, C.E., Whichard, L.P., Wall, M.E., Egley, G.H., Coggon, P., Luhan, P.A. *et al.*, Germination stimulants. II. The structure of strigol – a potent seed germination stimulant for witchweed (*Striga lutea* Lour.). *J. Am. Chem. Soc.*, 94, 6198–6199, 1972.
24. Siame, B.P., Weerasuriya, Y., Wood, K., Ejeta, G., Butler, L.G., Isolation of strigol, a germination stimulant for *Striga asiatica*, from host plants. *J. Agric. Food. Chem.*, 41, 9, 1486–1491, 1993.
25. Matsuura, H., Ohashi, K., Sasako, H., Tagawa, N., Takano, Y., Ioka, Y. *et al.*, Germination stimulant from root exudates of *Vigna unguiculata. Plant Growth Regul.*, 54, 31–36, 2008.
26. Xie, X., Yoneyama, K., Kusumoto, D., Yamada, Y., Takeuchi, Y., Sugimoto, Y. *et al.*, Sorgomol, a germination stimulant for root parasitic plants, is produced by Sorghum bicolor. *Tetrahedron Lett.*, 49, 2066–2068, 2008.
27. Xie, X., Yoneyama, K., Kusumoto, D., Yamada, Y., Yokota, T., Takeuchi, Y. *et al.*, Isolation and identification of alectrol as (+)-orobanchyl acetate, a novel germination stimulant for root parasitic plants. *Phytochem.*, 69, 427–431, 2008.
28. Rani, K., Zwanenburg, B., Sugimoto, Y., Yoneyama, K., Bouwmeester, H.J., Biosynthetic considerations could assist the structure elucidation of host plant-produced rhizosphere signalling compounds (strigolactones) for arbuscular mycorrhizal fungi and parasitic plants. *Plant Physiol. Biochem.*, 46, 7, 617–626, 2008.
29. Xie, X., Yoneyama, K., Kurita, J.Y., Harada, Y., Yamada, Y., Takeuchi, Y., Yoneyama, K., 7-Oxoorobanchyl acetate and 7-oxoorobanchol as germination stimulants for root parasitic plants from flax (*Linum usitatissimum*). *BBBIEJ*, *73*, 6, 1367–1370, 2009.
30. Mangnus, E.M., Van Vliet, L.A., Vandenput, D.A.L., Zwanenburg, B., Structural modifications of strigol analogs. Influence of the B and C rings on the bioactivity of the germination stimulant GR24. *J. Agric. Food. Chem.*, 40, 7, 1222–1229, 1992.
31. Xie, X., Kusumoto, D., Takeuchi, Y., Yoneyama, K., Yamada, Y., Yoneyama, K., Epi-orobanchol and solanacol, two unique strigolactones, germination stimulants for root parasitic weeds, produced by tobacco. *J. Agric. Food. Chem.*, 55, 20, 8067–8072, 2007.

32. Bouwmeester, H.J., Matusova, H., Sun, Z.K., Beale, M.H., Secondary metabolite signaling in host–parasitic plant interactions. *Curr. Opin. Plant Biol.*, 6, 4, 358–364, 2003.
33. Hristeva, T., Dekalska, T., Denev, I., Structural and functional biodiversity of microbial communities in the rhizosphere of plants infected with broomrapes (Orobanchaceae). *Biotechnol. Biotechnol. Equip.*, 27, 5, 4082–4086, 2013.
34. Rasmussen, A., Depuydt, S., Goormachtig, S., Geelen, D., Strigolactones fine-tune the root system. *Planta*, 238, 4, 615–626, 2013.
35. Parker, C., Observations on the current status of Orobanche and Striga problems worldwide. *Pest Manag. Sci.*, 65, 5, 453–459, 2009.
36. Parker, C., Parasitic weeds: a world challenge. *Weed Sci.*, 60, 2, 269–276, 2012.
37. Yoneyama, K., Awad, A.A., Xie, X., Yoneyama, K., Takeuchi, Y., Strigolactones as germination stimulants for root parasitic plants. *PCP*, 51, 7, 1095–1103, 2010.
38. Grenz, J.H. and Sauerborn, J., Mechanisms limiting the geographical range of the parasitic weed Orobanche crenata. *Agric. Ecosys. Environ.*, 122, 3, 275–281, 2007.
39. Kgosi, R.L., Zwanenburg, B., Mwakaboko, A.S., Murdoch, A.J., Strigolactone analogues induce suicidal seed germination of Striga spp. in soil. *Weed Res.*, 52, 3, 197–203, 2012.
40. Bebawi, F.F., Eplee, R.E., Harris, C.E., Norris, R.S., Longevity of witchweed (Striga asiatica) seed. *Weed Sci.*, 32, 4, 494–497, 1984.
41. Cardoso, C., Ruyter-Spira, C., Bouwmeester, H.J., Strigolactones and root infestation by plant-parasitic Striga, Orobanche and Phelipanche spp. *Plant Sci.*, 180, 3, 414–420, 2011.
42. Joel, D.M., Gressel, J., Musselman, L.J., Parasitic Orobanchaceae. *Parasitic mechanisms and control strategies*, Springer Berlin, Heidelberg, 978-3-642-38146-1, 2013.
43. Kim, H.I., Xie, X.N., Kim, H.S., Chun, J.C., Yoneyama, K., Nomura, T., Takeuchi, Y., Yoneyama, K., Structure-activity relationship of naturally occurring strigolactones in Orobanche minor seed germination stimulation. *J. Pest Sci.*, 35, 3, 344–347, 2010.
44. Nomura, S., Nakashima, H., Mizutani, M., Takikawa, H., Sugimoto, Y., Structural requirements of strigolactones for germination induction and inhibition of *Striga gesnerioides* seeds. *Plant Cell Rep.*, 32, 6, 829–838, 2013.
45. Yoneyama, K., Xie, X., Yoneyama, K., Takeuchi, Y., Strigolactones: structures and biological activities. *Pest Manag. Sci.*, 65, 5, 447–470, 2009.
46. Matusova, R., van Mourik, T., Bouwmeester, H.J., Changes in the sensitivity of parasitic weed seeds to germination stimulants. *Seed Sci. Res.*, 14, 4, 335–344, 2005.
47. Song, W.J., Zhou, W.J., Jin, Z.L., Cao, D.D., Joel, D.M., Takeuchi, Y., Yoneyama, K., Germination response of Orobanche seeds subjected to conditioning

temperature, water potential and growth regulator treatments. *Weed Res.*, 45, 6, 467–476, 2005.
48. Akiyama, K., Matsuzaki, K., Hayashi, H., Plant sesquiterpenes induce hyphal branching in arbuscular mycorrhizal fungi. *Nature*, 435, 824–827, 2005.
49. Bever, J.D., Schultz, P.A., Pringle, A., Morton, J.B., Arbuscular mycorrhizal fungi: fore diverse than meets the eye, and the ecological tale of why. *BioScience*, 51, 11, 923–932, 2001.
50. Smith, S.E., Jakobsen, I., Grønlund, M., Smith, F.A., Roles of arbuscular mycorrhizas in plant phosphorus nutrition: interactions between pathways of phosphorus uptake in arbuscular mycorrhizal roots have important implications for understanding and manipulating plant phosphorus acquisition. *Plant Physiol.*, 156, 3, 1050–1057, 2011.
51. Quilambo, O.A., The vesicular-arbuscular mycorrhizal symbiosis. *Afr. J. Biotechnol.*, *2*, 12, 539–546, 2003.
52. Yoshida, S., Kameoka, H., Tempo, M., Akiyama, K., Umehara, M., Yamaguchi, S., Hayashi, H., Kyozuka, J., Shirasu, K., The D3 F-box protein is a key component in host strigolactone responses essential for arbuscular mycorrhizal symbiosis. *New Phytol.*, 196, 4, 1208–1216, 2012.
53. Foo, E., Ross, J.J., Jones, W.T., Reid, J.B., Plant hormones in arbuscular mycorrhizal symbioses: an emerging role for gibberellins. *Ann. Bot.*, 111, 5, 769–779, 2013.
54. Gomez-Roldan, V., Fermas, S., Brewer, P.B., Puech-Pages, V., Dun, E.A., Pillot, J.-P. *et al.*, Strigolactone inhibition of shoot branching. *Nature*, 455, 189–194, 2008.

# 10

# Karrikin-Related Effects on Plant Development, Stress Tolerance, and Beyond

**Tamás Bodor[1,2], Gábor Fejes[1,2], Dóra Kondak[1], Selahattin Kondak[1,2], Réka Szőllősi[1] and Zsuzsanna Kolbert[1*]**

*[1]Department of Plant Biology, University of Szeged, Szeged, Hungary*
*[2]Doctoral School of Biology, Faculty of Science and Informatics, University of Szeged, Szeged, Hungary*

***Abstract***
Karrikins (KARs) are butenolide compounds formed during the combustion of cellulose and found in smoke. KARs share signal pathway with strigolactones, and plant species are able to respond to the presence of KARs by improved germination, seedling growth, and stress tolerance. According to the available literature, KARs have the potential to intensify plant tolerance against sub- or supraoptimal temperature, drought, heavy metal exposure, salinity, and bacterial pathogens. Beyond plants, some effects of KARs have been revealed on plants' ecological partners like arbuscular mycorrhiza fungi. Smoke water containing KARs has promising applications in plant cultivation due to its stimulating effects on plant growth and development. This chapter aims to collect and discuss the recent literature about the direct effects of KARs and smoke water on plant development and stress resilience and also about their indirect effects *via* the modulation of the ecological partners of plants. Future research directions for better exploring KARs' (phyto) effects are also suggested.

***Keywords***: Abiotic stress, arbuscular mycorrhiza, biotic stress, karrikins, plant development, smoke water

**Corresponding author*: ordogne.kolbert.zsuzsanna@szte.hu

Tariq Aftab and Kaiser Iqbal Wani (eds.) Strigolactones: Emerging Plant Hormones, (211–234) © 2025 Scrivener Publishing LLC

## 10.1 Introduction

The first research articles dealing with the post-fire effects on plants were published in the 1980s. These reports demonstrated that fire which naturally occurs in the Californian chaparrals, in the flora of Cape, in the vegetation of Western Australia, or in Mediterranean regions of Europe is deleterious for all plant species [1–4]. Indeed numerous species require fire and/or plant-derived smoke to develop successfully [5]. It is well known that certain plants with thick and tough seed coat or dry fruits possessing sclerenchymatous pericarp show higher germination rate after heat due to the breaking of physical seed dormancy [4]. Based on the experimental findings of De Lange and Boucher [6] about the positive effects of smoke and smoke water on seed germination, researchers of the 1990–2004 era focused on the beneficial influences of smoke water on germinative processes including breaking seed dormancy [7, 8]. In addition, commercially distributed liquid smoke proved to be promotive in the germination of red rice [9].

In the first years of the 1990s, several attitudes of the periodic fires were emphasized, such as weakening the hard seed coat or promoting the embryonic development [2]. It was also announced that ethylene (ET) and ammonia found in the smoke play a stimulative role in seed germination. Furthermore, plant-derived smoke or smoke extract proved to induce seed germination in case of several fynbos species without knowing the exact chemical composition of the smoke [2]. Baxter *et al.* [7] found in the case of *Themeda triandra* (grass family) that ET is not an active component of plant-derived smoke, but the end product of hemicellulose or cellulose exposed to heat (fire) is potentially responsible for better germination even at a suboptimal temperature. Later, only a few experimental data supported that not only the direct effects of heat but also the chemical products of fire (smoke) might be responsible for better germination rate depending on the seed type, the duration, and the form of smoke application [3].

In the above-mentioned era (1990–2004), many efforts were made to identify the main and most effective component(s) of plant-derived smoke and smoke water. It was recognized quite early by Keeley and Pizzorno [1] that this "stimulatory compound" derived from charred wood is water soluble, and it is produced at 160–200°C and becomes volatile above 200°C (reviewed by Van Staden *et al.* [8]). Based on the findings of Jäger *et al.* [10] and Keeley and Bond [11], the similarity of the active components of charred wood and smoke seemed to be probable.

The bioactive compound(s) of the smoke strongly absorb to the surface of soil particles or seeds and remain stable for years, especially in solved form [8]. Compared to the first extraction and characterization of strigolactones (SLs) in 1966 [12], smoke components were identified much later. Finally, in 2004, Flematti *et al.* [13] presented the structure of the compound which triggers seed germination of fire-dependent species.

This butenolide-type substance was 3-methyl-2H furo[2,3-c] pyran-2-one (Figure 10.1), which is the derivate of cellulose combustion, and therefore the active component of smoke water as well. Its common name is karrikinolide ($KAR_1$), and it has got five derivates methylated at various positions ($KAR_{2-6}$). All these together are the members of so-called *karrikins* (KARs) [14–16]. In the chemical sense, stigolactones are terpenoid lactones, and the currently known at least 20 SLs are synthetized from carotenoids in the plastids with the involvement of multiple but only partly known enzymes [17, 18]. In contrast, to date, there is no evidence for the existence of endogenous KARs in plants since the apocarotenoid origin of KARs has not been proven, and the biosynthetic routes have not been uncovered [19].

Beyond the identification of $KAR_1$, [20] also tested the effects of different $KAR_1$-derivates substituted at various positions on seed germination. It was assessed that the methyl substituent at C-3 found in $KAR_{1,3-4}$ is responsible for the germination-inducing action. Kulkarni *et al.* [21] investigated the potentially positive influences of smoke water and isolated $KAR_1$ on tomato plants at both vegetative and reproductive stages. They found that plants irrigated with smoke water (in 1:500 dilution, v:v) were much taller

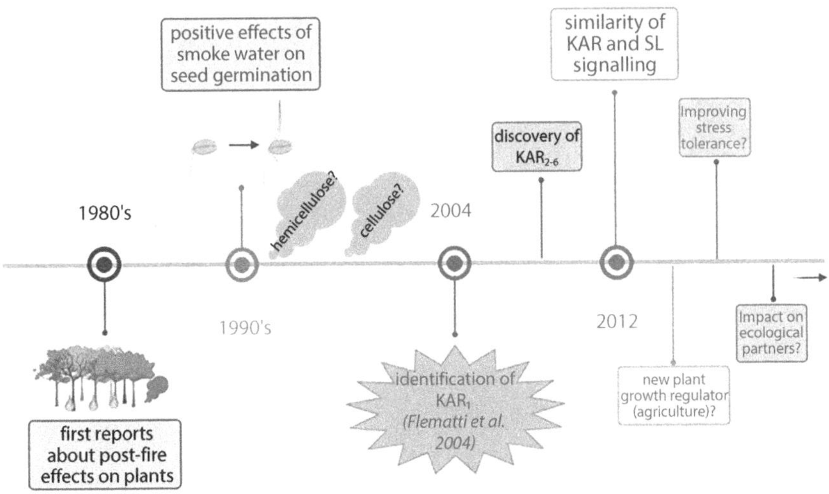


**Figure 10.1** Milestone discoveries about karrikins (KARs) (see details in the text). SL, strigolactones.

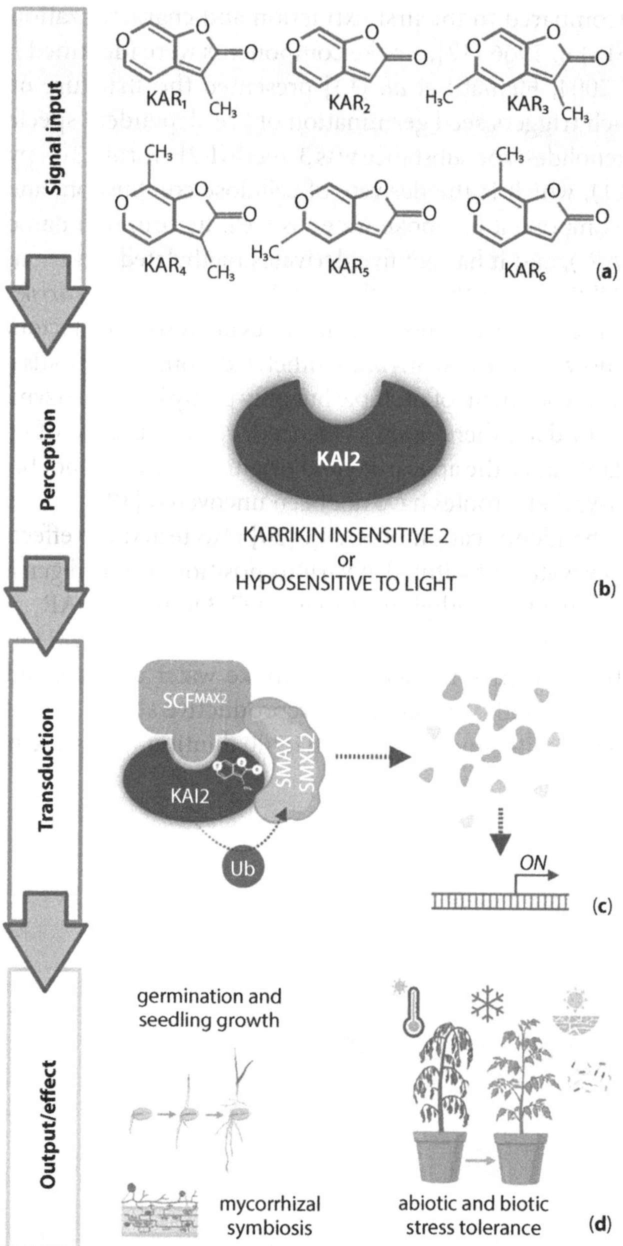


**Figure 10.2** Signal transduction of karrikins in plants. (a) Chemical structures of the $KAR_{1-6}$ identified in smoke water, (b) KAI2 receptor perceives KARs, (c) signal transduction involves ubiquitination of SMAX1/SMXL2 repressor, and (d) transcriptional changes lead to the following outputs/effects: induced seed germination and seedling growth, abiotic stress tolerance, and mycorrhizal symbiosis.

than control plants and yielded significantly larger-sized fruits. Plants treated with isolated butenolide ($KAR_1$) also produced more fruits, but it was not so remarkable. These and further experimental data (summarized by Kulkarni *et al.* [22]) suggested that smoke water could be beneficially used at low concentrations ($10^{-4}$–$10^{-9}$ M) as a new exogenous plant growth regulator in agricultural systems (Figure 10.1).

There are common components in the signal pathways of the terpenoid lactone SLs produced endogenously in plants and the smoke component butenolide KARs sensed by plants. The perception of SLs involves the DWARF14 (D14) protein having α/β fold hydrolase activity, while KARs are perceived by KARRIKIN INSENSITIVE2 (KAI2) also with α/β-hydrolase activity [23, 24]. Due to the restricted ecological frequency of fire, the primary role of KAI2 might not be to perceive KARs but to sense and respond to one or more unknown phytohormone-like metabolites, collectively termed "KAI2 ligand" (KL) [19]. The perceptions of SLs and KL are similar since in both cases the receptor (D14/KAI2) proteins suffer conformational change as the effect of ligand binding which in turn facilitates their interaction with MAX2. The common component in the SL and KAR signal transductions is MAX2 which is the F-box subunit of the SCF class of the E3 ubiquitin–protein ligase complex. Additionally, the receptor and/or MAX2 interact with one or more repressors from the SMXL family triggering their degradation. These events lead to alterations in the transcription of target genes the products of which participate in seed germination, hypocotyl elongation, development of root system architecture, defense responses against abiotic and biotic stressors, and the establishment of mycorrhizal symbiosis (Figure 10.2) [25].

## 10.2 Direct KAR-Associated Effects on Plants

### 10.2.1 Promotion of Seed Germination and Seedling Growth in Healthy Plants

It has been known in agriculture for ages that fire has an effect on vegetation. This knowledge based on observation has been often used by farmers. Researchers became curious about the scientific background of this phenomenon; thus, they conducted various experiments to study the effects of fire on several plant species—for instance, Walters *et al.* [26] found that fire has a role in inducing germination of some *Acacia* species, but it is not the sole driver of the process. Later, it seemed more promising not to focus on the fire itself but on the smoke and dust that it leaves behind. One of the most intriguing compound in plant-derived smoke is KAR that has a significant impact on seed germination.

According to the early results of Nelson *et al.* [15], KAR has a positive impact on breaking *Arabidopsis thaliana* seed dormancy even in low concentrations. However, KAR has a structure that no other hormone has; it has a ring in its chemical structure, which is similar to the ring of SLs. To evaluate whether the effects of the two compounds are similar, the impacts of $KAR_1$, $KAR_2$, $KAR_3$, and $KAR_4$ to the synthetic SL, GR24, on germination were compared. The results showed that $KAR_2$ is the most effective in germination induction, but the other types of KARs have a similar impactful effect. GR24 was also effective, but at 100 times higher concentration than the weakest KAR [15]. In general, we can state that KARs are effective in breaking seed dormancy and germination induction not just in *Arabidopsis thaliana*, but also, e.g., in *Lactuca sativa*, *Conostylis aculeate*, *Stylidium affine*, *Syncarpha vestita*, *Emmenanthe penduliflora*, *Nicotiana attenuate*, *Prunus armeniaca*, *Onopordum caricum*, *Arcopoterium spinosum*, and *Stachys cretica* [13, 19, 27–29].

Seed germination is regulated by a delicate balance maintained by germination-promoting gibberellic acids (GAs) and seed dormancy promoting abscisic acid (ABA). The seed germination promoting effect of KARs seem to manifest through the interplay with both of the germination regulating phytohormones. The signal transduced by KAI2 during *Arabidopsis* seed germination is aimed at the induction of WRKY33, which, in turn, induces CYP707A ABA catabolic enzyme, thus terminating ABA-associated dormancy [30]. On the other hand, KAR treatment has been found to require the biosynthesis of the germination inductor GA although $KAR_1$ is able to stimulate the germination of *sleepy* mutant deficient in GA signaling, reflecting the partial dependence of KAR effect on GA signaling [15]. Recently, Bunsick *et al.* [31] revealed that *smax1* seeds are capable of germinating in the presence of inhibitors of GA biosynthesis, which indicates the requirement of SMAX1 for germination limitation by the repressors of GA signaling.

Mutant analyses revealed an elongated hypocotyl and a shorter cotyledon size of *Arabidopsis kai2* mutant compared to the wild type (WT). However, *smax1/smxl2* double mutant shows a shorter hypocotyl and larger cotyledons compared to WT. The effects of the mutations are similar to the effects of KAR treatment of the WT [32]. Both the WT and *hy5* seedlings showed the same level of *KAI2* transcript accumulation when grown in continuous light or dark, suggesting that HY5 is needed for KAI2 induction only in light-to-dark transition [33]. Moreover, in *hy5* mutants, the hypocotyl response to KAR treatment is slight; however, mutations in *KAI2* and *MAX2* have an additive effect to the mutation in *HY5* [33]. These imply that the HY5-associated and the KAI2-associated signal pathways regulating hypocotyl elongation are parallel but mostly independent. Furthermore, Bursch *et al.* [34] revealed that BBX zinc finger proteins (known to interact with HY5 and promote light-responsive growth) contribute to KAI2-triggered shortening of hypocotyls but are not essential for this process. These results point in the direction that HY5 may be required to some extent for hypocotyl size reduction mediated by KAI2.

During the shift from darkness to light, there is an increase in SMAX1 accumulation in *Arabidopsis* seedlings. Conversely, in the absence of light, SMAX1 levels are minimal and diminish over time [35]. This observation implies that KAI2 signaling more significantly influences hypocotyl growth under illuminated conditions compared to darkness. In light, SMAX1 prohibits the accumulation of DELLA proteins in cell nuclei, which causes hypocotyl elongation. Supplementation of KAR leads to the accumulation of DELLA proteins which results in hypocotyl growth inhibition [35]. Another interesting effect shows up in *Arabidopsis* seedlings when transitioning from darkness to light happens. Hamon-Josse *et al.* [36] revealed that KAI2 triggers a diminution in PIN7 abundance in older tissues and induces an elevation of PIN1/PIN2 abundance in the root meristem. Furthermore, removing PIN3, PIN4, and PIN7 from *kai2* mutants or the pharmacological inhibition of auxin transport and synthesis repress the majority of *kai2* phenotypes. The results indicate that KAI2 is required for the appropriate changes in PIN protein abundance during photomorphogenesis. According to the results of Xu *et al.* [37], KAR signaling is involved in the regulation of shade avoidance by affecting auxin homeostasis in the hypocotyl. This is supported by the fact that SMAX1 is required by KAI2 and MAX2 for the limitation of hypocotyl growth associated with shade avoidance. Inhibition of auxin accumulation by l-kynurenine leads to the partial restoration of hypocotyl elongation in *kai2* mutant under simulated shade condition. Moreover, based on RNA-seq data, SMAX1-overexpression modulated the expression of various auxin homeostasis genes, especially under simulated shade.

In the work of Waters *et al.* [24], the phenotypes of *d14*, *kai2*, and *smax2* mutants were compared to each other, and it was found that *d14* and *smax2-1* exhibit similarly reduced petiole length, and both mutants are characterized by wider and more deeply lobed leaf shapes. In comparison, the leaves of *kai2* are elongated and have more curled margins. At the flowering stage, *d14* plants exhibit an increased branching phenotype, similar to the *max2* mutants. However, the *kai2* mutants display a branching pattern that is indistinguishable from the wild type. These observations suggest that KAI2 is not essential for branching control, and KARs cannot take over SLs' role in axillary bud growth [24]. In the *kai2* mutant of the model grass *Brachypodium distachyon*, the internode elongation increased, and the chlorophyll levels are decreased compared to the WT. Furthermore, *Bdkai2* shows intensified lateral root induction but inhibited the root hair growth and does not effectively facilitate the usual colonization of its roots by arbuscular mycorrhizal fungi [38].

KAI2-associated KAR signaling has been shown to greatly impact the root system. KARs modulate the primary root elongation, and the development of lateral roots and root hairs and even root skewing and waving. In *Arabidopsis kai2* mutants, the root hair length and density are decreased and this shows similarities with the previously mentioned *Brachypodium distachyon* [38, 39]. In *Lotus japonicus*, the protein SMAX1 plays a negative regulatory role, since it suppresses the expression of *1-AMINOCYCLOPROPANE CARBOXYLIC ACID (ACC) SYNTHASE 7 (ACS7)*, which is a key enzyme involved in ET biosynthesis. By doing so, SMAX1 fine-tunes the architecture of the root system. The *smax1* mutants have increased ET level and *ACS7* expression compared to WT, which lead to longer root hairs and suppressed length of primary roots. These phenotypes can be rescued by pharmacological modulation of ET levels and signaling [40]. Also in *Arabidopsis*, ACC supplementation complemented the short root hair phenotype of *kai2* and *acs7* mutants [36]. It was also evidenced that functional KAI2 is necessary for the optimization of the auxin level in epidermal cells, resulting in the promotion of root hair growth. In more detail, KAI2 contributes to the accumulation of AUXIN TRANSPORTER PROTEIN 1 (AUX1) and PIN-FORMED2 (PIN2) in different root tissues, and these alterations in auxin transport leads to root hair formation. Thus, AUX1 and PIN2 are downstream elements of KAR signaling during root hair development [36]. KAI2 and D14 are equally important in the regulation of the root system and seem to additively regulate the properties of the root system. In *Brachypodium distachyon*, a mutation in KAI2 results in a significant augmentation of lateral roots [38]. It is important to emphasize the connection between D14 and KAI2 as both signals triggering the degradation of SMXL2 repressor. As was mentioned before, SMXL2 has a crucial role in hypocotyl growth processes. This discovery reveals a significant convergent pathway in gene expression and hypocotyl elongation, regulated by both SLs and KARs, and at the same time reveals a new field of research [41].

### 10.2.2 Amelioration of Abiotic Stress-Related Damages

Plants are continually exposed to various stress factors during their growth, but unlike animals, plants cannot change location to defend themselves. Nowadays, there are several definitions of plant stress, but its common knowledge that plants need to exert physiological, biochemical, and molecular changes to avoid the damage to their growth, development, and productivity. Depending on the source of the stress, biotic (e.g., bacterial or fungal diseases) and abiotic factors (e.g., heat or salinity) stresses exist. Butenolide molecules such as KARs and SLs are reported not only to stimulate seed germination and regulate seedling growth but also to increase the seedling vigor during biotic and abiotic stress conditions. Two classical approaches have been applied for revealing stress amelioration of KARs: exogenous administration of KARs to stress-affected plants and evaluation and comparison stress-induced phenotypical, biochemical, and molecular alterations in KAR signaling mutants and the WT.

#### *10.2.2.1 Drought Stress*

Drought or water deficit is one of the major abiotic stressors that reduce plant yield *via* various types of damages to cellular, physiological, and biochemical processes [42]. Among the disadvantageous effects of water deficit, the increased seed germination time and delayed plant production can be emphasized [43, 44].

In one of the first studies, the drought tolerance/sensitivity of *kai2 Arabidopsis* mutant was evaluated [45]. The authors found that mutations in KAI2 result in hypersensitivity to water deprivation. As a result of drought stress, the mutant shows intensified water loss and drought-triggered cell damage due to more opened stomata and increased cuticular permeability. Moreover, drought-affected *kai2* plants have reduced anthocyanin biosynthesis and showed hyposensitivity to ABA regarding stomatal closure and cotyledon opening compared to WT. The comparison of SL defective *d14* and the *kai2* plants revealed that both butenolides contribute to plant drought adaptation; however, SLs do not affect the development of cuticle [45]. The germination of *kai2* seeds proved to be more sensitive to inhibition by mannitol-induced osmotic stress. Under mild osmotic stress, KAR2 treatment intensified the expression of drought-associated *DREB2A*, *WRKY33*, and *ERF5* genes, but not ABA signaling genes. These results support that the KAR–KAI2 signaling system can provide protection against abiotic stress [46].

In the in-depth analysis of Li *et al.* [47], various drought resistance-associated mechanisms of *d14* and *kai2* and the double-mutant *d14kai2* were compared. The results pointed out that functional KAI2 is more relevant in the maintenance of cell membrane integrity, cuticle structure, and ABA-induced leaf senescence but has a slighter role in drought-induced anthocyanin biosynthesis relative to the SL-associated D14. Furthermore, additive effects of D14 and KAI2 have been evidenced regarding cell membrane integrity and leaf cuticle development in drought-stressed plants.

Recently, another element has been added to the molecular map of KAR effects. The mutation in *KAR UPREGULATED F-BOX 1* (*KUF1*) enhanced drought tolerance in *Arabidopsis* [48]. It was found that *kuf1* plants shows less water loss due to reduced stomatal aperture and cuticular permeability. In addition, *kuf1* plants have increased sensitivity to ABA and have higher root/shoot ratios and root hair density compared to WT, reflecting that *KUF* mutation results in improved water absorption. Overall, KUF1 negatively regulates drought tolerance by modulating various morphological and physiological processes and ABA responses [48].

As for the pharmacological approach, a few studies have been conducted so far. Using *Sapium sebiferum* oil plant, $KAR_1$ treatment at a nanomolar concentration improves biomass production and upregulates antioxidative enzymes, such as superoxide dismutase (SOD), peroxidase (POD), catalase (CAT), and ascorbate peroxidase (APX), and lowers hydrogen peroxide ($H_2O_2$) levels and membrane damage [49]. Additionally, $KAR_1$ administration significantly increases organic acid and amino acid contents and intensifies the expression of several ABA signaling genes, such as *SNF1-RELATED PROTEIN KINASE2.3*, *SNF1-RELATED PROTEIN KINASE2.6*, *ABI3*, and *ABI5*, implying potential interactions between KARs and ABA signaling during drought stress responses [49].

Another study aimed to observe the drought stress-ameliorating effect of exogenous $KAR_1$ application on creeping bentgrass (*Agrostis stolonifera* cv. PennA4) [50]. All the measured parameters (relative water content, electrolyte leakage, chlorophyll content, and photosystem II effectiveness of leaves) in drought-stressed creeping bentgrass are significantly increased by exogenous $KAR_1$ (100 nM) compared to water treatment. Additionally, $KAR_1$ increases the activity of several participants of the enzymatic antioxidant defense such as POD, SOD, and APX. The $KAR_1$-induced beneficial effects on drought endurance may be due to the upregulation of KAR-responsive genes (*KAI2*, *MAX2*, and *AFL1*), water stress-associated transcription factors (*ABF3*, *bHLH148*, *MYB13*, and *DREB2A*), antioxidant genes (*Cu/Zn-SOD*, *APX2*, *CAT1*, and *POD2*), and downregulation of the expression of chlorophyll-degradation genes (*PPH* and *Chl-PRX*) [50].

#### 10.2.2.2 Temperature Stresses

Temperature stress in plants has three types, including heat, chilling, and freezing. Plants affected by high or low temperatures show low germination rates, limited growth, and inhibited photosynthesis. The responses of plants to temperature stress include alterations of membrane lipid composition, upregulation of stress-related transcription factors, metabolite synthesis, and detoxification mechanisms [51]. The role of temperature is extraordinary in the case of plants because it determines the distribution of plant species on the planet since each species has an optimal temperature ensuring growth [52].

*Eragrostis tef* is a C4 annual graminaceous species between tropical and temperate climate conditions. The seeds of this species were treated with smoke water or $KAR_1$ solution ($10^{-8}$ M) and studied under high temperature and low osmotic potentials [53]. In the case of increased temperature, both treatments enhance the percentage of germination, and there is a significant increase in seedling lengths as the effect of smoke water and $KAR_1$ treatments at all the studied temperatures (especially at 30/15°C alternating temperature). Similar to this, priming *Brassica napus* seeds with $KAR_1$ or smoke water extended the germination range from 10–35°C to 40°C, suggesting the beneficial effect of KARs under heat stress [54]. The exogenously applied $KAR_1$ efficiently improves the tolerance of *Brassica alboglabra* against multiple stressors, including Cd, heat, and persistent organic pollutant by promoting enzymatic antioxidant defense, increasing membrane stability, and the removal of pollutants [55].

Beyond exogenous KAR treatments, the genetic approach has also been applied in order to reveal the involvement and signaling of KARs under heat stress. Wang *et al.* [46] found that the germination of *kai2* seeds is more sensitive to elevated temperature than that of the WT, which implies the role of KAI2-associated signaling in the heat stress response of *Arabidopsis*. In order to reveal the mechanism of KAI2 action in heat tolerance, experiments at the level of phenotype, physiology, and gene expression were performed in *kai2* mutant and WT *Arabidopsis* under control and heat stress conditions [56]. The *kai2* mutant was found to be more sensitive to high-temperature stress than the WT due to its increased membrane damage and leaf surface temperature. The increased heat sensitivity of the *kai2* mutant is associated with the downregulation of several HSP- and HSF-encoding genes (those are involved in many heat stress-responsive pathways, e.g., heat shock protein binding and protein folding chaperone) compared to WT plants under both control and heat stress conditions. There are additional stress-responsive genes (WRKY- and NAC-encoding genes) which are downregulated by heat stress in the *kai2* mutant [56].

Low temperature is a major environmental condition that seriously impacts on plant development and affects crop productivity. Plants have evolved a series of physiological and molecular mechanisms ensuring acclimation to cold stress [57].

It has been found recently that KAI2 is not only involved in seed germination, seedling development, and alleviation of heat-associated damages but can also regulate responses to freezing as well [58]. The *kai2 Arabidopsis* mutant shows increased mortality as the effect of -20°C treatment, while *SsKAI2* overexpressing (OE) *Sapium sebiferum KAI2* has improved recovery after freezing [58]. The OE plants contain a higher amount of proline, soluble sugars, and soluble protein, show an increased expression of cold stress-related genes, and are hypersensitive to ABA during cold stress. Additionally, cold stress affects ABA signaling genes in *SsKAI2* OEs, suggesting a potential interaction between *SsKAI2*- and ABA-associated signaling cascades. The freezing-triggered electrolyte leakage, $H_2O_2$, and malondialdehyde contents are lower in the *SsKAI2* OE line compared to WT due to increased antioxidant activities like that of APX, CAT, PD, and SOD and antioxidant contents such as total glutathione level [58].

### *10.2.2.3 Salinity*

The minerals and nutrients in the soil are crucial for plants' life, although the unnecessary presence of soluble salts could cause osmotic and ionic stress *via* reducing the plants' water uptake capacity.

In the case of *Sapium sebiferum*, a nanomolar concentration of $KAR_1$ causes significantly increased recovery in seed germination under salt and osmotic stress conditions (sodium chloride and mannitol, respectively) compared to the untreated seeds [49]. Moreover, $KAR_1$ increases seedling biomass production, the taproot length, and the lateral root number under salt stress. The $KAR_1$-treated, salt-exposed seedlings possess promoted activity of antioxidative enzymes, such as SOD, POD, CAT, and APX, compared to the plants exposed to salinity alone. $KAR_1$ treatment results in a lower level of $H_2O_2$, malondialdehyde, and electrolyte leakage. Additionally, $KAR_1$ administration notably increases the level of organic and amino acids of salt-exposed plants. Interestingly, under salt stress, $KAR_1$ causes no increment of ABA level but modifies the expression of various ABA signaling genes, suggesting potential interactions between KARs and ABA signaling [49].

As for the genetic approach, the salt responses of *KAI2*-deficient line were compared to those of the WT, and it was revealed that *KAI2* mutation results in delayed and inhibited cotyledon opening, phenotypic aberrations, biomass reduction, intensified water loss, and oxidative damage under salt stress [59]. The lower $K^+/Na^+$ ratio in *kai2* seedlings compared to the WT implies severe ion toxicity. Accordingly, in *kai2* plants, lower expression levels of genes associated with $Na^+$ homeostasis have been detected. In *kai2 Arabidopsis*, salt stress leads to lower glutathione level and antioxidant enzyme activities than in the WT. Furthermore, *KAI2* deficiency resulted in a lower expression of genes involved in the biosynthesis of SLs, salicylic acid, and jasmonic acid and the signaling of ABA and SLs than those of WT plants, indicating an interplay between KAI2 signaling with other hormone signaling pathways in modulating plant responses to salinity.

#### *10.2.2.4 Heavy Metals*

Heavy metals can be accumulated in soil due to anthropogenic activities such as mining, exhaust emissions, sewage irrigation, and the use of metal products. Heavy metal exposure of agrosystems has increased during the last decades globally [60]. Among HMs, cadmium (Cd) is a harmful metal that causes physiological, structural, and metabolic disturbances due to transcriptional changes in plants [61].

The responses against Cd stress and combined stresses have been studied in *Brassica alboglabra* by [55] who found that the administration of $KAR_1$ ameliorates the deleterious effects of Cd by reducing *in planta* Cd contents and improving photosynthetic parameters, antioxidant defense mechanisms, proline, and beta-carotin contents as well as growth attributes. In the case of *Coriandrum sativum*, seed priming with $KAR_1$ ($10^{-6}$ M) results in promoted germination rate and subsequent improvement of photosynthesis and biomass production of Cd-exposed seedlings [62]. Moreover, Cd+$KAR_1$-treated plants have higher amount of proline, higher relative water content, leaf osmotic potential, improved gas exchange parameters, and better membrane stability than plants treated with Cd alone. Additionally, $KAR_1$ treatment limited the Cd-triggered $H_2O_2$ and MDA accumulation and electrolyte leakage due to the upregulation of antioxidant enzymes (SOD, POD, and CAT).

### 10.2.3 Amelioration of Biotic Stress-Induced Damages

Biotic stress can be caused by almost every type of organisms, and the damage can affect all levels of organization of plants from molecules to whole populations [63].

It was discovered that KARs can affect bacterial group behavior especially in quorum sensing (QS) system. QS is the mechanism that can help microorganisms to sense each other *via* secretion and detection of small, diffusible molecules in order to communicate that lead to synchronized gene expression and make it possible for them to act like a single organism [64]. Mandabi *et al.* [65] discovered that $KAR_1$ and $KAR_2$ modulate the QS in *A. tumefaciens*, *P. aeruginosa*, and *V. harveyi*. However, $KAR_1$ does not prevent the infection of *Arabidopsis thaliana* (WT) by *P. aeruginosa*, and a significant reduction in loss of leaves was observed.

In a recent molecular study, *kai2 Arabidopsis* mutant proved to be sensitive *Pseudomonas syringae* pv. tomato relative to the WT and SL-deficient mutants, suggesting that the KAR-associated signaling, not the SL-associated pathway, upregulates the immune response in this pathosystem [66]. The exogenous supplementation of $KAR_2$ promotes resistance against *P. syringae* depending on salicylic acid (SA) signaling. Furthermore, deficiency in KAR signaling delays SA-regulated defense responses and limits pathogen-triggered protein biosynthesis. These results imply the possibility that KAR regulates plant pathogen defense, but further studies are needed to support this.

## 10.3 Indirect KAR-Associated Effects on Plants

Though there are many experimental data about the direct effects of KARs on seed germination and root development, both *in vitro* and pot (soil) systems, we still have little knowledge of how smoke, smoke water, or its active compounds, namely, KARs, shape the ecological partners of plants, especially the rhizomicrobiome. Additionally, it is also a question whether smoke, smoke water, or the KARs exert their physiological (negative or positive) effects on plants' rhizobiome directly or *via* the plants' KAR/SL signaling pathway. For more than 10 years, it is clear that due to their structural similarities SLs and KARs use similar signaling cascades inside the plants sharing MAX2 [67].

Since almost 80% of the land plant species have mycorrhizal partners, we may hypothesize that KARs play an important role not only in the early development (e.g., seed germination, root hair formation, and lateral root emergence) of plants but also in the root colonization of mycorrhizal fungi [68, 69]. In the past few years, researchers have focused on the exploration of root exudates (e.g., SLs) which promote the formation of arbuscular mycorrhiza within the roots (summary by Lanfranco *et al.* [68]). They pronounced that the KAR receptor complex, KAI2, is potentially involved in sensing the fungal partner. This was confirmed in monocots by Meng *et al.* [38], who demonstrated that arbuscular symbiosis between grass

*Brachypodium distachyon* and its fungal partner *Rhizophagus irregularis* depends on the presence of KAI2 at the host side. Moreover, Bonanomi *et al.* [70] presented that smoke water treatments may differently affect the growth of the hyphae of soil dweller pathogenic *Rhizoctonia solani* and *Sclerotium rolfsii* or *Trichoderma harzianum* which is often used as fungicide. Depending on the degree of dilution and the origin of the smoke water (i.e., prepared from cellulose or alfalfa hay), pathogenic fungi proved to be less sensitive compared to *Trichoderma* strains. Though the exact composition the smoke water types applied in this study was not identified, the results suggest that fungi found in rhizobiome may be differently influenced by KARs derived from smoke in fire-prone vegetations.

## 10.4 Smoke Water as a Promising Agent for Agricultural Applications

Throughout history, fire and smoke have been integral components of traditional agricultural practices. Biologically active compounds extracted from smoke show promise for applications in agriculture and horticulture. Smoke technology is cost-effective and could help mitigate the adverse effects of chemical fertilizers, pesticides, fungicides, and herbicides on the soil and the environment [71, 72] by reducing their usage. In fact, plant-derived smoke has been discovered to regulate somatic embryogenesis, flowering, photosynthesis, root development, and crop yield in plants. These discoveries have led to the expansion of smoke technology into various agricultural sectors, including commercially cultivated crops, to alleviate the stresses caused by auxin, boron, and salinity [73]. The possible advantages of utilizing smoke in agriculture and horticulture can be realized through different application methods, such as using "crude" smoke, aerosol smoke, or smoke extracts. One of the most straightforward smoke treatment methods involves directly applying aerosol smoke. In this approach, seeds are exposed directly to smoke produced by burning plant material [22]. The utilization of smoke water in agriculture is safe and does not pose any toxic effects on crops [74]. Compared to aerosol smoke formation, smoke extract treatments are advantageous because the active compounds present in smoke readily dissolve in water. The application of smoke water can be done through seed treatment, soil drenching, foliar spraying, irrigation systems, or a combination thereof [75]. Employing smoke water treatments has shown promising results in improving seed germination, seedling growth, and crop yield. Furthermore, the positive influence of smoke water on crop flowering and fruiting is notable, particularly given the growing recognition of karrikins found in smoke as promising regulators of plant growth.

De Lange and Boucher [6] were the first to report that biologically active compounds present in smoke readily dissolve in water, and when this smoke extract is used as a diluted solution, seeds of many plant species show significant improvement in germination. They generated smoke by burning in a drum, and using compressed air, the smoke was bubbled through distilled water. Based on this method, various plant materials have been used to prepare aqueous smoke extracts. A concentrated aqueous smoke extract can be mixed with water at ratios of 1:250, 1:500, 1:1,000, 1:1,500, and 1:2,000 (v/v). Dilutions within this range typically prove effective in stimulating germination, though effectiveness may vary depending on the species.

Multiple crop species have been studied regarding the impacts of smoke and smoke solutions. Kulkarni *et al.* [76] investigated the effects of foliar application of smoke water and the smoke compound butenolide on various growth parameters of *Abelmoschus esculentus* and *Lycopersicon esculentum* seedlings under greenhouse conditions. Both seedling species' shoot/root length, shoot fresh/dry weight, number of leaves, total leaf area, and absolute growth rate per week were stimulated by the applied treatments. In 2021, El-Nour and colleagues described the application of four different concentrations (0.5%, 1%, 1.5%, and 2%) of smoke water extract as soil drenching during seed germination and seedling growth of vegetables such as pepper, tomato, carrot, onion, celery, and lettuce, cultivated in a greenhouse, or in combination with foliar spray to study their effects on the productivity of bean (*Phaseolus vulgaris*) in open field conditions. The results indicated that all concentrations of smoke water encouraged seed germination percentage and rate of pepper, celery, and lettuce. However, tomato and onion seeds positively responded to 1.0% concentration of smoke water compared to higher concentrations (1.5% and 2%), which caused inhibitory effects. Meanwhile, the low level of smoke water (0.5%) was sufficient for superior germination values of carrot seeds compared to other levels. Additionally, the results showed that the high concentration of smoke water extract (2%) was adequate to support the growth parameters of pepper, carrot, and lettuce seedlings, such as height, fresh and dry weight, number of leaves, and chlorophyll content. On the other hand, all levels of smoke water had negative effects on tomato, onion, and celery seedling growth. The maximum results of snap bean vegetative growth, green pod yield, and the accumulation of plant dry matter were achieved using 1.5% smoke water as a foliar spray combined with soil drenching. However, the high level of smoke water (2%) led to a decrease in snap bean vegetative growth, total green pod yield (ton/fed), pod characteristics, and pod chemical contents in the two seasons.

Gupta and colleagues described in 2024 [77] the effects of standardized smoke water (at a volume ratio of 1:2,500) foliar application on the seedling growth of radish (*Raphanus sativus* L.), spinach (*Spinacia oleracea* L.), and turnip (*Brassica rapa* L.) and determined the mineral elemental composition of leaves. The foliar treatment with smoke water significantly improved the length and weight of seedlings compared to seedlings sprayed with water (control). Among the mineral elements, potassium and zinc showed notable increases in the leaves of all smoke water-treated seedlings [77].

Biostimulants are substances or microorganisms that stimulate natural processes in plants, thereby increasing yield and quality. They reduce the need for synthetic fertilizers, enhance plant growth, and improve stress tolerance. Effective in low concentrations, they offer new opportunities for regulating plant physiological processes. Smoke water is also referred to as a biostimulant nowadays [78]. Nemahunguni *et al.* [79] proposed plant-derived smoke as the most effective organic biostimulant for potentially enhancing seed germination and promoting the growth of *Cleome gynandra* (African cabbage). Plant-derived smoke, acting as an agent with biostimulant properties, has been demonstrated in numerous studies to promote abiotic stress tolerance. This includes alleviating salt stress in rice and maize [80, 81] heat stress in tomato [82], flooding stress in soybean [83], and reversing ABA stress in lettuce [84].

The efficacy of plant-derived smoke and smoke water in agriculture and horticulture comes with numerous advantages. These substances stimulate biological processes in plants, thus increasing yield and quality. By reducing the use of chemical fertilizers, insecticides, fungicides, and herbicides, they contribute to mitigating the harmful effects on soil and the environment. Plant-derived smoke and smoke water can be used with various application methods and have been proven to improve seed germination, seedling growth, and yield. Additionally, they have shown positive effects on plant flowering and fruiting. Smoke water is safe and has no toxic effects on plants. The method and concentration of application can vary depending on the plant species, effectively enhancing plant growth and stress tolerance. Furthermore, due to the benefits mediated by biostimulants, there is potential for further dissemination and utilization of plant-derived smoke and smoke water technology in various agricultural sectors to help alleviate various stressors.

## 10.5 Conclusions and Perspectives

In the past 20 years, several aspects of the effects of plant-derived smoke components, KARs, have been characterized. Regarding their plant perception and signaling, KAI2 receptor is well characterized, although its ligand is still unknown. It is also a relevant and still uncovered question: whether KARs are synthetized by plant cells similar to SLs? The discovery of a wide range of effects is underway, and practice-oriented research is also performed in relation to smoke water and $KAR_{1/2}$. In connection with the latter, the characterization of the growth and stress tolerance of signaling mutants and the pharmacological modification of KAR signaling are the two major approaches used in plants. Interactions of KARs with other growth-regulating signals and hormones including SLs remain to be discovered and represent a putative research direction. KAR-associated effects can be divided into direct effects on plants (e.g., seed germination and induction of hypocotyl elongation) and indirect effects on the plants' symbiont partners. The putative indirect effects are poorly known; therefore, a promising future research direction is the further investigation of the involvement of KAR signaling on mycorrhizae as well as on plant growth-promoting rhizobacteria (including N-fixing symbioses) and pollinators.

Overall, research related to KARs is an excitingly developing area in which basic research, *Arabidopsis* genetics, and practice-oriented (smoke water and crops) research efforts are combined.

## Acknowledgments

This work was supported by the "Lendület" MOMENTUM project of the Hungarian Academy of Sciences under grant no. LP2023-14/2023 (Zs.K.).

## References

1. Keeley, S.C. and Pizzorno, M., Charred wood stimulated germination of two fire-following herbs of the California chaparral and the role of hemicellulose. *Am. J. Bot.*, *73*, 9, 1289–1297, 1986.
2. Brown, N.A.C., Promotion of germination of fynbos seeds by plant-derived smoke. *New Phytol.*, *123*, 3, 575–583, 1993.
3. Dixon, K.W., Roche, S., Pate, J.S., The promotive effect of smoke derived from burnt native vegetation on seed germination of Western Australian plants. *Oecologia*, *101*, 185–192, 1995.

4. Moreira, B. and Pausas, J.G., Tanned or burned: the role of fire in shaping physical seed dormancy. *PLoS One*, *7*, 12, e51523, 2012.
5. Mojzes, A. and Kalapos, T., The role of smoke produced during the burning of vegetation in the regeneration of plants. *Landsc. Ecol.*, *10*, 2, 247–270, 2012.
6. De Lange, J.H. and Boucher, C., Autecological studies on *Audouinia capitata* (Bruniaceae). I. Plant-derived smoke as a seed germination cue. *S. Afr. J. Bot.*, *56*, 6, 700–703, 1990.
7. Baxter, B.J.M., Van Staden, J., Granger, J.E., Brown, N.A.C., Plant-derived smoke and smoke extracts stimulate seed germination of the fire-climax grass Themeda triandra. *Environ. Exp. Bot.*, *34*, 2, 217–223, 1994.
8. Staden, J.V., Brown, N.A., Jäger, A.K., Johnson, T.A., Smoke as a germination cue. *Plant Species Biol.*, *15*, 2, 167–178, 2000.
9. Doherty, L.C. and Cohn, M.A., Seed dormancy in red rice (*Oryza sativa*). XI. Commercial liquid smoke elicits germination. *Seed Sci. Res.*, *10*, 4, 415–421, 2000.
10. Jäger, A.K., Light, M.E., Van Staden, J., Effects of source of plant material and temperature on the production of smoke extracts that promote germination of light-sensitive lettuce seeds. *Environ. Exp. Bot.*, *36*, 4, 421–429, 1996.
11. Keeley, J.E. and Bond, W.J., Convergent seed germination in South African fynbos and Californian chaparral. *Plant Ecol.*, *133*, 153–167, 1997.
12. Cook, C.E., Whichard, L.P., Turner, B., Wall, M.E., Egley, G.H., Germination of witchweed (*Striga lutea* Lour.): isolation and properties of a potent stimulant. *Sci.*, *154*, 3753, 1189–1190, 1966.
13. Flematti, G.R., Ghisalberti, E.L., Dixon, K.W., Trengove, R.D., A compound from smoke that promotes seed germination. *Sci.*, *305*, 5686, 977–977, 2004.
14. Chiwocha, S.D., Dixon, K.W., Flematti, G.R., Ghisalberti, E.L., Merritt, D.J., Nelson, D.C., Stevens, J.C., Karrikins: a new family of plant growth regulators in smoke. *Plant Sci.*, *177*, 4, 252–256, 2009.
15. Nelson, D.C., Riseborough, J.A., Flematti, G.R., Stevens, J., Ghisalberti, E.L., Dixon, K.W., Smith, S.M., Karrikins discovered in smoke trigger Arabidopsis seed germination by a mechanism requiring gibberellic acid synthesis and light. *Plant Physiol.*, *149*, 2, 863–873, 2009.
16. Light, M.E., Burger, B.V., Staerk, D., Kohout, L., Van Staden, J., Butenolides from plant-derived smoke: natural plant-growth regulators with antagonistic actions on seed germination. *J. Nat. Prod.*, *73*, 2, 267–269, 2010.
17. Alder, A., Jamil, M., Marzorati, M., Bruno, M., Vermathen, M., Bigler, P., Al-Babili, S., The path from β-carotene to carlactone, a strigolactone-like plant hormone. *Sci.*, *335*, 6074, 1348–1351, 2012.
18. Brewer, P.B., Yoneyama, K., Filardo, F., Meyers, E., Scaffidi, A., Frickey, T., Beveridge, C.A., LATERAL BRANCHING OXIDOREDUCTASE acts in the final stages of strigolactone biosynthesis in Arabidopsis. *PNAS*, *113*, 22, 6301–6306, 2016.

19. Kamran, M., Melville, K.T., Waters, M.T., Karrikin signalling: impacts on plant development and abiotic stress tolerance. *J. Exp. Bot.*, *75*, 4, 1174–1186, 2024.
20. Flematti, G.R., Goddard-Borger, E.D., Merritt, D.J., Ghisalberti, E.L., Dixon, K.W., Trengove, R.D., Preparation of 2 H-furo [2, 3-c] pyran-2-one derivatives and evaluation of their germination-promoting activity. *J. Agric. Food Chem.*, *55*, 6, 2189–2194, 2007.
21. Kulkarni, M.G., Ascough, G.D., Van Staden, J., Smoke-water and a smoke-isolated butenolide improve growth and yield of tomatoes under greenhouse conditions. *Hort. Technol.*, *18*, 3, 449–454, 2008.
22. Kulkarni, M.G., Light, M.E., Van Staden, J., Plant-derived smoke: old technology with possibilities for economic applications in agriculture and horticulture. *S. Afr. J. Bot.*, *77*, 4, 972–979, 2011.
23. Sun, X.D. and Ni, M., HYPOSENSITIVE TO LIGHT, an alpha/beta fold protein, acts downstream of ELONGATED HYPOCOTYL 5 to regulate seedling de-etiolation. *Mol. Plant*, *4*, 1, 116–126, 2011.
24. Waters, M.T., Nelson, D.C., Scaffidi, A., Flematti, G.R., Sun, Y.K., Dixon, K.W., Smith, S.M., Specialisation within the DWARF14 protein family confers distinct responses to karrikins and strigolactones in Arabidopsis. *Dev.*, *139*, 7, 1285–1295, 2012.
25. Waters, M.T. and Nelson, D.C., Karrikin perception and signalling. *New Phytol.*, *237*, 5, 1525–1541, 2023.
26. Walters, M., Midgley, J.J., Somers, M.J., Effects of fire and fire intensity on the germination and establishment of *Acacia karroo, Acacia nilotica, Acacia luederitzii* and *Dichrostachys cinerea* in the field. *BMC Ecol.*, *4*, 1–13, 2004.
27. Nelson, D.C., Flematti, G.R., Ghisalberti, E.L., Dixon, K.W., Smith, S.M., Regulation of seed germination and seedling growth by chemical signals from burning vegetation. *Annu. Rev. Plant Biol.*, *63*, 1, 107–130, 2012.
28. Hürkan, Y.K. and Akı, C., Effects of Plant-derived Smoke, Karrikin, and Salinity Stress on *Prunus armeniaca* cv. Şalak seeds and seedlings: A Morphological, Biochemical, and Molecular Approach. *J. Agric. Sci.*, *30*, 2, 273–283, 2024.
29. Çatav, S.S., Akbaş, K., Tavşanoğlu, C., Küçükakyüz, K., Germination response of three smoke-and karrikinolide-responsive species to gibberellic acid and reactive oxygen species. *Ann. Bot. Fenn.*, 61, 1, 31–40, 2024, January. Finnish Zoological and Botanical Publishing Board.
30. Brun, G., Thoiron, S., Braem, L., Pouvreau, J.B., Montiel, G., Lechat, M.M., Delavault, P., CYP707As are effectors of karrikin and strigolactone signalling pathways in *Arabidopsis thaliana* and parasitic plants. *Plant Cell Environ.*, *42*, 9, 2612–2626, 2019.
31. Bunsick, M., Toh, S., Wong, C., Xu, Z., Ly, G., McErlean, C.S., Lumba, S., SMAX1-dependent seed germination bypasses GA signalling in Arabidopsis and Striga. *Nat. Plants*, 6, 6, 646–652, 2020.

32. Stanga, J.P., Morffy, N., Nelson, D.C., Functional redundancy in the control of seedling growth by the karrikin signaling pathway. *Planta, 243*, 1397–1406, 2016.
33. Waters, M.T. and Smith, S.M., KAI2-and MAX2-mediated responses to karrikins and strigolactones are largely independent of HY5 in Arabidopsis seedlings. *Mol. Plant*, *6*, 1, 63–75, 2013.
34. Bursch, K., Niemann, E.T., Nelson, D.C., Johansson, H., Karrikins control seedling photomorphogenesis and anthocyanin biosynthesis through a HY5-BBX transcriptional module. *TPJ*, *107*, 5, 1346–1362, 2021.
35. Kim, J.Y., Park, Y.J., Lee, J.H., Park, C.M., SMAX1 integrates karrikin and light signals into GA-mediated hypocotyl growth during seedling establishment. *Plant Cell Physiol.*, 63, 7, 932–943, 2022.
36. Villaécija-Aguilar, J.A., Hamon-Josse, M., Carbonnel, S., Kretschmar, A., Schmidt, C., Dawid, C., Gutjahr, C., SMAX1/SMXL2 regulate root and root hair development downstream of KAI2-mediated signalling in Arabidopsis. *PLos Genet.*, *15*, 8, e1008327, 2019.
37. Xu, P., Jinbo, H., Cai, W., Karrikin signaling regulates hypocotyl shade avoidance response by modulating auxin homeostasis in Arabidopsis. *New Phytol.*, *236*, 5, 1748–1761, 2022.
38. Meng, Y., Varshney, K., Incze, N., Badics, E., Kamran, M., Davies, S.F., Waters, M.T., KARRIKIN INSENSITIVE2 regulates leaf development, root system architecture and arbuscular-mycorrhizal symbiosis in *Brachypodium distachyon*. *TPJ*, *109*, 6, 1559–1574, 2022.
39. Varshney, K. and Gutjahr, C., KAI2 can do: karrikin receptor function in plant development and response to abiotic and biotic factors. *PCP*, *64*, 9, 984–995, 2023.
40. Carbonnel, S., Das, D., Varshney, K., Kolodziej, M.C., Villaécija-Aguilar, J.A., Gutjahr, C., The karrikin signaling regulator SMAX1 controls *Lotus japonicus* root and root hair development by suppressing ethylene biosynthesis. *PNAS*, *117*, 35, 21757–21765, 2020.
41. Wang, L., Xu, Q., Yu, H., Ma, H., Li, X., Yang, J., Wang, B., Strigolactone and karrikin signaling pathways elicit ubiquitination and proteolysis of SMXL2 to regulate hypocotyl elongation in Arabidopsis. *Plant Cell*, *32*, 7, 2251–2270, 2020.
42. Lei, S. and Huang, B., Metabolic regulation of α-Ketoglutarate associated with heat tolerance in perennial ryegrass. *PPB*, *190*, 164–173, 2022.
43. Kaya, M.D., Okçu, G., Atak, M., Çıkılı, Y., Kolsarıcı, O., Seed treatments to overcome salt and drought stress during germination in sunflower (*Helianthus annuus* L.). *Eur. J. Agron.*, *24*, 4, 291–295, 2006.
44. Li, Y., Zhao, H., Duan, B., Korpelainen, H., Li, C., Effect of drought and ABA on growth, photosynthesis and antioxidant system of *Cotinus coggygria* seedlings under two different light conditions. *Environ. Exp. Bot.*, *71*, 1, 107–113, 2011.

45. Li, W., Nguyen, K.H., Chu, H.D., Ha, C.V., Watanabe, Y., Osakabe, Y., Tran, L.S.P., The karrikin receptor KAI2 promotes drought resistance in *Arabidopsis thaliana*. *PLos Genet.*, *13*, 11, e1007076, 2017.
46. Wang, L., Waters, M.T., Smith, S.M., Karrikin-KAI2 signalling provides Arabidopsis seeds with tolerance to abiotic stress and inhibits germination under conditions unfavourable to seedling establishment. *New Phytol.*, *219*, 2, 605–618, 2018.
47. Li, G., Zhang, C., Zhang, G., Fu, W., Feng, B., Chen, T., Fu, G., Abscisic acid negatively modulates heat tolerance in rolled leaf rice by increasing leaf temperature and regulating energy homeostasis. *Rice*, *13*, 1–16, 2020.
48. Tian, H., Watanabe, Y., Nguyen, K.H., Tran, C.D., Abdelrahman, M., Liang, X., Li, W., KARRIKIN UPREGULATED F-BOX 1 negatively regulates drought tolerance in Arabidopsis. *Plant Physiol.*, *190*, 4, 2671–2687, 2022.
49. Shah, F.A., Wei, X., Wang, Q., Liu, W., Wang, D., Yao, Y., Wu, L., Karrikin improves osmotic and salt stress tolerance via the regulation of the redox homeostasis in the oil plant *Sapium sebiferum*. *Front. Plant Sci.*, 11, 216, 2020.
50. Tan, Z.Z., Wang, Y.T., Zhang, X.X., Jiang, H.Y., Li, Y., Zhuang, L.L., Yang, Z.M., Karrikin1 enhances drought tolerance in creeping bentgrass in association with antioxidative protection and regulation of stress-responsive gene expression. *Agron.*, *13*, 3, 675, 2023.
51. Abera, B.B., Rice genotypic variation on phenological development and yield performance in cold prone high altitude cropping systems, 2021, (Doctoral dissertation, Dissertation, Hohenheim, Universität Hohenheim, 2021).
52. Criddle, R.S., Hopkin, M.S., McArthur, E.D., Hansen, L.D., Plant distribution and the temperature coefficient of metabolism. *Plant Cell Environ.*, *17*, 3, 233–243, 1994.
53. Ghebrehiwot, H.M., Kulkarni, M.G., Kirkman, K.P., Van Staden, J., Smoke-water and a smoke-isolated butenolide improve germination and seedling vigour of Eragrostis Tef (Zucc.) Trotter under high temperature and low osmotic potential. *J. Agron. Crop Sci.*, *194*, 4, 270–277, 2008.
54. Moyo, M., Amoo, S.O., Van Staden, J., Seed priming with smoke water and karrikin improves germination and seedling vigor of *Brassica napus* under varying environmental conditions. *J. Plant Growth Regul.*, *97*, 2, 315–326, 2022.
55. Ahmad, A., Shahzadi, I., Mubeen, S., Yasin, N.A., Akram, W., Khan, W.U., Wu, T., Karrikinolide alleviates BDE-28, heat and Cd stressors in Brassica alboglabra by correlating and modulating biochemical attributes, antioxidative machinery and osmoregulators. *Ecotoxicol. Environ. Saf.*, *213*, 112047, 2021.
56. Abdelrahman, M., Mostofa, M.G., Tran, C.D., El-Sayed, M., Li, W., Sulieman, S., Tran, L.S.P., The karrikin receptor KARRIKIN INSENSITIVE2 positively regulates heat stress tolerance in *Arabidopsis thaliana*. *PCP*, *63*, 12, 1914–1926, 2022.

57. Ding, Y., Shi, Y., Yang, S., Advances and challenges in uncovering cold tolerance regulatory mechanisms in plants. *New Phytol.*, *222*, 4, 1690–1704, 2019.
58. Shah, F.A., Ni, J., Yao, Y., Hu, H., Wei, R., Wu, L., Overexpression of karrikins receptor gene *Sapium sebiferum* KAI2 promotes the cold stress tolerance via regulating the redox homeostasis in Arabidopsis thaliana. *Front. Plant Sci.*, *12*, 657960, 2021.
59. Mostofa, M.G., Abdelrahman, M., Rahman, M.M., Tran, C.D., Nguyen, K.H., Watanabe, Y., Tran, L.S.P., Karrikin receptor KAI2 coordinates salt tolerance mechanisms in Arabidopsis thaliana. *PCP*, *63*, 12, 1927–1942, 2022.
60. Feng, D., Wang, R., Sun, X., Liu, P., Tang, J., Zhang, C., Liu, H., Heavy metal stress in plants: Ways to alleviate with exogenous substances. *Sci. Total Environ.*, 897, 165397, 2023.
61. Li, Y., Ding, L., Zhou, M., Chen, Z., Ding, Y., Zhu, C., Transcriptional regulatory network of plant cadmium stress response. *Int. J. Mol. Sci.*, *24*, 5, 4378, 2023.
62. Sardar, R., Ahmed, S., Yasin, N.A., Seed priming with karrikinolide improves growth and physiochemical features of Coriandrum sativum under cadmium stress. *Env. Adv.*, *5*, 100082, 2021.
63. Peterson, J.D., Umayam, L.A., Dickinson, T., Hickey, E.K., White, O., The comprehensive microbial resource. *Nucleic Acids Res.*, *29*, 1, 123–125, 2001.
64. Fuqua, W.C., Winans, S.C., Greenberg, E.P., Quorum sensing in bacteria: the LuxR-LuxI family of cell density-responsive transcriptional regulators. *J. Bacteriol.*, *176*, 2, 269–275, 1994.
65. Mandabi, A., Ganin, H., Krief, P., Rayo, J., Meijler, M.M., Karrikins from plant smoke modulate bacterial quorum sensing. *Chem Comm.*, *50*, 40, 5322–5325, 2014.
66. Zheng, X., Liu, F., Yang, X., Li, W., Chen, S., Yue, X., Sun, X., The MAX2-KAI2 module promotes salicylic acid-mediated immune responses in Arabidopsis. *J. Integr. Plant Biol.*, *65*, 6, 1566–1584, 2023.
67. Antala, M., Sytar, O., Rastogi, A., Brestic, M., Potential of karrikins as novel plant growth regulators in agriculture. *Plants*, *9*, 1, 43, 2019.
68. Lanfranco, L., Fiorilli, V., Gutjahr, C., Partner communication and role of nutrients in the arbuscular mycorrhizal symbiosis. *New Phytol.*, *220*, 4, 1031–1046, 2018.
69. Xue, L. and Wang, E., Arbuscular mycorrhizal associations and the major regulators. *Front. Agric. Sci. Eng.*, *7*, 3, 296–306, 2020.
70. Bonanomi, G., Jesu, G., Zotti, M., Idbella, M., d'Errico, G., Laudonia, S., Abd-ElGawad, A., Biochar-derived smoke-water exerts biological effects on nematodes, insects, and higher plants but not fungi. *Sci. Total Environ. 750*, 142307, 2021.
71. Ye, X.J., Wang, Z.Q., Li, Q.S., The ecological agriculture movement in modern China. *AGR. ECOSYST. ENVIRON.*, *92*, 2–3, 261–281, 2002.
72. Aeron, A., Kumar, S., Pandey, P., Maheshwari, D.K., Emerging role of plant growth promoting rhizobacteria in agrobiology, in: *Bacteria in agrobiology:*

*crop ecosystems*, pp. 1–36, Springer Berlin Heidelberg, Berlin, Heidelberg, 2011.

73. Akeel, A., Khan, M.M.A., Jaleel, H., Uddin, M., Smoke-saturated water and karrikinolide modulate germination, growth, photosynthesis and nutritional values of carrot (*Daucus carota* L.). *J. Plant Growth Regul.*, *38*, 1387–1401, 2019.
74. Verschaeve, L., Maes, J., Light, M.E., van Staden, J., Genetic toxicity testing of 3-methyl-2H-furo [2, 3-c] pyran-2-one, an important biologically active compound from plant-derived smoke. *Mutat. Res. Genet. Toxicol. Environ. Mutagen.*, *611*, 1–2, 89–95, 2006.
75. Gupta, S., Kulkarni, M.G., White, J.F., Stirk, W.A., Papenfus, H.B., Doležal, K., Van Staden, J., Categories of various plant biostimulants–mode of application and shelf-life, in: *Biostimulants for Crops from Seed Germination to Plant Development*, pp. 1–60, Academic Press, London, United Kingdom, 2021.
76. Kulkarni, M.G., Ascough, G.D., Van Staden, J., Effects of foliar applications of smoke-water and a smoke-isolated butenolide on seedling growth of okra and tomato. *HJHSAR*, *42*, 1, 179–182, 2007.
77. Gupta, S., Kulkarni, M., Doležalova, I., Novák, O., Van Staden, J., Doležal, K., Foliar smoke-water application improves growth and changes the leaf mineral composition of radish, spinach and turnip seedlings. *S. Afr. J. Bot.*, *167*, 527–534, 2024.
78. Khatoon, A., Rehman, S.U., Aslam, M.M., Jamil, M., Komatsu, S., Plant-derived smoke affects biochemical mechanism on plant growth and seed germination. *Int. J. Mol. Sci.*, *21*, 20, 7760, 2020.
79. Nemahunguni, N.K., Gupta, S., Kulkarni, M.G., Finnie, J.F., Van Staden, J., The effect of biostimulants and light wavelengths on the physiology of Cleome gynandra seeds. *J. Plant Growth Regul.*, *90*, 467–474, 2020.
80. Jamil, M., Kanwal, M., Aslam, M.M., Khan, S.U., Malook, I., Tu, J., ur Rehman, S., Effect of plant-derived smoke priming on physiological and biochemical characteristics of rice under salt stress condition. *Aust. J. Crop Sci.*, *8*, 2, 159–170, 2014.
81. Waheed, M.A., Jamil, M., Khan, M.D., Shakir, S.K., Rehman, S.U., Effect of plant-derived smoke solutions on physiological and biochemical attributes of maize (Zea mays L.) under salt stress. *Pak. J. Bot.*, *48*, 5, 1763–1774, 2016.
82. Jain, N., Kulkarni, M.G., van Staden, J., A butenolide, isolated from smoke, can overcome the detrimental effects of extreme temperatures during tomato seed germination. *J. Plant Growth Regul.*, *49*, 263–267, 2006.
83. Li, X., Yamaguchi, H., Hitachi, K., Tsuchida, K., Yamaguchi, T., Komatsu, S., Proteomic analysis of the effect of plant-derived smoke on soybean during recovery from flooding stress. *J. Proteom.*, *181*, 238–248, 2018.
84. Kamran, M., Imran, Q., Khatoon, A., Lee, I., Rehman, S., Effect of plant extracted smoke and reversion of abscisic acid stress on lettuce. *Pak. J. Bot.*, *45*, 5, 1541–1549, 2013.

# 11

# Strigolactones: Key Phytohormones in Plant–Microbe Interactions and Development

**Asif Hussain Hajam and Gausiya Bashri***

*Department of Botany, Faculty of Life Sciences, Aligarh Muslim University, Aligarh, India*

***Abstract***

Strigolactones (SLs) are a class of phytohormones derived from carotenoids and are crucial in shaping how plants grow and react to changes in their surroundings. SLs are synthesized in roots, especially when there is a lack of phosphate, and are then released into the surrounding soil, promoting the expansion of fungal hyphae in arbuscular mycorrhizal (AM) fungi and promoting symbiosis and nutrient uptake. SL biosynthesis involves carotenoid cleavage dioxygenases (CCDs) and 9-cis-epoxy-carotenoid cleavage dioxygenases (NCEDs), with D27 potentially involved in C13/C14 apocarotenoid biosynthesis. In recent years, their involvement in establishing and regulating symbiotic relationships, particularly with arbuscular mycorrhizal (AM) fungi, has garnered significant attention. In addition to their role in AM symbiosis, emerging evidence suggests that SLs also participate in other plant-microbe interactions, including rhizobial and parasitic plant interactions, root development, and nodule formation in leguminous plants. SLs also influence the plant's capacity to withstand both biotic and abiotic challenges, indicating their potential role as biocontrol agents against various challenges posed by the environment. This chapter describes the complex interactions between SLs and mycorrhizal symbionts, highlighting their dual role as signaling molecules that mediate plant–fungal interactions and modulators of plant growth and development. Understanding the molecular mechanisms of SLs and their roles in symbiotic relationships can provide insights into optimizing agricultural techniques and practices, enhancing nutrient use efficiency, and increasing crop quantity and quality.

**Corresponding author*: gausiya.bashri@gmail.com

Tariq Aftab and Kaiser Iqbal Wani (eds.) Strigolactones: Emerging Plant Hormones, (235–260) © 2025 Scrivener Publishing LLC

***Keywords*:** Strigolactones, legumes, nodules, rhizobia, AM fungi, roots, nutrition

## 11.1 Introduction

Strigolactones (SLs), a set of phytohormones derived from carotenoids, are produced in both plastids and the cytosol of plants [1]. They serve as signaling molecules that aid plants in interacting with their environment. These potent compounds are primarily synthesized in roots and subsequently released into the rhizosphere; however, small quantities of these compounds have been detected in other plant tissues [2].

Arbuscular mycorrhizal (AM) symbiosis is a mutually beneficial relationship between certain plant roots and fungi belonging to phylum Glomeromycota. In this symbiosis, fungi colonize the roots of plants, creating branched structures known as arbuscules and vesicles within the root cells. These specialized structures promote the exchange of nutrients between the fungus and the plant. The fungus receives sugars from the plant, which it needs for growth, and the plant hosts them and acquires better access to nutrients, such as phosphorous (P) and nitrogen (N), from the soil, which the fungus can absorb more efficiently [3–5]. AM symbiosis is prevalent across the plant kingdom and plays a vital role in nutrient uptake and sustaining plant health.

The beneficial interactions of SLs with microbes encompass a range of collaborative relationships between plants and microorganisms, mainly fungi and bacteria. SLs can act as signaling molecules that facilitate symbiotic relationships, such as AM symbiosis and interactions with rhizosphere bacteria [6]. SLs initiate and control the establishment of AM symbiosis, thereby participating in signaling pathways that attract and stimulate the growth of mycorrhizal fungi toward plant roots. This symbiotic association enhances plant nutrient acquisition, stress tolerance, and ecosystem resilience, making them integral to plant–microbe interactions under diverse environmental conditions. This association also affects the makeup and variety of microorganisms in the rhizosphere. SLs regulate interactions among microbes by altering the biosynthesis of secondary compounds and impacting communication systems among microorganisms [7, 8]. This adjustment of the microbial community fosters a conducive rhizosphere environment, which promotes productive interactions between plants and microbes while mitigating the detrimental effects of stress factors. Secondary metabolites, such as flavonoids, were discovered to be under the control of SLs, thereby bolstering plant defense mechanisms against osmotic stresses [9].

## 11.2 Identification of SLs as Signaling Molecules in AM Symbiosis

The discovery of SLs as signaling molecules in AM symbiosis significantly advances the understanding of the molecular mechanisms involved in plant–fungal interactions. This breakthrough came from studies investigating the factors involved in recognizing and establishing AM symbioses between plants and fungi. Researchers have shown that certain compounds released by plant roots can attract fungal hyphae to the roots, initiating symbiotic associations; these compounds were later identified as SLs [10].

The key milestones in the discovery of SLs as signaling molecules in AM symbiosis include the recognition of root exudate-containing compounds (SLs) that can stimulate the growth of mycorrhizal fungi toward plant roots, suggesting the presence of signaling molecules involved in the recognition process [11]. Additionally, bioassays and chemical analysis techniques were employed to isolate and characterize the active compounds in root exudates responsible for attracting fungal hyphae, and these efforts led to the identification of SLs as potent signaling molecules [12]. Moreover, genetic and molecular approaches have helped elucidate the biosynthetic pathways involved in plant SL production. Mutant studies and gene expression analyses furnished additional evidence reinforcing the significance of SLs in AM symbiosis [10]. Furthermore, functional studies, including studies on the exogenous application of synthetic SLs and genetic manipulation of SL biosynthesis pathways, have demonstrated their essential role in promoting AM fungal colonization and symbiotic development in plant roots [6].

In the 1960s, researchers explored the mechanisms of root parasitism and observed that certain compounds present in the root exudates of host plants stimulated the germination of parasitic plants such as *Striga* and *Orobanche* [13]. These parasitic plants depend on host plants for nutrients and latch onto their roots to extract water and carbohydrates. Following the discovery of SLs as elicitors for germinating seeds from root parasitic plants, another crucial role of SLs in the rhizosphere emerged. This discovery sheds light on why plants have continued to evolve mechanisms for secreting compounds that promote the germination of potentially harmful parasites.

## 11.3 SL Perception by AM Fungi

The effectiveness of SLs on AM fungi can be recognized at remarkably low concentrations, as evidenced by their activity even at concentrations as low as $10^{-13}$ M in certain AM species [14]. This underscores the existence of an exceedingly sensitive perception mechanism in these fungi. However, despite this sensitivity, no receptors of the SL signaling pathway in the fungus have been conclusively identified to date, and identifying such receptors presents a considerable challenge due to several factors.

AM fungi are not readily amenable to classical genetic manipulation, and obtaining significant quantities of fungal biomass for biochemical investigations is difficult [15]. Although searches within AM genomic sequence data have been conducted, they have yet to reveal apparent homologs of the α/β-fold hydrolases that are recognized as receptors for plant SLs [16]. Regarding the characteristics of a ligand, the structural prerequisites for a compound resembling SLs to influence hyphal branching notably contrast with the criteria reported for its capacity to inhibit shoot branching in pea plants [17]. The presence of the D-ring appears to be crucial for the diverse biological activities of SLs, including parasitic seed germination, hyphal branching in AM fungi, and shoot suppression in flowering plants [18]. However, the structural requirements of the ABC moiety vary among these target organisms; these observations suggest potential divergence between the corresponding receptors, although it remains plausible that AM fungal receptors may still belong to the α/β-fold hydrolase family [19]. On the other hand, receptors in fungi and plants could also independently evolve via convergent evolution and may not share structural similarities.

## 11.4 Influence of SLs on AM Fungi at the Cellular and Molecular Levels

AM fungi induce hyphal growth, spore germination, and mitosis in SLs; these effects are accompanied by increased respiration and mitochondrial biogenesis, suggesting a metabolic shift toward an active state conducive to root colonization [20]. At the cellular level, SLs influence the growth, development, and hyphal branching of AM fungi and act as chemoattractants, guiding fungal hyphae toward potential host plant roots [21]. Once in proximity to the roots, SLs promote hyphal penetration and colonization of the root cortex, aiding in the exchange of nutrients between the plant and the fungus. SLs contribute to the formation of specialized structures known as arbuscules and vesicles within root cells that serve as sites for nutrient transfer, allowing the fungus to deliver P and other essential nutrients to the host plant in return for carbon compounds [22]. SLs influence structures specialized for nutrient exchange and symbiotic interactions within host plants. Additionally, SL treatment leads to a swift elevation in cytosolic calcium levels within the hyphae of AM fungi [23].

At the molecular level, SLs elicit changes in gene expression, including the increased expression of genes associated with respiration and the upregulation of genes encoding potential secreted proteins [24]. One of these proteins, SIS1, has been identified as crucial for symbiosis. Moreover, SLs prompt the release of short chitin oligosaccharides by AM fungi, which are integral components of signals emitted by fungi that are necessary for recognition by host plants [25]. This finding emphasizes the pivotal importance of SLs in facilitating molecular communication within the mutualistic relationship between plant and fungal partners in AM symbiosis. Certain AM fungal species harbor endosymbiotic bacteria, and SL treatment has been associated with enhanced bacterial replication and the accumulation of transcripts from specific bacterial genes [26]. However, it remains unclear whether these effects of SLs are directly mediated or indirectly influenced by the fungus. Furthermore, SLs exert diverse effects on AM fungal physiology and molecular processes, orchestrating metabolic changes, gene expression patterns, and the release of signaling molecules critical for symbiotic interactions with host plants.

## 11.5 SLs and AM Fungi Mediate Root Development

Several novel properties of SLs have been discovered, including their role as AM fungal branching factors, regulators of plant structure, and responses to abiotic factors [10]. SLs act as plant hormones that modulate aboveground plant architecture by inhibiting shoot branching and bud outgrowth [27]. SLs stimulate AM fungal hyphal growth toward host plant roots, facilitating colonization and the establishment of symbiotic structures such as arbuscles and vesicles. Several studies have provided evidence indicating that SLs participate in controlling the growth and development of plant roots. Mitra *et al.* [6] demonstrated that SLs play a role in regulating the growth of plant roots and influence the elongation of both primary and adventitious roots while inhibiting the formation and development of lateral roots. Sun *et al.* [28] demonstrated that the effect of SLs on root growth and development can be reversed by the addition of GR24, an active synthetic SL compound, to SL-deficient plants and normal plants but not to mutants insensitive to SLs. Thus, SLs regulate root branching and elongation, aiding in the establishment of a robust root system for nutrient acquisition, and their influence on AM fungal colonization further enhances nutrient uptake and the exploration of soil resources (Figure 11.1) [29]. In rice, SLs positively regulate the formation of adventitious roots. Conversely, it was observed that SL mutants in rice produced fewer adventitious roots during the initial growth phase of the plant and a reduced quantity per tiller during maturation [30]. The addition of GR24 complemented the deficiency in adventitious root formation in the synthetic SL mutant (*d10*) but not in the SL signaling mutant (*d3*) [18].

Additionally, the symbiotic interactions between plants and AM fungi involve complex signaling networks in which SLs act as key signaling molecules. SL perception by plants and fungi triggers cascades of molecular responses that orchestrate root development and symbiotic establishment [31]. These responses involve the activation of specific genes, signaling pathways, and physiological processes that facilitate symbiotic interactions and nutrient exchange. Moreover, SLs play an essential role in adaptive reactions to P and N deficiencies by promoting symbiotic relationships with AM fungi, inducing branching of hyphae (Figure 11.1), and adjusting the shoot structure to cope with nutrient deficiencies [32]. Studies have shown that SLs control plant perception and react to P deficiency in soils, particularly by influencing root growth in rice plants under P and N deficiencies [33]. Furthermore, SLs induce the germination of parasites and

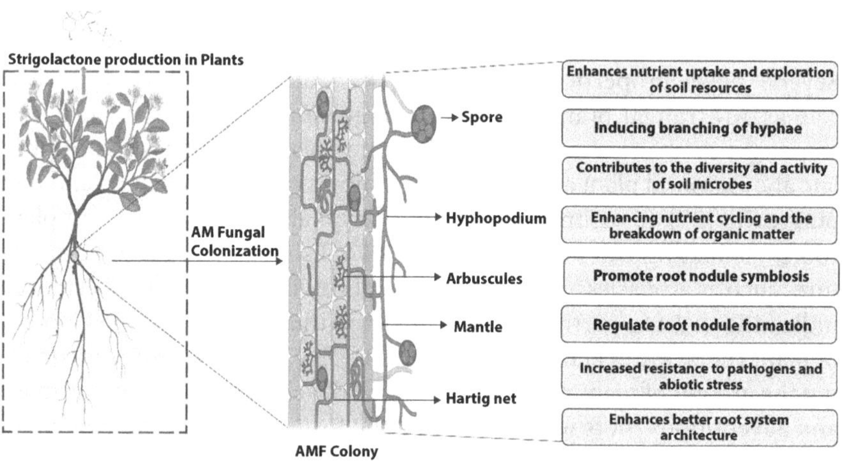


**Figure 11.1** Impact of strigolactones on AM fungal colonization for mediating the growth and tolerance of plants.

recruit AM fungal minerals and nutrients, suggesting that the management of parasites can be achieved by regulating AM fungal colonization [34]—for instance, *Striga hermonthica* germination significantly decreases in *Zea mays* and *Sorghum bicolor* varieties inoculated with AM fungi [35].

Guillotin [36] revealed that AM fungal inoculation reduces *Striga* germination partially due to SL downregulation, possibly as a mechanism of autoregulation to avoid excessive colonization. Structural variations in SLs critically impair AM fungal hyphal branching and stimulatory activity [37]—for example, in *Gigaspora margarita*, 5-deoxystrigol, lacking a hydroxyl group, is 30 times more potent than sorgomol, which contains a hydroxyl group in the initial tricyclic lactone ring. *Striga*-susceptible cultivars primarily secrete 5-deoxystrigol, while resistant cultivars produce orobanchol, indicating a role in field resistance to *Striga* [38]. Genetic analysis of *Striga*-resistant sorghum cultivars revealed a sulfotransferase gene mutation that reduces *S. hermonthica* germination by promoting orobanchol production more than 5-deoxystrigol [39]. *Striga* seeds can determine host diversity based on SL composition, suggesting that selecting high orobanchol producers could enhance P acquisition and AMF responsiveness while reducing *Striga* germination activity.

## 11.6 Impact of AM Fungi on Nutrient Acquisition (Particularly Phosphate) and Plant Growth

AM fungi have been recognized for their efficacy in enhancing nutrient absorption, especially P, and promoting biomass accumulation in numerous crops grown in P-deficient soils [40]. AM fungi facilitate nutrient transfer from one host plant to another. Mycorrhizal hyphae can extend from one host plant to infiltrate the roots of one or multiple adjacent plants. In AM fungi, P can be absorbed through two pathways: directly at the "soil–root interface" via the root epidermis and root hairs and through the "mycorrhizal" pathway facilitated by external AM hyphae in the soil [41]. These hyphae take up P and transport it swiftly to AM within the roots, including hyphae that exist between cells, coils of hyphae inside cells, and arbuscules within the cells, releasing it to the interfacial apoplast located adjacent to the cortical cells of the root [42]. Estimations of fungal involvement in overall P uptake typically consider the collaborative functioning of both direct uptake pathways and mycorrhizal pathways. However, direct uptake from the soil might diminish when the concentration of P in the soil solution decreases in proximity to the roots. The ability of hyphae to stretch beyond this depleted area and transport P to roots is considered to be the foundation of their beneficial impacts on P absorption and plant development [5, 43]. When there is no observable improvement in P uptake or plant growth as a result of symbiosis, it is presumed that the mycorrhizal uptake pathway is not operational. In such instances, all P uptake by the plant is assumed to occur via P transporters in the root hairs and root epidermis. Any increase in the growth of AM fungi is attributed to other advantages of symbiosis, including the suppression of pathogens.

Javeria [44] demonstrated that assimilates, such as $^{14}C$ photosynthate and $^{32}P$, are primarily transported between plants via hyphae of AM instead of via root exudation from donor plants. Mycorrhizal treatments positively influence the uptake of P and Zn, and analyses of P concentrations in plant tissues have revealed variations among mycorrhizal species at different inoculation times [45]. However, consistency between different mycorrhizal species and the timing of inoculation is limited because of the dilution effect caused by plant growth. Since P uptake integrates the P concentration achieved through mycorrhizal inoculation, the P concentration serves as a crucial indicator. Generally, plants inoculated with mycorrhizae exhibit a greater percentage of P concentrations than noninoculated plants across various inoculation times in multiple experiments [46]. AM fungal colonization occurs behind the root apex, where AM fungi develop an extensive hyphal network in the soil that absorbs P a few centimeters from the surface of roots, successfully extending the depletion zone [47]. The rapid translocation of P to the roots, likely in the form of polyphosphate, overcomes the gradual diffusion in the soil solution. Compared with roots, fungal hyphae have smaller diameters, allowing them to enter narrower soil pores and expanding the area of soil explored [48]. Specialized interfaces form between AM fungi and plant roots within root cortical cells and are characterized by elaborate fungal structures such as arbuscules and coiled hyphae. These formations are enveloped by the plasma membrane, creating specific membranes for both the plant and fungus, with an apoplastic region separating them. P is crucial for controlling nutrient transfer between symbionts; however, the mechanisms of P release from fungi to the interfacial apoplast are not fully understood, and P uptake into plants is well documented [49]. AM-inducible plant phosphate transporter (*PiT*) genes, which are distinct from those in the direct pathway, are expressed in cortical cells that are colonized [50]. These *PiTs* aid in the absorption of P released by fungi and are found in all plant species capable of forming AM associations, irrespective of their responsiveness to colonization by AM fungi. Furthermore, proton pumps ($H^+$-ATPases) stimulate the plant plasma membrane to enclose intracellular fungal structures, enabling active P uptake. Overall, AM symbiosis tends to have a favorable impact on the growth, development and well-being of plants, particularly under challenging environmental conditions. However, neutral or adverse impacts on the growth of plants, which are linked to P starvation and excessive carbon utilization by the AM fungus, have been reported.

The adverse plant reaction to AM fungal colonization has been suggested to be linked to diminished P absorption via the "direct pathway" triggered by symbiosis as well as reduced P uptake capacity by AM fungal infection via the "AM pathway" [40]. Therefore, finding the optimal "dance partner" is essential for establishing a mutually adventitious association between plants and AM fungi. This suggests that not all plant-fungus interactions result in positive outcomes and that the effectiveness of AM symbiosis can be altered by factors such as plant species, soil conditions, and environmental stresses. Hence, understanding the dynamics of plant-fungus interactions and optimizing symbiotic relationships are essential for enhancing the benefits derived from AM symbiosis in agricultural and ecological contexts.

## 11.7 AM Symbiosis: Serving as a Biofertilizer and Biocontrol Agent

The agricultural revolution following World War II led to soil overexploitation and excessive use of agrochemicals, raising concerns about their environmental impact. Consequently, there is growing interest in environmentally friendly agriculture, with AM symbiosis emerging as a potential solution [51]. As biofertilizer AM fungi symbiosis enhances nutrient uptake, particularly P, by forming symbiotic associations with plant roots, fungi stretch their hyphae into the soil, increasing the volume of soil exploring and absorbing minerals and nutrients from a wider area, which improves plant nutrient availability, resulting in higher crop yields and enhanced soil fertility.

Additionally, AM fungi can enhance the uptake of other essential nutrients, such as N, K, and micronutrients, further enhancing plant growth and productivity [52]. AM symbiosis promotes soil aggregation and stability, which enhances soil structure and reduces erosion. The extensive hyphal network formed by AM fungi improves soil aeration and water infiltration, leading to better soil moisture retention and drought tolerance [4]. Moreover, AM fungal symbiosis contributes to soil microbial diversity and activity, enhancing nutrient cycling and the breakdown of organic matter (Figure 11.1). As biocontrol AM fungi can inhibit pathogens that reside in the soil and pests through various mechanisms, they compete with pathogens for root colonization sites and nutrients, reducing pathogen proliferation [53]. AM symbiosis also induces systemic resistance in plants, activating defense mechanisms that enhance plant immunity against pathogens [54]. Furthermore, some AM fungi produce antifungal compounds that inhibit the growth of pathogenic organisms in the soil by improving nutrient uptake and enhancing plant resilience to biotic and abiotic stresses, and AM symbiosis reduces the reliance on chemical fertilizers and pesticides [5, 55]. This promotes environmentally friendly agricultural practices, minimizes negative impacts on soil and water quality, and contributes to sustainable farming systems.

SLs play a crucial role in AM symbiosis establishment, and breeding for cultivars with high SL production or the exogenous application of SLs can enhance mycorrhizal colonization. Controlled stress conditions that do not overly harm plants may also promote AM symbiosis, but caution is needed due to the role of SLs as stimulants for triggering germination in parasitic plant roots and their involvement in various plant functions. Understanding the structure–activity relationships of SLs is crucial before their application. Research has shown differences in the effects of SLs on the germination of parasitic weeds, branching of AM fungi and branching of shoots [20]. The roles of specific SLs in host plant–AM fungal interactions, including transport and exudation, require further investigation to optimize AM symbiosis in agroecosystems.

## 11.8 Interactions of SL with Non-AM Fungi

Limited data exist on the interplay of SLs with nonarbuscular mycorrhizal (non-AM) fungi. Dor and Poveda [56, 57] separately conducted experiments with GR24, a synthetic SL, and observed no significant variations in the growth or branching patterns of ectomycorrhizal fungi such as *Trichoderma* and *Piriformospora indica*, as well as soil-borne tomato pathogens such as *Fusarium oxysporum* f. sp. *lycopersici*, *Botrytis cinerea*, and *Cladosporium* sp. Despite these findings, indirect evidence suggests that SLs may indeed affect pathogenic fungi—for instance, the growth pattern of the pathogen *Sporisorium reilianum* was affected by maize root exudates containing compounds similar to SLs.

SLs exhibit antifungal properties against a variety of non-AM fungi; they can disrupt fungal cell membrane integrity and inhibit fungal spore germination and hyphal growth. These antifungal properties suggest that SLs may have potential applications in controlling fungal diseases in agricultural and horticultural crops [58]. Moreover, SLs released by plants may also influence the composition and activity of soil microbial communities, including non-AM fungi. SLs in the rhizosphere can change the richness and variety of soil fungi, potentially affecting nutrient cycling and soil ecosystem functioning. Furthermore, a fraction of maize root exudates was found to activate or inhibit fungal cell proliferation and induce the transition between yeast and hyphal forms in the pathogen [56]. Recent research has also shown that the application of GR24 to *S. reilianum* induces the rapid expression of genes related to cell division and respiration, similar to observations with AM fungi [59]. These results indicated that SLs might play a broader role in signaling within plant-microbe interactions occurring in the soil, although more data on such roles are needed to fully understand their impact.

## 11.9 Strigolactones and Root Nodule Symbiosis

Plants need N compounds such as nitrate or ammonia to grow, but these compounds are often scarce in natural soil, leading to the slow growth of plants. Agriculture relies on chemical N fertilizers for compensation, but this process is unsustainable and pollutes the environment. Diazotrophic bacteria possess the Nase enzyme and convert inert N gas into ammonia, which plants can use. Certain bacteria facilitate the transfer of ammonia to plant roots, facilitating plant growth in exchange for carbon from plants. This symbiotic relationship helps plants thrive in N-deficient soils. An instance of nitrogen-fixing symbiosis occurs between leguminous plants and rhizobia, and this association results in the formation of specialized root structures called nodules, where rhizobia efficiently convert atmospheric nitrogen into a usable form [60].

SLs serve as signal molecules that attract beneficial microbes, including rhizobia bacteria, which form symbiotic relationships with legume roots to establish root nodule symbiosis [61]. SLs not only promote root nodule symbiosis (Figure 11.2) but also induce AM symbiosis in non-leguminous plants [62, 63], suggesting that SLs play a broader role in facilitating beneficial interactions between plants and soil microbes. [64] demonstrated that SLs can modulate the nodulation process in leguminous plants and may act as signaling molecules that influence the initiation and development of root nodules by regulating the expression of Nod genes and interacting with other signaling pathways involved in nodulation. When plants detect N deficiency, they trigger molecular communication with neighboring rhizobia by secreting a blend of secondary metabolites, primarily flavonoids and isoflavonoids [65]. These compounds activate Nod genes within rhizobia, producing Nod factors and lipochitooligosaccharides with strain-specific chemical modifications [19]. In the presence of Nod, plant roots initiate nodule organ development and allow rhizobial colonization. This process exhibits strong host specificity, with only certain legumes forming symbiotic associations with specific rhizobial strains.

Evidence supporting the involvement of SLs in the establishment of the rhizobium–legume symbiosis emerged from a study demonstrating that the synthetic SL analog GR24 enhanced nodule formation in *Medicago truncatula* plants inoculated with *Sinorhizobium meliloti* [66], while other legumes with symbionts such as *Azorhizobium caulinodans* have *Sesbania rostrata* [67]. Interestingly, this positive effect on nodulation was not associated with an increase in the growth of bacteria or the expression of nod genes, but this specificity arises from the molecular dialog between rhizobia and legumes, where rhizobial strains activate Nod genes in response to specific flavonoids, and legumes initiate nodulation upon perceiving Nod factors with particular structures [68]. Rhizobia typically enter roots through root hairs, which curl to entrap them, and an infection thread forms via the inward folding of the cytoplasmic membrane in the root, directing bacteria toward the cortical cells of the root [69]. Concurrently, cortical cells divide to form nodule primordia, where rhizobia are released into symbiosomes surrounded by a host-derived membrane for N fixation. Mature nodules may be determinate or indeterminate, with determinate nodules being round-shaped and fully occupied by infected cells, while indeterminate nodules are elongated, with a meristem continuously providing new cells for infection.

SLs interact with other phytohormones and signaling molecules involved in plant-microbe interactions, such as cytokinins, auxins, and ethylene. The cross-talk between SL signaling pathways and other pathways may fine-tune the plant's response to symbiotic signals and regulate the establishment and maintenance of root nodule symbiosis [70]. Understanding the role of SLs in root nodule symbiosis has implications for agricultural practices, particularly in enhancing N fixation in legume crops. Manipulating SL biosynthesis or signaling pathways may offer strategies to enhance the efficiency of N fixation and enhance crop yields in sustainable agriculture.

## 11.10 Effect of SLs on Nodule Number (Quantity)

Studies involving legume plants with modified levels of endogenous SLs revealed that while these phytohormones are not indispensable for functional nodule development, they may also influence optimal nodule numbers (Figure 11.2)—for instance, in pea, the SL-deficient mutant *rms1* formed approximately 40% fewer nodules (Figure 11.2(b)) than did the wild type (Figure 11.2(a)), which was partially restored by GR24 application [71]. In *Medicago truncatula*, GR24 had a positive effect on the number of nodules at low rac-GR24 concentrations, as shown in Figure 11.2(c), but had a negative effect at high rac-GR24 concentrations (Figure 11.2(d)) [72]. Similarly, *Lotus japonicus* plants with silenced *LjCCD7*, resulting in 80% reduction in SL, exhibited a 20% decrease in the number of nodules (Figure 11.3(b)) [73]. However, the overall understanding of this phenomenon may not be straightforward, as data obtained from various legume species show conflicting results—for example, contrary to expectations, there was an increase in the number of nodules observed in the SL-insensitive ramosus4 (*rms4*) mutant; this mutant is characterized by a mutation in the ortholog of the *Arabidopsis thaliana MORE AXILLARY BRANCHES 2* (*MAX2*), which is a component of the SL receptor complex [74]. Therefore, these findings indicated that the modulation of the nodulation process by SLs may vary depending on the concentration of SLs and the particular sensitivity of legume species to SLs.

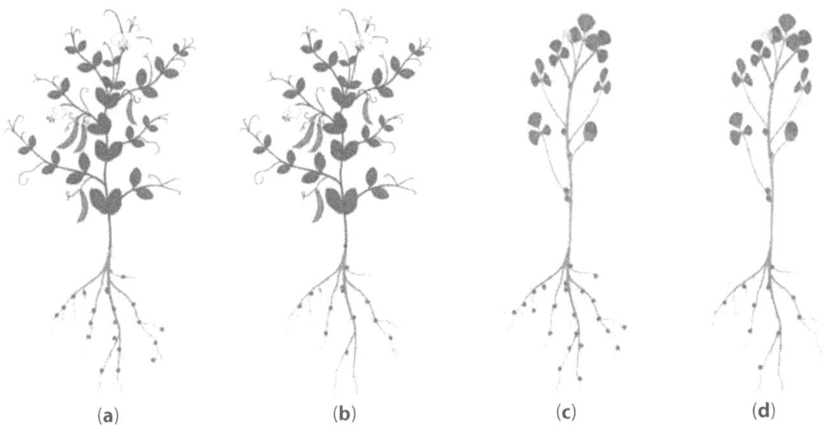


**Figure 11.2** Wild type pea plant showing normal nodule numbers (a), SL deficient mutant *rms*1 formed approximately 40% less nodules than wild type (b), In *Medicago truncatula* low concentration of *rac*-GR24 increase nodule number (c), while high concentration of *rac*-GR24 decreases nodule number (d).

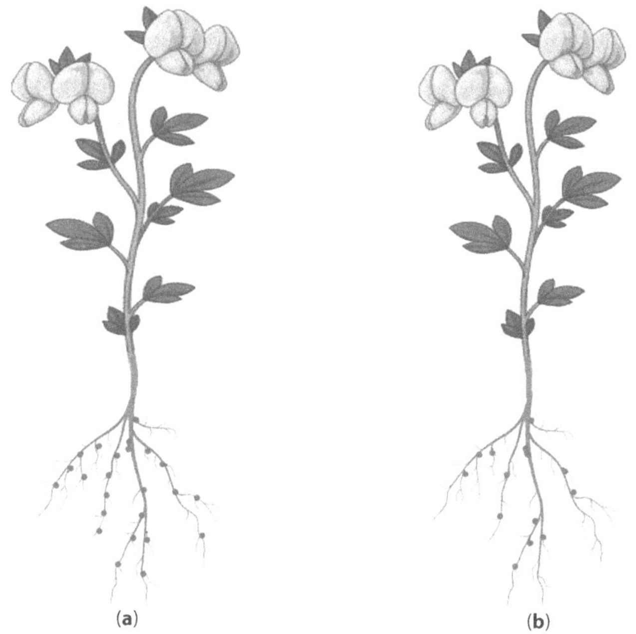


**Figure 11.3** Wild type *Lotus japonicus* shows normal number of root nodules (a), *Lotus japonicus LjCCD7*-silenced plants with an 80% SL reduction carried 20% fewer nodules (b).

Notably, *MAX2/RMS4* is also implicated in recognizing smoke-derived karrikin and an unidentified endogenous ligand in conjunction with *KARRIKIN-INSENSITIVE 2* (*KAI2*) [19]. Therefore, it remains plausible that other carotenoid-derived compounds recognized by a *KAI2* and *MAX2/RMS4* complex might contribute to the nodulation process, and rac-GR24, comprising two Dia stereoisomers, appears to be acknowledged by *KAI2* in *Arabidopsis* [75]. Hence, future research endeavors should meticulously examine the roles of both the SL and karrikin signaling pathways, integrating thorough metabolite studies to establish precise correlations among plant mutants, metabolites, and their functional significance. Moreover, the SL biosynthesis genes *MtD27*, *MtCCD7*, and *MtCCD8* were expressed in root hairs, meristematic cells, and nodule infection zones in response to *Rhizobium* and Nod [76], highlighting the multifaceted role of SLs in the *Rhizobium*–legume interaction.

An important clue for comprehending the role of SLs in nodule formation lies in examining the specific patterns of expression of SL-related genes within tissues during nodulation. Studies have revealed the upregulation of

SL biosynthesis genes during nodulation, which is governed by the Nod factor signaling pathway [77]. Promoter activity analyses indicate that these genes exhibit activity in nodule meristems, nodule primordia and the early infection zone of mature indeterminate nodules in *Medicago truncatula* [78]. This spatial pattern of activity at cell division sites suggests a potential role for auxin and cytokinin in regulating cell division. Despite these observations, no significant alterations in nodule structure have been noted in diverse SL mutants across various legume species, suggesting that this role may not be pivotal. Another interpretation of this expression pattern is its potential involvement in autoregulation of nodulation (AON), whereby legume plants control the quantity of nodules per plant to manage N fixation [79]; however, investigations with AON mutants have yet to verify this hypothesis. Consequently, further research is warranted to clarify the precise role of SLs in nodule formation and their potential participation in AON mechanisms.

SLs likely influence the early stages of root nodule symbiosis, as observed in *Medicago truncatula,* where rac-GR24 hindered infection thread development and related gene expression, as indicated by an ethylene perception mutant [80]. Conversely, a pea SL biosynthesis mutant displayed reduced infection threads. The conflicting results between rac-GR24 treatment and mutant analysis suggest the involvement of the Karrikin receptor complex in nodulation or the necessity of precise control over SL concentrations for desired outcomes in legume-rhizobia interactions [81]. Different legumes may have varying sensitivities to SLs, and the locations where hormone biosynthesis occurs and where their effects take place may not necessarily overlap. Thus, while SLs may be produced in dividing cells, they might exert their effects at the level of rhizobial infection. Further research, including detailed, cell-specific expression analysis of genes related to SL perception, is necessary to uncover the precise role of SLs in modulating nodule quantity and to elucidate their significance in plant-microbe interactions and agricultural practices.

## 11.11 Impact of SLs on Rhizobia

Apart from their endogenous functions, SLs may also serve as rhizosphere signals, acting outside plant cells after secretion into the soil, while SLs may not play as significant a role in rhizosphere signaling for nodulation as they do for AM formation; their presence in the rhizosphere could still impact rhizobia, albeit challenging to observe under *in vitro* conditions [82].

The symbiosis between legumes and *Rhizobium* bacteria, dating back more than 60 million years, facilitates N acquisition by the host plant by converting atmospheric N into ammonia within specialized root nodules [83]. Various studies have suggested that SLs may affect rhizobial motility, particularly swarming behavior, although their precise function is not fully understood [84–86]. Treatment with rac-GR24, a synthetic SL analog, did not affect nodulation factor production, bacterial growth, or bacterial calcium spiking, suggesting that rac-GR24 functions primarily as a hormone, in contrast to its role as a rhizosphere signaling molecule in AM fungal interactions [87].

Nevertheless, it is notable that the crystal structure of the rice SL receptor D14 is similar to that of the *Bacillus subtilis* RsbQ protein, indicating that SL potentially influences rhizobial cellular activity differently [88]. Indeed studies suggest a role for SL in chemical communication between plants and rhizobia; extracts from the moss *Physcomitrella patens* stimulate *Rhizobium leguminosarum* swarming motility, whereas this effect diminishes with extracts from SL-deficient strains [76]. Additionally, rac-GR24 triggers *Sinorhizobium meliloti* swarming motility in a dose-dependent manner, and these findings, coupled with increased flagellin *flaA* gene expression, suggest that rhizobia perceive and respond to SL [89]. SLs can act as signals for rhizobia detection by the host in the rhizosphere, potentially enhancing bacterial swarming motility and influencing the symbiotic relationship between rhizobia and legumes [90]. Furthermore, essential genes involved in SL biosynthesis, such as *D27*, *CCD7*, and *CCD8*, are activated during interactions with *Rhizobium*, possibly through elements of the common SYM pathway [8]. However, the ecological significance of this swarming behavior is not well understood and warrants further investigation to improve our understanding of nodulation.

## 11.12 Future Directions and Conclusion

One of the primary hurdles facing agriculture in the foreseeable future involves identifying sustainable and eco-friendly alternatives to reduce reliance on chemical fertilizers and pesticides while also safeguarding the yield and quality of crops [5]. A promising approach involves harnessing interactions with beneficial soil microorganisms such as AM fungi and nitrogen-fixing rhizobia; these interactions have the potential to reduce the need for fertilizers while also bolstering crop resilience to diverse stresses encountered in field conditions. Indeed these microorganisms are now being marketed as biofertilizers and bio protection agents by numerous companies in the agri-food sector. SLs, which are natural plant-derived compounds, play crucial roles in facilitating the establishment of AM and rhizobial symbiosis [91]. Future investigations may uncover additional roles for SLs in other mutualistic associations. However, emerging evidence suggests that SL production and release from plant roots orchestrate the symbiotic relationships between plants and AM fungi and between *Rhizobium* bacteria and legumes, ultimately enhancing plant growth and development under normal and stressful conditions [92]. SLs are crucial for facilitating the recognition of plants by symbiotic fungi, particularly AM fungi, which foster a robust root system to enhance mineral and water absorption, while AM fungi ensure a steady supply of reduced carbon [93].

Further research is needed to explore the potential involvement of SLs in diseases caused by various organisms, including nematodes, bacteria, fungi, viruses and insect pests. Understanding whether SLs directly affect microorganisms or induce defense mechanisms in host plants by interacting with other phytohormones is crucial. Additionally, SLs appear to operate within the regulatory network governing phosphate metabolism and root morphogenesis, processes influenced to some extent by AM symbiosis [94]. Understanding the biological significance of each component within this intricate network and their interactions presents a formidable challenge for future research endeavors. Moreover, exploring the involvement of SLs in root-to-shoot communication during interactions between plants and microbes is justified, considering their mobility within plants from roots to shoots. The regulation of nodulation by SLs and the specific signaling pathways involved, mainly whether D14 or KAI2 mediates the impact of rac-GR24 application on infection thread formation and nodulation, remain unclear [95]. Understanding this mechanism awaits the availability of legume mutants in each receptor gene. Altering SL concentrations in roots through genetic or pharmacological means has produced partly divergent results in *M. truncatula* and pea [96, 97], suggesting potential species-specific sensitivity to SLs. Alternatively, differences in growth conditions such as light intensity or nutrient concentration, which are known to affect SL biosynthesis and other hormone pathways, could account for variations in SL effects. Hence, discrepancies in the effects of SL on nodulation may stem from differences in the balance and interactions of various hormones. A systematic comparison of the effects of SL on nodulation across various growth conditions could support this hypothesis.

Understanding how various environmental factors, particularly the nutritional status of plants, impact the synthesis of SLs is essential. It is crucial to explore how different stresses, whether abiotic or biotic, affect the production of SLs to determine their role in plant interactions. Furthermore, it is vital to distinguish the specific functions of different SLs produced by the same plant, given that each plant produces a unique combination of SLs. Future research efforts should prioritize elucidating the specific functions of individual SLs to gain deeper insights into their involvement in plant–microbe interactions, nodulation processes, nodule quantity, and prospective use in sustainable farming practices.

## References


1. Al-Babili, S. and Bouwmeester, H.J., Strigolactones, a novel carotenoid-derived plant hormone. *Annu. Rev. Plant Biol.*, *66*, 161–186, 2015.
2. Samynathan, R., Venkidasamy, B., Shariati, M.A., Muthuramalingam, P., Thiruvengadam, M., Biosynthesis, functional perspectives, and agricultural applications of strigolactones. *Braz. J. Bot.,* 47, 751–770, 2024.
3. Behie, S.W. and Bidochka, M.J., Nutrient transfer in plant–fungal symbioses. *Trends Plant Sci.*, 19, 11, 734–740, 2014.
4. Wahab, A., Muhammad, M., Munir, A., Abdi, G., Zaman, W., Ayaz, A., Reddy, S.P.P., Role of arbuscular mycorrhizal fungi in regulating growth, enhancing productivity, and potentially influencing ecosystems under abiotic and biotic stresses. *Plants*, 12, 17, 3102, 2023.
5. Hajam, A.H., Ali, M.S., Singh, S.K., Bashri, G., Dubey, G., Implications of strigolactones in plant biology: Achievements, future perspectives, and challenges. *Strigolactones*, 221–236, 2024.
6. Mitra, D., Rad, K.V., Chaudhary, P., Ruparelia, J., Sagarika, M.S., Boutaj, H., Panneerselvam, P., Involvement of strigolactone hormone in root development, influence and interaction with mycorrhizal fungi in plant: Mini-review. *Curr. Res. Microb. Sci.*, 2, 100026, 2021.
7. Soliman, S., Wang, Y., Han, Z., Pervaiz, T., El-Kereamy, A., Strigolactones in Plants and Their Interaction with the Ecological Microbiome in Response to Abiotic Stress. *Plants*, 11, 24, 3499, 2022.
8. Begum, N., Zhao, T., Antwi-Boasiako, A., Zhang, L., Ashraf, M., Role of strigolactones in rhizobiology: Plant-microbe interactions, in: *Phytohormones and Stress Responsive Secondary Metabolites*, pp. 13–25, Academic Press, United Kingdom, 2023.
9. Ahmad, S., Belwal, V., Punia, S.S., Ram, M., Rajput, S.S., Mohamed, H. I., Role of plant secondary metabolites and phytohormones in drought tolerance: a review. *Gesunde Pflanzen*, 75, 4, 729–746, 2023.
10. Boyno, G., Rezaee Danesh, Y., Demir, S., Teniz, N., Mulet, J.M., Porcel, R., The Complex Interplay between Arbuscular Mycorrhizal Fungi and Strigolactone: Mechanisms, Sinergies, Applications and Future Directions. *Int. J. Mol. Sci.*, 24, 23, 16774, 2023.

11. Koltai, H., Matusova, R., Kapulnik, Y., Strigolactones in root exudates as a signal in symbiotic and parasitic interactions. *Secretions Exudates Biol. Sys.*, 12, 49–73, 2012.
12. Cavar, S., Zwanenburg, B., Tarkowski, P., Strigolactones: occurrence, structure, and biological activity in the rhizosphere. *Phytochem. Rev.*, 14, 691–711, 2015.
13. Cardoso, C., Ruyter-Spira, C., Bouwmeester, H.J., Strigolactones and root infestation by plant-parasitic Striga, *Orobanche* and *Phelipanche spp. Plant Sci.*, 180, 3, 414–420, 2011.
14. Besserer, A., Puech-Pagès, V., Kiefer, P., Gomez-Roldan, V., Jauneau, A., Roy, S., Séjalon-Delmas, N., Strigolactones stimulate arbuscular mycorrhizal fungi by activating mitochondria. *PLoS Biol.*, 4, 7, e226, 2006.
15. Fitter, A.H., Darkness visible: reflections on underground ecology. *J. Ecol.*, 93, 2, 231–243, 2005.
16. de Saint Germain, A., Clavé, G., Badet-Denisot, M.A., Pillot, J.P., Cornu, D., Le Caer, J.P., Boyer, F.D., An histidine covalent receptor and butenolide complex mediates strigolactone perception. *Nat. Chem. Biol.*, 12, 10, 787–794, 2016.
17. Dun, E.A., de Saint Germain, A., Rameau, C., Beveridge, C.A., Dynamics of strigolactone function and shoot branching responses in *Pisum sativum. Mol. Plant Pathol.*, 6, 1, 128–140, 2013.
18. Pandey, A., Sharma, M., Pandey, G.K., Emerging roles of strigolactones in plant responses to stress and development. *Front. Plant Sci.*, 7, 434, 2016.
19. Rochange, S., Goormachtig, S., Lopez-Raez, J.A., Gutjahr, C., The role of strigolactones in plant–microbe interactions. *Strigolactones-Biol. Appl.*, 121–142, 2019.
20. Kountche, B.A., Novero, M., Jamil, M., Asami, T., Bonfante, P., Al-Babili, S., Effect of the strigolactone analogs methyl phenlactonoates on spore germination and root colonization of arbuscular mycorrhizal fungi. *Heliyon.*, 4, 11, e00936, 2018.
21. Ogawa, S., Cui, S., White, A.R., Nelson, D.C., Yoshida, S., Shirasu, K., Strigolactones are chemoattractants for host tropism in Orobanchaceae parasitic plants. *Nat. Commun.*, 13, 1, 4653, 2022.
22. Fusconi, A., Regulation of root morphogenesis in arbuscular mycorrhizae: what role do fungal exudates, phosphate, sugars and hormones play in lateral root formation? *Ann. Bot.*, 113, 1, 19–33, 2014.
23. Moscatiello, R., Sello, S., Ruocco, M., Barbulova, A., Cortese, E., Nigris, S., Navazio, L., The hydrophobin HYTLO1 secreted by the biocontrol fungus *Trichoderma longibrachiatum* triggers a NAADP-mediated calcium signaling pathway in Lotus japonicus. *Int. J. Mol. Sci.*, 19, 9, 2596, 2018.
24. Lanfranco, L., Fiorilli, V., Venice, F., Bonfante, P., Strigolactones cross the kingdoms: plants, fungi, and bacteria in the arbuscular mycorrhizal symbiosis. *J. Exp. Bot.*, 69, 9, 2175–2188, 2018.

25. Zhai, X., Jia, M., Chen, L., Zheng, C.J., Rahman, K., Han, T., Qin, L.P., The regulatory mechanism of fungal elicitor-induced secondary metabolite biosynthesis in medical plants. *Crit. Rev. Microbiol.*, 43, 2, 238–261, 2017.
26. Salvioli, A., Ghignone, S., Novero, M., Navazio, L., Venice, F., Bagnaresi, P., Bonfante, P., Symbiosis with an endobacterium increases the fitness of a mycorrhizal fungus, raising its bioenergetic potential. *ISME J.*, 10, 1, 130–144, 2016.
27. Faizan, M., Faraz, A., Sami, F., Siddiqui, H., Yusuf, M., Gruszka, D., Hayat, S., Role of strigolactones: Signaling and crosstalk with other phytohormones. *Open Life Sci.*, 15, 1, 217–228, 2020.
28. Sun, H., Tao, J., Gu, P., Xu, G., Zhang, Y., The role of strigolactones in root development. *Plant Signal Behav.*, 11, 1, e1110662, 2016.
29. Andreo-Jimenez, B., Ruyter-Spira, C., Bouwmeester, H.J., Lopez-Raez, J.A., Ecological relevance of strigolactones in nutrient uptake and other abiotic stresses, and in plant-microbe interactions below-ground. *Plant Soil*, 394, 1–19, 2015.
30. Sun, H., Tao, J., Hou, M., Huang, S., Chen, S., Liang, Z., Zhang, Y., A strigolactone signal is required for adventitious root formation in rice. *Ann. Bot.*, 115, 7, 1155–1162, 2015.
31. Gutjahr, C., Phytohormone signaling in arbuscular mycorhiza development. *Curr. Opin. Plant Biol.*, 20, 26–34, 2014.
32. Miyata, K. and Umehara, M., Roles of Arbuscular Mycorrhizal Fungi for Essential Nutrient Acquisition Under Nutrient Deficiency in Plants, in: *Arbuscular Mycorrhizal Fungi and Higher Plants: Fundamentals and Applications*, pp. 123–148, Springer Nature Singapore, Singapore, 2024.
33. Lu, H., Wang, F., Wang, Y., Lin, R., Wang, Z., Mao, C., Molecular mechanisms and genetic improvement of low-P tolerance in rice. *PLCEDV.*, 46, 4, 1104–1119, 2023.
34. Sharma, S., Bhuvaneswari, V., Saikia, B., Karthik, R., Rajeshwaran, B., Shree Naveena, P., Gayithri, M., Multitrophi c Reciprocity of AMF with Plants and Other Soil Microbes in Relation to Biotic Stress, in: *Microbial Symbionts and Plant Health: Trends and Applications for Changing Climate*, pp. 329–366, Springer Nature Singapore, Singapore, 2023.
35. Othira, J.O., Omolo, J.O., Wachira, F.N., Onek, L.A., *Effectiveness of arbuscular mycorrhizal fungi in protection of maize (Zea mays L.) against witchweed (Striga hermonthica Del Benth) infestation*, 2012, http://repository.seku.ac.ke/handle/123456789/3523.
36. Guillotin, B., *Spatiotemporal regulation of the arbuscular mycorrhiza symbiosis establishment*, 2016, (Doctoral dissertation, Université Paul Sabatier-Toulouse III).
37. Tominaga, T., Ueno, K., Saito, H., Egusa, M., Yamaguchi, K., Shigenobu, S., Kaminaka, H., Monoterpene glucosides in Eustoma grandiflorum roots promote hyphal branching in arbuscular mycorrhizal fungi. *Plant Physiol.*, 193, 4, 2677–2690, 2023.

38. Abdelhalim, T., Jannoura, R., Joergensen, R.G., Mycorrhiza response and P acquisition efficiency of sorghum cultivars differing in strigolactone composition. *Plant Soil*, 437, 55–63, 2019.
39. Yoda, A., Mori, N., Akiyama, K., Kikuchi, M., Xie, X., Miura, K., Nomura, T., Strigolactone biosynthesis catalyzed by cytochrome P450 and sulfotransferase in sorghum. *New Phytol.*, 232, 5, 1999–2010, 2021.
40. Etesami, H., Jeong, B.R., Glick, B.R., Contribution of arbuscular mycorrhizal fungi, phosphate-solubilizing bacteria, and silicon to P uptake by plant. *Front. Plant Sci.*, 12, 699618, 2021.
41. Smith, S.E. and Smith, F.A., Roles of arbuscular mycorrhizas in plant nutrition and growth: new paradigms from cellular to ecosystem scales. *Annu. Rev. Plant Biol.*, 62, 227–250, 2011.
42. Smith, S.E., Dickson, S., Smith, F.A., Nutrient transfer in arbuscular mycorrhizas: how are fungal and plant processes integrated? *Funct. Plant Biol.*, 28, 7, 685–696, 2001.
43. Smith, S.E., Jakobsen, I., Grønlund, M., Smith, F.A., Roles of arbuscular mycorrhizas in plant P nutrition: interactions between pathways of P uptake in arbuscular mycorrhizal roots have important implications for understanding and manipulating plant P acquisition. *Plant Physiol.*, 156, 3, 1050–1057, 2011.
44. Javeria, S., Kumar, V., Sharma, P., Prasad, L., Kumar, M., Varma, A., Mycorrhizal symbiosis: ways underlying plant–fungus interactions. *Mycorrhiza-ecophysiology, Secondary Metabolites, Nanomater.*, 183–207, 2017.
45. Yu, B.G., Chen, X.X., Cao, W.Q., Liu, Y.M., Zou, C.Q., Responses in zinc uptake of different mycorrhizal and nonmycorrhizal crops to varied levels of P and zinc applications. *Front. Plant Sci.*, 11, 606472, 2020.
46. Rubio, R., Borie, F., Schalchli, C., Castillo, C., Azcón, R., Occurrence and effect of arbuscular mycorrhizal propagules in wheat as affected by the source and amount of P fertilizer and fungal inoculation. *Appl. Soil Ecol.*, 23, 3, 245–255, 2003.
47. Thonar, C., Schnepf, A., Frossard, E., Roose, T., Jansa, J., Traits related to differences in function among three arbuscular mycorrhizal fungi. *Plant Soil*, 339, 231–245, 2011.
48. Hammer, E.C., Arellano-Caicedo, C., Mafla-Endara, P.M., Kiers, E.T., Shimizu, T., Ohlsson, P., Aleklett, K., Hyphal exploration strategies and habitat modification of an arbuscular mycorrhizal fungus in microengineered soil chips. *Fungal Ecol.*, 67, 101302, 2024.
49. Chiu, C.H. and Paszkowski, U., Mechanisms and impact of symbiotic phosphate acquisition. *Cold Spring Harbor Perspect. Biol.*, 11, 6, a034603, 2019.
50. Chen, A., Hu, J., Sun, S., Xu, G., Conservation and divergence of both phosphate- and mycorrhiza-regulated physiological responses and expression patterns of phosphate transporters in solanaceous species. *New Phytol.*, 173, 4, 817–831, 2007.

51. Kapoor, R., Evelin, H., Mathur, P., Giri, B., Arbuscular mycorrhiza: approaches for abiotic stress tolerance in crop plants for sustainable agriculture, in: *Plant acclimation to environmental stress*, pp. 359–401, Springer New York, New York, NY, 2012.
52. Bhantana, P., Rana, M.S., Sun, X.C., Moussa, M.G., Saleem, M.H., Syaifudin, M., Hu, C.X., Arbuscular mycorrhizal fungi and its major role in plant growth, zinc nutrition, phosphorous regulation and phytoremediation. *Symbiosis*, 84, 19–37, 2021.
53. Jain, P. and Pundir, R.K., Biocontrol of soil phytopathogens by arbuscular mycorrhiza–a review. Mycorrhiz. *Pedogen*, 221–237, 2019.
54. Jung, S.C., Martinez-Medina, A., Lopez-Raez, J.A., Pozo, M.J., Mycorrhiza-induced resistance and priming of plant defenses. *J. Chem. Ecol.*, 38, 651–664, 2012.
55. Begum, N., Qin, C., Ahanger, M.A., Raza, S., Khan, M., II, Ashraf, M., Zhang, L., Role of arbuscular mycorrhizal fungi in plant growth regulation: implications in abiotic stress tolerance. *Front. Plant Sci.*, 10, 1068, 2019.
56. Dor, E., Joel, D.M., Kapulnik, Y., Koltai, H., Hershenhorn, J., The synthetic strigolactone GR24 influences the growth pattern of phytopathogenic fungi. *Planta*, 234, 419–427, 2011.
57. Poveda, J., Abril-Urias, P., Escobar, C., Biological control of plant-parasitic nematodes by filamentous fungi inducers of resistance: Trichoderma, mycorrhizal and endophytic fungi. *Front. Microbiol.*, 11, 992, 2020.
58. Lastochkina, O., Aliniaeifard, S., SeifiKalhor, M., Bosacchi, M., Maslennikova, D., Lubyanova, A., Novel approaches for sustainable horticultural crop production: Advances and prospects. *Hortic*, 8, 10, 910, 2022.
59. SABBAGH, S., Mazaheri, M., Panjehkeh, N., SALARI, M., Transcriptomic analysis of Sporisorium reilianum in response to the strigolactone analog GR24. *Phytopathol. Mediterr.*, 51, 2, 283–291, 2012.
60. Yang, J., Lan, L., Jin, Y., Yu, N., Wang, D., Wang, E., Mechanisms underlying legume–rhizobium symbioses. *J. Integr. Plant Biol.*, 64, 2, 244–267, 2022.
61. Kalia, V.C., Gong, C., Patel, S.K., Lee, J.K., Regulation of plant mineral nutrition by signal molecules. *Microorganisms*, 9, 4, 774, 2021.
62. Yoneyama, K., Xie, X., Kim, H., II, Kisugi, T., Nomura, T., Sekimoto, H., Yoneyama, K., How do nitrogen and P deficiencies affect strigolactone production and exudation? *Planta*, 235, 1197–1207, 2012.
63. Sindhu, S.S., Sharma, R., Sindhu, S., Sehrawat, A., Soil fertility improvement by symbiotic rhizobia for sustainable agriculture. *Soil Fertil. Manage. Sustain. Dev.*, 101–166, 2019.
64. Liu, H., Zhang, C., Yang, J., Yu, N., Wang, E., Hormone modulation of legume-rhizobial symbiosis. *J. Integr. Plant Biol.*, 60, 8, 632–648, 2018.
65. Hassan, S. and Mathesius, U., The role of flavonoids in root–rhizosphere signaling: opportunities and challenges for improving plant–microbe interactions. *J. Exp. Bot.*, 63, 9, 3429–3444, 2012.

66. Menéndez, E., Martínez-Hidalgo, P., Silva, L.R., Velázquez, E., Mateos, P.F., Peix, A., Recent advances in the active biomolecules involved in rhizobia-legume symbiosis. *Microbes Legume Improv.*, 45–74, 2017.
67. Rinaudo, G., Orenga, S., Fernandez, M.P., Meugnier, H., Bardin, R., DNA homologies among members of the genus Azorhizobium and other stem- and root-nodulating bacteria isolated from the tropical legume Sesbania rostrata. *Int. J. Syst. Evol. Microbiol.*, 41, 1, 114–120, 1991.
68. Cooper, J.E., Early interactions between legumes and rhizobia: disclosing complexity in a molecular dialog. *J. Appl. Microbiol.*, 103, 5, 1355–1365, 2007.
69. Tsyganova, A.V., Brewin, N.J., Tsyganov, V.E., Structure and development of the legume-rhizobial symbiotic interface in infection threads. *Cell*, 10, 5, 1050, 2021.
70. Mitra, D., BE, G.S., Khoshru, B., De Los Santos Villalobos, S., Belz, C., Chaudhary, P., Mohapatra, P.K.D., Impacts of arbuscular mycorrhizal fungi on rice growth, development, and stress management with a particular emphasis on strigolactone effects on root development. *Commun. Soil Sci. Plant Anal.*, 52, 14, 1591–1621, 2021.
71. Foo, E. and Davies, N.W., Strigolactones promote nodulation in pea. *Planta*, 234, 1073–1081, 2011.
72. De Cuyper, C., Fromentin, J., Yocgo, R.E., De Keyser, A., Guillotin, B., Kunert, K., Goormachtig, S., From lateral root density to nodule number, the strigolactone analog GR24 shapes the root architecture of *Medicago truncatula*. *J. Exp. Bot.*, 66, 1, 137–146, 2015.
73. Liu, J., Novero, M., Charnikhova, T., Ferrandino, A., Schubert, A., Ruyter-Spira, C., Cardinale, F., Carotenoid cleavage dioxygenase 7 modulates plant growth, reproduction, senescence, and determinate nodulation in the model legume Lotus japonicus. *J. Exp. Bot.*, 64, 7, 1967–1981, 2013.
74. Umehara, M., Hanada, A., Magome, H., Takeda-Kamiya, N., Yamaguchi, S., Contribution of strigolactones to the inhibition of tiller bud outgrowth under phosphate deficiency in rice. *PCP*, 51, 7, 1118–1126, 2010.
75. Flematti, G.R., Scaffidi, A., Waters, M.T., Smith, S.M., Stereospecificity in strigolactone biosynthesis and perception. *Planta*, 243, 1361–1373, 2016.
76. Peláez-Vico, M.A., Bernabéu-Roda, L., Kohlen, W., Soto, M.J., López-Ráez, J.A., Strigolactones in the Rhizobium-legume symbiosis: Stimulatory effect on bacterial surface motility and downregulation of their levels in nodulated plants. *Plant Sci.*, 245, 119–127, 2016.
77. Limpens, E. and Bisseling, T., Nod factor signal transduction in the Rhizobium–legume symbiosis. *Root Hairs*, 12, 249–276, Springer, Berlin, Heidelberg, 2009.
78. Moreau, S., Verdenaud, M., Ott, T., Letort, S., De Billy, F., Niebel, A., Gamas, P., Transcription reprogramming during root nodule development in *Medicago truncatula*. *PLoS One*, 6, 1, e16463, 2011.

79. Isidra-Arellano, M.C., Pozas-Rodríguez, E.A., del Rocío Reyero-Saavedra, M., Arroyo-Canales, J., Ferrer-Orgaz, S., del Socorro Sánchez-Correa, M., Valdés-López, O., Inhibition of legume nodulation by Pi deficiency is dependent on the autoregulation of nodulation (AON) pathway. *TPJ*, 103, 3, 1125–1139, 2020.
80. Xie, X., Yoneyama, K., Yoneyama, K., The strigolactone story. *Annu. Rev. Phytopathol.*, 48, 93–117, 2010.
81. Richmond, B., *Investigating the role of strigolactones in root system architecture*, 2023, (Doctoral dissertation, University of Warwick).
82. Antunes, P.M. and Goss, M.J., Communication in the tripartite symbiosis formed by arbuscular mycorrhizal fungi, rhizobia and legume plants: a review. *Roots Soil Manage.: Interact. Between Roots Soil.*, 48, 199–222, 2005.
83. Schwember, A.R., Schulze, J., Del Pozo, A., Cabeza, R.A., Regulation of symbiotic nitrogen fixation in legume root nodules. *Plants*, 8, 9, 333, 2019.
84. Bais, H.P., Weir, T.L., Perry, L.G., Gilroy, S., Vivanco, J.M., The role of root exudates in rhizosphere interactions with plants and other organisms. *Annu. Rev. Plant Biol.*, 57, 233–266, 2006.
85. Faizan, M., Cheng, S.H., Tonny, S.H., Robab, M., II, Specific roles of strigolactones in plant physiology and remediation of heavy metals from contaminated soil. *PPB*, 192, 186–195, 2022.
86. Clark, J. and Bennett, T., Cracking the enigma: understanding strigolactone signaling in the rhizosphere. *J. Exp. Bot.*, 75, 4, 1159–1173, 2024.
87. Gavrin, A. and Schornack, S., *Medicago truncatula* as a model organism to study conserved and contrasting aspects of symbiotic and pathogenic signaling pathways. The Model Legume. *Medicago Truncatula*, 317–330, 2019.
88. Wang, Q., Smith, S.M., Huang, J., Origins of strigolactone and karrikin signaling in plants. *Trends Plant Sci.*, 27, 5, 450–459, 2022.
89. De Cuyper, C. and Goormachtig, S., Strigolactones in the rhizosphere: friend or foe? *Mol. Plant-microbe Interact.*, 30, 9, 683–690, 2017.
90. Sanchez-Contreras, M., Bauer, W.D., Gao, M., Robinson, J.B., Allan Downie, J., Quorum-sensing regulation in rhizobia and its role in symbiotic interactions with legumes. *Philos. Trans. R. Soc. B: Biol. Sci.*, 362, 1483, 1149–1163, 2007.
91. Liao, D., Wang, S., Cui, M., Liu, J., Chen, A., Xu, G., Phytohormones regulate the development of arbuscular mycorrhizal symbiosis. *Int. J. Mol. Sci.*, 19, 10, 3146, 2018.
92. Choi, J., Summers, W., Paszkowski, U., Mechanisms underlying establishment of arbuscular mycorrhizal symbioses. *Annu. Rev. Phytopathol.*, 56, 135–160, 2018.
93. Chauhan, S., Mahawar, S., Jain, D., Udpadhay, S.K., Mohanty, S.R., Singh, A., Maharjan, E., Boosting sustainable agriculture by arbuscular mycorrhiza under stress condition: Mechanism and future prospective. *Biomed Res. Int.*, 2022, 2022.

94. Rui, W., Mao, Z., Li, Z., The roles of P and nitrogen nutrient transporters in the arbuscular mycorrhizal symbiosis. *Int. J. Mol. Sci.*, 23, 19, 11027, 2022.
95. Ahmad, M.Z., Rehman, N.U., Yu, S., Zhou, Y., Haq, B.U., Wang, J., Zhao, J., GmMAX2–D14 and–KAI interaction-mediated SL and KAR signaling play essential roles in soybean root nodulation. *TPJ*, 101, 2, 334–351, 2020.
96. Charron, D., Pingret, J.L., Chabaud, M., Journet, E.P., Barker, D.G., Pharmacological evidence that multiple phospholipids signaling pathways link Rhizobium nodulation factor perception in *Medicago truncatula* root hairs to intracellular responses, including $Ca^{2+}$ spiking and specific ENOD gene expression. *Plant Physiol.*, 136, 3, 3582–3593, 2004.
97. Ligerot, Y., de Saint Germain, A., Waldie, T., Troadec, C., Citerne, S., Kadakia, N., Rameau, C., The pea branching RMS2 gene encodes the PsAFB4/5 auxin receptor and is involved in an auxin-strigolactone regulation loop. *PLoS Genet.*, 13, 12, e1007089, 2017.

# 12

# Strigolactones and Control of Parasitic Weeds

**Sonal Jain[1], Rahul Kumar[2], Divya Gunsola[2], Sanja Živković[3], Tanja Vasić[3], Sourav Chattaraj[4], Somya Sinha[1], Prateek Gururani[1], Shraddha Bhaskar Sawant[5], Guerra Sierra B.E.[6], Wiem Alloun[7] and Debasis Mitra[2]***

[1]*Department of Biotechnology, Graphic Era (Deemed to be University), Dehradun, Uttarakhand, India*
[2]*Department of Microbiology, Graphic Era (Deemed to be University), Dehradun, Uttarakhand, India*
[3]*Faculty of Agriculture, University of Niš, Kruševac, Serbia*
[4]*Centre for Industrial Biotechnology Research, School of Pharmaceutical Science, Siksha 'O' Anusandhan University, Bhubaneswar, Odisha, India*
[5]*Department of Plant Pathology, Bihar Agricultural University, Bhagalpur, Bihar, India*
[6]*Universidad de Santander-Facultad de Ciencias Exactas Naturales Y Agropecuarias, Bucaramanga, Colombia*
[7]*Department of Applied Biology, University of Constantine 1, Constantine, Algeria*

***Abstract***
Parasitic weeds such as *Striga* and *Orobanche* species pose a significant threat to agricultural productivity by attaching to the roots of host plants, depriving them of essential water, nutrients, and energy derived from photosynthesis. The seeds of these parasites require specific chemical signals, known as germination stimulants, released by host plant roots to initiate their germination process. These stimulants mainly consist of isoprenoids categorized as strigolactones (SLs), a class of plant hormones crucial for regulating various aspects of plant growth and development. The structural diversity of SLs is attributed to their biosynthetic pathway originating from carotenoids. SLs mediate interactions between plants and other organisms, particularly parasitic plants. They serve a dual role in these interactions: while facilitating the germination of parasitic weed seeds, they also influence root development, shoot branching, and plant responses to environmental cues.

**Corresponding author*: debasismitra3@gmail.com

Tariq Aftab and Kaiser Iqbal Wani (eds.) Strigolactones: Emerging Plant Hormones, (261–280) © 2025 Scrivener Publishing LLC

Leveraging this understanding, researchers aim to develop potent SL agonists and small-molecule SL agonists that induce suicidal germination of parasitic weeds at low concentrations. Such strategies could effectively control these harmful pests, enhancing crop yields sustainably. In summary, insights into the function of SLs in plant–parasitic weed relationships provide a foundation for developing innovative approaches to manage these agricultural threats and improve overall crop productivity.

***Keywords***: Strigolactones, host–parasite interaction, biosynthetic pathway, GR24, sustainable agriculture

## 12.1 Introduction

A recent class of plant hormones known as SLs is essential for many aspects of root architecture and plant development, particularly for controlling endogenous plant responses to environmental stressors and interactions with other rhizospheric organisms [1, 2]. On the other hand, a significant interaction with parasitic weeds was noted. It is well known that parasitic weeds are plant species that get all or part of their nourishment from other live plants, sometimes known as hosts. Through specialized structures called haustoria, these weeds attach to the host plant and enable the extraction of nutrients, water, and signaling chemicals. Significant agricultural losses may arise from this parasitic relationship's primary effects on the host plants' development and output. SLs have become essential in weeds' parasitic associations with their hosts. According to Akiyama and Hayashi [3], they specifically serve as signaling molecules that promote the development of parasitic weed seeds in the soil. Increases in SL concentration in the soil cause parasitic weed seeds to germinate, resulting in the appearance of seedlings and their attachment to the host plant. According to Zwanenburg and Pospíšil [4] and Smith [5], the main functions of these SLs are to promote the germination of parasitic weed seeds, induce AM fungus to branch, and suppress shoot branching and bud development. Due to the crucial function of SLs in germinating parasitic weed seeds, weed management techniques have been developed. One strategy to prevent the promotion of parasitic weed seed germination is to target the manufacture or perception of SLs in the host plant [6]. By interfering with SL-mediated signaling pathways, parasitic weeds can be prevented from spreading, and their adverse effects on agricultural yields can be reduced. Another effective method for controlling parasitic weeds is the application of SLs, which inhibit the germination of seeds from parasitic plants.

## 12.2 What are SLs?

SLs are a class of small organic compounds derived from carotenoids (Figure 12.1). These compounds were initially identified because of their role in promoting seed germination in parasitic plants, such as *Striga* and *Orobanche* species, which led to their naming as "strigolactones" (Figure 12.2). Nevertheless, additional research has uncovered their involvement in various other physiological processes in plants, including the following.

### 12.2.1 Stimulation of Seed Germination

SLs are well known for their ability to encourage the germination of dormant seeds in parasitic plants by releasing chemical signals from host plant roots into the soil.

### 12.2.2 Inhibition of Shoot Branching

SLs suppress the outgrowth of axillary buds, regulate shoot architecture, and prevent excessive branching while promoting resource allocation to the main shoot.

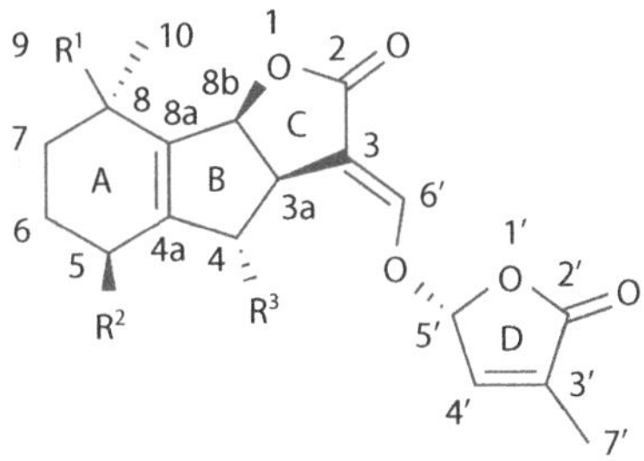


**Figure 12.1** General structure of SLs.

**Figure 12.2** Parasitic plant: *Striga* species.

### 12.2.3 Influence on Root Development

SLs play a significant role in shaping root architecture, affecting both primary and lateral root growth, root branching, and interactions with symbiotic mycorrhizal fungi.

### 12.2.4 Response to Stress

SLs help plants respond to various environmental stressors like salt, drought, and nutrient shortage. This allows plants to modify their development and resource allocation in response to changing environmental conditions. The characteristic structure of SLs comprises three annulated rings called the ABC scaffold. An enol ether unit connects the ABC scaffold to a butenolide ring (D-ring) (Figure 12.1). By altering this basic structure, new SL derivatives with distinct biological activity can be produced. Based on their chemical structure, SLs are divided into canonical and non-canonical SLs. Thus far, natural SLs with C-2 configurations have only been found in (2′R) configurations. The tricyclic lactone structure of canonical SLs consists of three rings (ABC rings) joined to a butenolide group (D-ring) via an enol ether bridge. The D-ring and enol-ether bridge portions are essential for their biological functions, especially in plants [7–9] (Figure 12.1). Because canonical SLs have β- and α-oriented C-rings, their stereochemistry allows further classification into strigol and orobanchol (ORO) kinds. Non-canonical SLs, on the other hand, have D-ring moieties and an enol-ether bridge but lack conventional ABC rings. Because the structure of these natural SLs is too complex for practical use, SL analogues with a much simpler structure that retains vital bioactivity have been created. SLs work in plant cells by attaching to specific receptor proteins. These receptors recognize SL signals and initiate intracellular signaling pathways that control gene expression and physiological reactions [10]. Additionally, they are released from the roots into the rhizosphere, where they promote cooperative relationships with arbuscular mycorrhizal fungi (AMF) and hyphae branching. AMF give their host plants inorganic nutrients, especially phosphate. Consequently, different species respond to phosphate and nitrogen deficits by increasing the production and extrusion of AMF into the soil [6, 11]. SLs interact with various plant hormones, including gibberellins, cytokinins, and auxins, to control plant growth and development. The interplay between many hormonal pathways is often involved in this interaction, allowing plants to combine multiple signals and precisely modify their responses. Plant hormones typically have two main effects: they can act locally, at or close to the biosynthetic site, or move and affect various tissues. The diverse and intricate mechanisms of hormone cross-talk are numerous.

## 12.3 Phenomena of Host–Parasitic Plant Interaction

Root parasitic plants have similar lifecycles, though they differ slightly depending on the host they parasitize worldwide. These include the germination of seeds in response to a stimulus from the root host, the growth of the radicle towards the host root, and the attachment and penetration of the plant through the haustorium (Figure 12.3). The parasitic plant draws nutrients and water from the host plant once a vascular connection is made, which hurts crop productivity and plant fitness [6]. Once these plants break through the ground, they blossom and release more mature seeds, which break and add to the growth of the seed bank [6] (Figure 12.3). Because most of their life cycles occur underground, parasitic weeds are challenging to eradicate. As a result, new control techniques that concentrate on the early phases of the host–parasite relationship must be developed [6]. The seeds are usually present in the soil, as shown in Figure 12.3, where they detect SLs—stimulants secreted by the host plant's roots—and start to germinate. After sprouting, the seeds form an apparatus known as a haustorium that allows them to cling to the host plant's root system and form a xylem–xylem link. The host plant's growth and crop yield are reduced as the mature branches of the parasitic plant poke out of the earth. The parasitic plant eventually yields ripe seeds, which are then added to soil and affect subsequent crops.

The hallmark of parasitic plants is a partial or whole loss of photosynthetic capacity. They also rely on their hosts for water, minerals, and photosynthesis. Hemiparasites and holoparasites are the two main groups of parasitic plants. Hemiparasites take water and mineral nutrients from their host while also containing chlorophyll and can photosynthesize it.

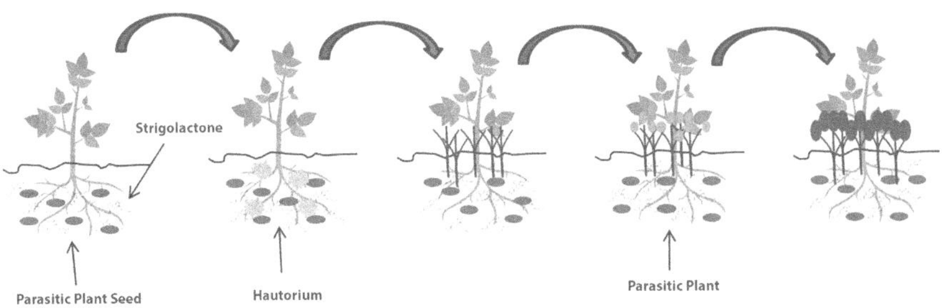


**Figure 12.3** Life cycle of root parasitic plants and interaction of SLs.

Conversely, holoparasites are obligatory parasitic species that do not use photosynthesis and depend on their host. Notably, the holoparasite group includes only about 400 species of parasitic plants [12, 13]. The development of the haustorium is essential for a successful invasion. For both parasites, attachment to a host must happen within the first few days of germination to ensure survival. The capacity of parasitic plants to develop the haustorium, a specialized invasive organ, is one of their most essential characteristics. The haustoria of parasitic plants are multicellular organs that perform the functions of host attachment, host tissue invasion, and the creation of a vascular connection between the parasite and host for material transfer. This contrasts with the unicellular haustoria found in pathogenic fungi that infect plants. An organ that helps parasitic plants infiltrate their host is the haustorium. Based on where they originated, there are two primary types of haustoria: the lateral haustorium, which is present in both facultative and obligate parasites, and the terminal haustorium, which is only present in obligatory parasites of the Orobanchaceae family. The terminal haustorium emerges from the root apical meristem of the parasitic plant, whereas the lateral haustorium grows from the differentiated tissues of the parasite's stem or root [14].

While haustoria is a valuable tool for transferring materials between species, little is known about their internal composition and physiological processes. Whether the parasitic plant is a stem or root parasite determines how haustorial linkages form in the host. Whereas root parasites make connections in the cortex or pith of the host root, stem parasites establish haustorial connections near the base of the host stem. These parasites limit crop productivity and cause financial losses, making them a severe agricultural danger; for example, in many regions, including semi-arid Africa, southern and eastern Europe, the Middle East, North Africa, *Striga* spp., *Orobanche*, and *Phelipanche* spp. pose a serious threat to crops like maize, sorghum, upland rice, millet, legumes, tomatoes, tobacco, rapeseed, and sunflowers [15].

Parasitic angiosperms, like non-parasitic ones, release a large number of dormant seeds. In non-parasitic angiosperms, the germination of dormant seeds takes place under optimal environmental conditions, while in parasitic plants, seed germination necessitates the presence of chemical signaling molecules that indicate the nearness of a host [16]. *Striga* and *Orobanche*, two parasitic species, employ low concentrations of chemical signaling molecules exuded by plant roots to facilitate seed germination, making their isolation, purification, and structure elucidation extremely difficult. The majority of previously identified germination stimulants are SLs [17]. Once connected, parasitic plants draw nutrients from the host plant, causing a significant decline in the host's growth, vitality, and reproductive success. Additionally, parasitic plants may secrete enzymes or chemicals that aid in the uptake of nutrients from the host's tissues.

## 12.4 SL–Carotenoid/Biosynthetic Pathway for Stimulation

Through genetic screening of shoot branching mutants, the enzymes involved in the production of SL have been identified. These enzymes include cytochrome P450 monooxygenases, carotenoid isomerase D27, and carotenoid cleavage dioxygenases CCD7 and CCD8.

### 12.4.1 The Role of Carotenoid

SLs originate from the carotenoids and pigments present in plants and algae. The biosynthesis of SLs commences with the transformation of a particular carotenoid precursor, often all-trans-β-carotene, into carlactones.

### 12.4.2 Isomerization by D27

The fundamental process of SL biosynthesis commences with all-trans-β-carotene, which is subsequently transformed into 9-cis-β-carotene through the action of D27 isomerase.

### 12.4.3 Cleavage by CCD7 and CCD8

It was first discovered that two chromoplast-localized carotenoid cleavage dioxygenases (CCD7/MAX3 and CCD8/MAX4) in *Arabidopsis* catalyzed consecutive processes converting 9-cis-β-carotene to carlactone (CL). CL was first found in an *Escherichia coli in vitro* system and has an A- and D-ring structure. It has been discovered to stimulate the germination of root parasite plants and hinder rice plant tillers [18]. Subsequent research has also demonstrated that in rice and *Arabidopsis thaliana*, CL is an endogenous precursor of both canonical and non-canonical SLs.

### 12.4.4 Oxidation by Cytochrome P450 Enzymes

Carlactonoic acid (CLA) is produced when cytochrome P450 MORE AXILLARY GROWTH (MAX)1 (CYP711A1) transforms CL. An unknown methyltransferase changes CLA into methyl carlactonoate (MeCLA) [19].

### 12.4.5 LBO and Unknown Enzymes

Reverse genetics and transcriptomics were used to investigate the 2-oxoglutarate and Fe (II)-dependent dioxygenase known as lateral branching oxidoreductase (LBO). Under some circumstances, such as removing the shoot tip and treating it with an auxin transport inhibitor that affects shoot branching, it was discovered to be strongly co-expressed with genes involved in producing short-chain lipoxygenase (SL) products, including MeCLA. The LBO enzyme converts MeCLA into hydroxymethyl carlactonoate (1′-OH–MeCLA) (recombinant LBO *in vitro*); however, the process also yields CLA. LBO can directly generate CLA by the spontaneous conversion of 1′-OH–MeCLA to CLA or indirectly by the enzymatic demethylation of MeCLA. Fe (II)-dependent dioxygenases and 2-oxoglutarate can both cause demethylation. If additional processing is not performed on 1′-OH–MeCLA, CLA may be produced. Determining whether LBO-dependent CLA biosynthesis occurs in plants is essential [20–22].

## 12.5 Karrikins

Karrikins are compounds found in burned plant matter that have been demonstrated to promote seed germination in fire-following plant species that would otherwise stay latent in the soil. Chemically speaking, karrikins and SLs are related in that they share enol-ether and lactone (butenolide) moieties. Furthermore, long-dormant seeds in the soil can be roused to germinate by SLs; however, these seeds are usually associated with parasitic plant species such as *Orobanche*, *Phelipanche* spp., *Aletra* spp., and *Striga* (witchweed). Studies have shown that karrikins control photomorphogenesis in young seedlings and seed germination in *Arabidopsis* after high-temperature treatment. This involves increasing seedling vigor, encouraging cotyledon extension and greening, and suppressing hypocotyl elongation. Several thorough evaluations of karrikins have been published in other places [23].

## 12.6 Exploring Various Strategies for Controlling Parasitic Weed Infestations

### 12.6.1 Involvement of SL Biosynthesis Inhibitors

Lowering their stimulation can prevent the host plant's SL synthesis and restrict the growth and attachment of parasitic weeds. Promising candidates for blocking SL biosynthesis are nitrogen-containing heterocyclic compounds like imidazole and triazole groups, which have been shown to exhibit inhibitory activities against CYPs in plants, microorganisms, and mammals. At doses ranging from 0.1 to 10 μM, triflumizole, for instance, has been demonstrated to lower the levels of 4DO in rice roots and root exudates. It has also been discovered that hydroxyamic acid and its TIS108 derivatives impede the production of SL. The Os900-mediated conversion of CL to CLA and 4DO was decreased by triflumizole treatment, suggesting that Os900 is a target protein of triflumizole and that its inhibition lowers the levels of 4DO in rice roots and root exudates. The rice enzyme Os900 produces 10 CL derivatives by oxidizing the SL precursor CL to 4DO. Triflumizole could be an advantageous instrument in examining the impact of SL biosynthesis inhibition and managing harm to root-parasitic plants. It has been discovered that the CL compounds block Os900 and lower the 4DO levels in rice. CL compounds have demonstrated a relatively modest inhibitory effect against SL biosynthesis, whereas recognized inhibitors of SL biosynthesis display inhibitory activity at doses less than 10 μM [24]. Similarly, it has been demonstrated that the diacylhydrazine insect growth regulator chromafenozide lowers the endogenous concentration of 4DO, a significant SL in rice. Furthermore, administration of methoxyfenozide, a member of the same class of insect growth regulators, has led to a decrease in 4DO levels in exudates from rice roots [25]. Preventing SL production in host plants presents a viable strategy for managing parasitic weeds and lessens their adverse effects on agricultural output.

### 12.6.2 Genetic Manipulation

Limiting the synthesis and exudation of SL is one strategy to reduce the stimulants that promote parasite germination in the rhizosphere. Thus, PDR1 and SL biosynthetic genes could be targets for RNA-guided gene editing or selective breeding to modify crops. Mutations resulting in loss of function in these genes have the potential to lower parasite germination; however, this advantage could be countered by pleiotropic effects on plant development or crop yield-reducing AMF connections [26]. Recently, CRISPR/Cas9 gene-editing tools have been utilized to change genes in the trigolactone biosynthetic pathway. These genetic alterations can improve a plant's ability to withstand abiotic stress and interact with AMF by affecting biological processes such as root formation, water use, and nutrient uptake—for example, using CRISPR/Cas9-mediated gene manipulation, the carotenoid cleavage dioxygenase 8 (CCD8) genes (*SbCCD8a* and

*SbCCD8b*), which are involved in SL biosynthesis in sorghum, were shown to improve microbial activity and weed control [27–29].

### 12.6.3 Application of Synthetic SLs, Mimics, and Analogues/ Suicidal Germination Strategy for Parasitic Weed Control

SLs (SLs) are vital plant hormones that control growth and react to biotic and abiotic stimuli [30]. They have great promise as valid targets in agriculture due to their in-plant hormonal actions and ex-plant signaling effects (Figure 12.4). *Striga* is a problematic parasite to control because it produces a lot of seeds, which can survive in the soil for decades and build up a sizable seed bank. It is essential to deplete this seed bank. Since parasitic plants require a host to survive after germination, using seed germination boosters to cause suicide germination and remove the parasites' seed bank could help manage parasitic weeds in agriculture. The "suicidal germination" method utilizes artificial germination stimulants to induce *Striga* seed germination in the absence of a host. According to studies, MP3, MP16, and the GR series—which includes Nijmegen-1 GR5 and GR7—are potent SL analogues that promote the germination of *Striga* seeds [31]. It is possible to accomplish this chemically or biologically. SL-emitting trap crops that prevent parasitization are already used in crop rotation and intercropping plans. Increasing SL emissions in trap crops could make them

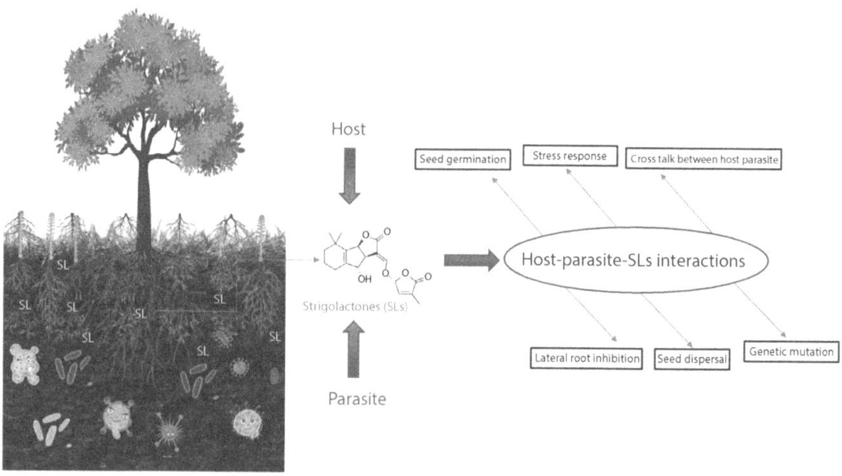


**Figure 12.4** Effects of SL analogs on the plant's growth and signaling pathway in relation to plant biological characteristics and soil and rhizosphere microbiome.

more efficient. Compound instability in soil and synthetic complexity and expense have been obstacles in creating chemical stimulants for parasite germination, like SL analogues, which are helpful for field application [32]. A well-formulated SL analogue increases its stability and shelf-life following field application and effectiveness. In a previous study, we discovered a novel chemical-based suicidal germination strategy (fluridone + thidiazuron) that can be used in infested fields to induce *Striga* suicidal germination without the need for SL signaling or the hyposensitive to LIGHT–SL interaction [33].

## 12.7 Impact of SL Application on Future Scientific Research and Agriculture

Environmental variables hinder plant growth and development, significantly lowering agricultural output worldwide. Since they are sessile by nature, plants are subject to various environmental stressors, including salt, drought, low temperatures, nutrient shortages, and biotic soil environments. To deal with these challenges, plants have evolved intricate signaling systems that include a variety of components, such as phytohormones. SLs have been a driving force in regulating plant growth and development, and they are essential in mitigating the effects of environmental stress. Future scientific studies utilizing SL will focus on upregulating SL signaling to create crop varieties resistant to drought. To illustrate, it was shown that transgenic rice that overexpressed the OsNAC14 transcription factor increased the expression of genes related to SL biosynthesis and genes related to plant defense, stress response, and DNA damage repair. Compared to non-transgenic controls, these transgenic plants exhibited higher survival rates and chlorophyll fluorescence during droughts. SLs assist plants to respond to a variety of abiotic challenges by regulating vegetative growth and development (Figure 12.5). This hormone, which is present in plants, is essential for controlling branch development, secondary growth, and root structure. On the other hand, little is currently known about SLs' actual use in large-scale farming. This can be explained by the high expense of manufacturing SLs and a deficiency of knowledge on possible environmental hazards and adverse effects from their breakdown products. Targeted research is required to close this knowledge gap to comprehend the unique effects of various SLs, the best time to apply them, any possible risks above or below ground at different dosage levels, and the community-level impact of SLs on soil microbiota. Like other hormones, SLs play a crucial function that should be acknowledged and emphasized in research where their effects are clearly shown. Practical cooperation between governmental, commercial, and research institutions is essential to the large-scale application of SL products. Furthermore, regulatory lobbying and public awareness may be required to give legislators, farmers, and the public correct information regarding SLs so they can make well-informed decisions (Figure 12.5).

Future research can focus on developing ways to use naturally occurring plant and microbial signaling molecules, such as SLs and AMF, to accomplish environmental protection, productivity, and sustainability goals in ecosystem management and agriculture. As part of these strategies, the symbiotic relationship between AMF and plants is strengthened, soil fertility is increased by encouraging phosphorus uptake using SLs, soil erosion is decreased, water retention capacity is improved, plant nutrient content is enhanced; resistance to malnutrition is developed, and the use of chemical fertilizers and pesticides is minimized through the use of biocontrol methods and natural nutrient delivery. Successfully executing these tactics necessitates considering factors

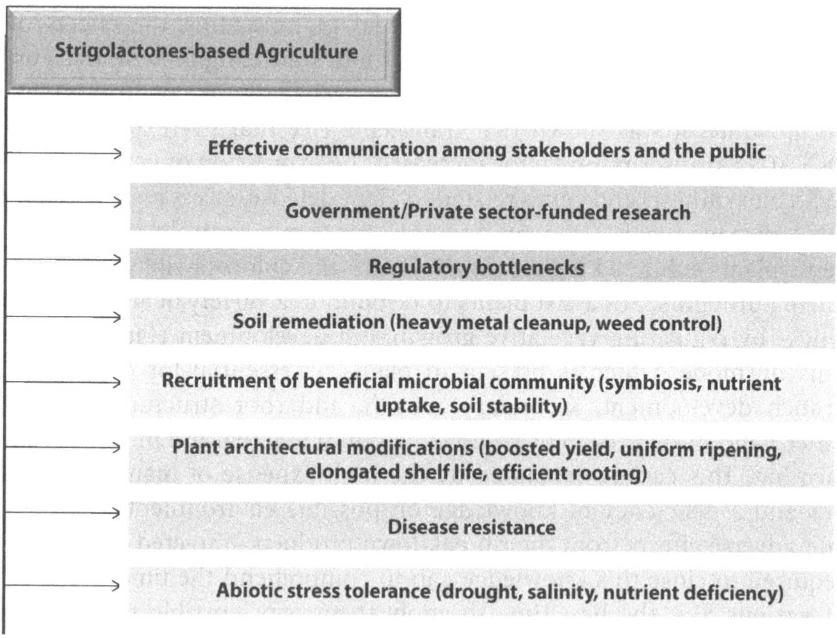


**Figure 12.5** Potential application benefits of SLs in agriculture.

like farmer education, scientific dissemination, and incorporation of technical breakthroughs. Pollock *et al.* [34] first documented the anti-cancer effects of SLs in breast cancer cells, finding that these natural compounds can selectively stop the growth of cancer cells and cause apoptosis, leaving non-cancerous cells unaffected. Promising outcomes for SLs as a possible treatment for other types of tumors and advanced prostate cancer have been observed in several studies. Validating and expanding the findings on HFF will require more study on physiologically relevant targets, such as monocytes, endothelial and epithelial cells, and cells that do not progress to a lytic infection. Studying other HCMV proteins, including anti-apoptotic proteins like vMIA, cICA, UL38, and IE2, as well as other viruses in the Herpesviridae family and newly developing viruses, is also crucial. These viruses are currently at the top of the medical research priority list [35].

By pursuing these future research directions (shown in Figure 12.6), we will fully realize the promise of the AMF–SL interaction for sustainable agriculture and understand it more deeply. Our evaluation work will aid in creating creative solutions that maximize crop yield, optimize nutrient utilization, and increase resilience to stress while reducing adverse environmental effects [36].

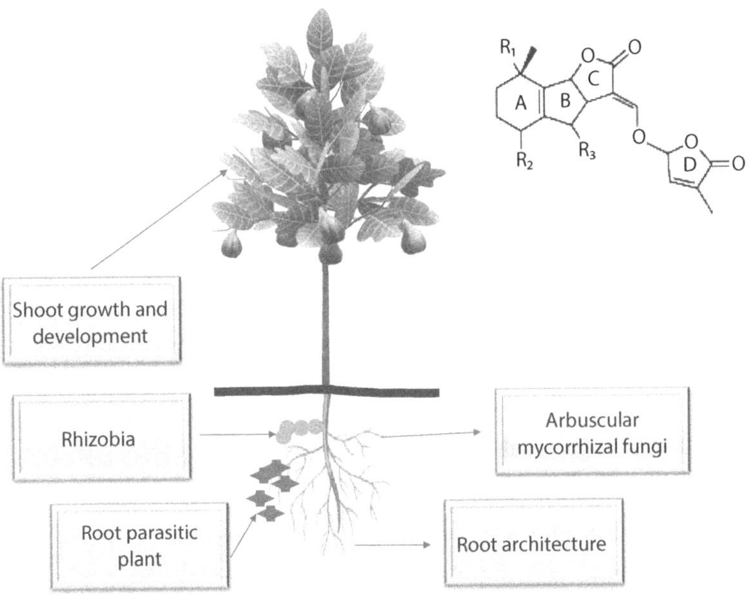


**Figure 12.6** Future research directions and application of SL for crop productivity.

**Table 12.1** Impacts of SLs on various plant species' functionalities.

| Functions | Plant species | Reference |
|---|---|---|
| Offers defense against bacterial infections caused by *Pseudomonas syringae*, *Pectobacterium carotovorum*, and *Rhadococcus fascians* | *Arabidopsis thaliana* | [39] |
| Raises the levels of nitric oxide and $H_2O_2$ | *Arabidopsis thaliana* | [40] |
| Lessens the impact of salinity and drought | *Arabidopsis thaliana* | [41] |
| Improves the responsiveness of stomata | *Arabidopsis thaliana*, *Solanum lycopersicum* | [42] |
| Enhances nodulation | *Glycine max* | [43] |
| Contributes favorably to nematode defense | *Solanum lycopersicum* | [44] |
| Improves tolerance to salt | *Sesbania cannabina* | [45] |
| Speed up the withering of leaves | *Bambusoideae* | [46] |
| Preventing tillering and drilling | *Oryza sativa* | [47] |
| Increases the mobility of the surface | *Sinorhizobium meliloti* | [48] |
| Controls freezing tolerance in a positive way | *Arabidopsis thaliana*, *Pisum sativum* | [49] |
| ABA-dependent stomatal closure is delayed | *Lotus japonicus* | [50] |
| Lengthens the root hair more | *Arabidopsis thaliana* | [51] |
| Encourages the germination of seeds | *Arabidopsis thaliana* | [52] |
| Bolsters defenses against abiotic stress | *Arabidopsis thaliana* | [53] |
| Suppresses the growth of lateral roots | *Arabidopsis thaliana* | [54] |
| Increases the level of chlorophyll | *Oryza sativa* | [55] |

## 12.8 SL Functions

SLs are a type of phytohormones that play an important part in various developmental processes, such as plant growth and development, symbiotic mycorrhizal relationships, and identification of host plants by parasitic weeds and mycorrhizal fungi. These hormones act as signaling molecules, moving from below to aboveground plant components to influence shoot branching and control the growth of plants. They also play a role in abiotic stress reactions and host identification signals for parasitic weeds and mycorrhizal fungi. Research has shown that SLs are involved in many crops and plants, as demonstrated in studies by Jia *et al.* [37] and Gomez-Roldan *et al.* [38] (Table 12.1).

## 12.9 Conclusion

The discovery and understanding of SLs has revolutionized approaches to control parasitic weeds in agriculture. By elucidating the intricate signaling pathways involved in host-parasite interactions, SLs have provided valuable targets for intervention. Researchers have leveraged this knowledge to devise innovative weed management strategies by inducing suicidal germination and developing SL inhibitors. The main application domains of SLs in sustainable agricultural methods are (i) controlling parasitic weeds, (ii) mitigating drought, and (iii) improving crop development and nutrient assimilation efficiency. However, the effectiveness of SL-based control methods lies not only in their specificity and potency but also in their integration with complementary practices such as crop rotation, intercropping, and biological control. Sustainability considerations and regulatory frameworks must guide the implementation of these approaches to ensure long-term viability. Ultimately, the continued exploration of SLs and their applications holds promise for mitigating the detrimental impacts of parasitic weeds on crop yields, fostering agricultural sustainability, and advancing our understanding of plant-microbe interactions in agroecosystems.

## References

1. Pandey, A., Sharma, M., Pandey, G.K., Emerging roles of SLs in plant responses to stress and development. *Front. Plant Sci.*, 7, 165491, 2016.
2. Mitra, D., Rad, K.V., Chaudhary, P., Ruparelia, J., Sagarika, M.S., Boutaj, H., Mohapatra, P.K.D., Panneerselvam, P., Involvement of SL hormone in root development, influence and interaction with mycorrhizal fungi in plant: Mini-review. *Curr. Res. Microb. Sci.*, 2, 100026, 2021.
3. Akiyama, K. and Hayashi, H., SLs: chemical signals for fungal symbionts and parasitic weeds in plant roots. *Ann. Bot.*, 97, 6, 925–931, 2006.
4. Zwanenburg, B. and Pospíšil, T., Structure and activity of SLs: new plant hormones with a rich future. *Mol. Plant*, 6, 1, 38–62, 2013.
5. Smith, S.M., Q&A: What are SLs and why are they important to plants and soil microbes? *BMC boil.*, 12, 1–7, 2014.
6. López-Ráez, J.A., Charnikhova, T., Gómez-Roldán, V., Matusova, R., Kohlen, W., De Vos, R., Verstappen, F., Puech-Pages, V., Bécard, G., Mulder, P., Bouwmeester, H., Tomato SLs are derived from carotenoids and their biosynthesis is promoted by phosphate starvation. *New Phytol.*, 178, 4, 863–874, 2008.
7. Mangnus, E.M., Dommerholt, F.J., De Jong, R.L., Zwanenburg, B., Improved synthesis of strigol analog GR24 and evaluation of the biological activity of its diastereomers. *J. Agric. Food. Chem.*, *40*, 7, 1230–1235, 1992.
8. Wu, F., Gao, Y., Yang, W., Sui, N., Zhu, J. Biological functions of strigolactones and their crosstalk with other phytohormones. *Front. Plant Sci.*, 13, 821563, 2022.
9. Umehara, M., Cao, M., Akiyama, K., Akatsu, T., Seto, Y., Hanada, A., Yamaguchi, S., Structural requirements of strigolactones for shoot branching inhibition in rice and Arabidopsis. *Plant Cell Physiol.*, *56*, 6, 1059–1072, 2015.
10. Wani, K. I, Zehra, A., Choudhary, S., Naeem, M., Khan, M.M.A., Castroverde, C.D.M., Aftab, T., Mechanistic insights into strigolactone biosynthesis, signaling, and regulation during plant growth and development. *J. Plant Growth Regul.*, 40, 1836–1852, 2021.
11. Yoneyama, K., Xie, X., Kim, H.I., Kisugi, T., Nomura, T., Sekimoto, H., Yokota, T., Yoneyama, K., How do nitrogen and phosphorus deficiencies affect SL production and exudation? *Planta*, 235, 1197–1207, 2012.
12. Bungard, R.A., Photosynthetic evolution in parasitic plants: insight from the chloroplast genome. *BioEssays*, *26*, 3, 235–247, 2004.
13. Nickrent, D.L. and Musselman, L.J., Introduction to parasitic flowering plants. *Plant Health Instr.*, 4 2004.
14. Yoshida, S., Cui, S., Ichihashi, Y., Shirasu, K., The haustorium, a specialized invasive organ in parasitic plants. *Annu. Rev. Plant Biol.*, *67*, 643–667, 2016.

15. Jhu, M.Y. and Sinha, N.R., Parasitic plants: an overview of mechanisms by which plants perceive and respond to parasites. *Annu. Rev. Plant Biol.*, *73*, 433–455, 2022.
16. Bouwmeester, H., Li, C., Thiombiano, B., Rahimi, M., Dong, L., Adaptation of the parasitic plant lifecycle: germination is controlled by essential host signaling molecules. *Plant Physiol.*, *185*, 4, 1292–1308, 2021.
17. Rani, K., Zwanenburg, B., Sugimoto, Y., Yoneyama, K., Bouwmeester, H.J., Biosynthetic considerations could assist the structure elucidation of host plant produced rhizosphere signalling compounds (SLs) for arbuscular mycorrhizal fungi and parasitic plants. *Plant Physiol. Biochem.*, *46*, 7, 617–626, 2008.
18. Yoneyama, K. and Brewer, P.B., SLs, how are they synthesized to regulate plant growth and development? *Curr. Opin. Plant Bio.*, *63*, 102072, 2021.
19. Abe, S., Sado, A., Tanaka, K., Kisugi, T., Asami, K., Ota, S., Kim, H.I., Yoneyama, K., Xie, X., Ohnishi, T., Seto, Y., Carlactone is converted to carlactonoic acid by MAX1 in Arabidopsis and its methyl ester can directly interact with AtD14 in vitro. *Proc. Nati. Acad. Sci.*, *111*, 50, 18084–18089, 2014.
20. Brewer, P.B., Yoneyama, K., Filardo, F., Meyers, E., Scaffidi, A., Frickey, T., Akiyama, K., Seto, Y., Dun, E.A., Cremer, J.E., Kerr, S.C., LATERAL BRANCHING OXIDOREDUCTASE acts in the final stages of SL biosynthesis in Arabidopsis. *Proc. Nati. Acad. Sci.*, *113*, 22, 6301–6306, 2016.
21. Yoneyama, K., Akiyama, K., Brewer, P.B., Mori, N., Kawano-Kawada, M., Haruta, S., Nishiwaki, H., Yamauchi, S., Xie, X., Umehara, M., Beveridge, C.A., Hydroxyl carlactone derivatives are predominant SLs in Arabidopsis. *Plant Direct*, *4*, 5, e00219, 2020.
22. Hagel, J.M. and Facchini, P.J., Dioxygenases catalyze the O-demethylation steps of morphine biosynthesis in opium poppy. *Nat. Chem. Bio.*, *6*, 4, 273–275, 2010.
23. Wang, B., Wang, Y., Li, J., Institute of Genetics and Developmental Biology, Chinese Academy of Sciences, Beijing, China, 2017.
24. Kawada, K., Saito, T., Onoda, S., Inayama, T., Takahashi, I., Seto, Y., Nomura, T., Sasaki, Y., Asami, T., Yajima, S., Ito, S., Synthesis of Carlactone Derivatives to Develop a Novel Inhibitor of SL Biosynthesis. *ACS omega*, *8*, 15, 13855–13862, 2023.
25. Kawada, K., Sasaki, Y., Asami, T., Yajima, S., Ito, S., Insect growth regulators with hydrazide moiety inhibit SL biosynthesis in rice. *J. Pest. Sci.*, *47*, 1, 43–46, 2022.
26. Khosla, A. and Nelson, D.C., SLs, super hormones in the fight against Striga. *Curr. Opin. Plant Bio.*, *33*, 57–63, 2016.
27. Butt, H., Jamil, M., Wang, J.Y., Al-Babili, S., Mahfouz, M., Engineering plant architecture via CRISPR/Cas9-mediated alteration of SL biosynthesis. *BMC. Plant Bio.*, *18*, 1–9, 2018.
28. Bari, V.K., Nassar, J.A., Kheredin, S.M., Gal-On, A., Ron, M., Britt, A., Steele, D., Yoder, J., Aly, R., CRISPR/Cas9-mediated mutagenesis of CAROTENOID

CLEAVAGE DIOXYGENASE 8 in tomato provides resistance against the parasitic weed Phelipanche aegyptiaca. *Sci. Rep.*, *9*, 1, 11438, 2019.

29. Hao, J., Yang, Y., Futrell, S., Kelly, E.A., Lorts, C.M., Nebie, B., Runo, S., Yang, J., Alvarez, S., Lasky, J.R., Schachtman, D.P., CRISPR/Cas9-mediated mutagenesis of carotenoid cleavage dioxygenase (CCD) genes in sorghum alters SL biosynthesis and plant biotic Interactions. *Phytobiomes J.*, *7*, 3, 339–351, 2023.
30. Jiang, W., Fei Lu, C., Xu, X., Riaz, M.W., Aimin, L.V., Shao, Q., SLs: Biosynthetic regulation, hormonal interaction, and their involvement in abiotic stress adaption. *Sci. Hort.*, 325, 112689, 2024.
31. Jamil, M., Wang, J.Y., Yonli, D., Ota, T., Berqdar, L., Traore, H., Margueritte, O., Zwanenburg, B., Asami, T., Al-Babili, S., Striga hermonthica suicidal germination activity of potent SL analogs: evaluation from laboratory bioassays to field trials. *Plants*, *11*, 8, 1045, 2022.
32. Fernández-Aparicio, M., Reboud, X., Gibot-Leclerc, S., Broomrape weeds. Underground mechanisms of parasitism and associated strategies for their control: a review. *Front. Plant Sci.*, *7*, 171714, 2016.
33. Jamil, M., Wang, J.Y., Berqdar, L., Alagoz, Y., Behisi, A., Al-Babili, S., Cytokinins as an alternative suicidal Striga germination compound. *Weed Res.*, 63, 1–8, 2023.
34. Pollock, C.B., Koltai, H., Kapulnik, Y., Prandi, C., Yarden, R.I., SLs: a novel class of phytohormones that inhibit the growth and survival of breast cancer cells and breast cancer stem-like enriched mammosphere cells. *Breast Cancer Res. Treat.*, *134*, 1041–1055, 2012.
35. Dell'Oste, V., Spyrakis, F., Prandi, C., SLs, from plants to human health: achievements and challenges. *Molecules*, *26*, 15, 4579, 2021.
36. Boyno, G., Rezaee Danesh, Y., Demir, S., Teniz, N., Mulet, J.M., Porcel, R., The Complex Interplay between Arbuscular Mycorrhizal Fungi and SL: Mechanisms, Sinergies, Applications and Future Directions. *Int. J. Mol. Sci.*, *24*, 23, 16774, 2023.
37. Jia, K.P., Baz, L., Al-Babili, S., From carotenoids to SLs. *J. exp. Bot.*, *69*, 9, 2189–2204, 2018.
38. Gomez-Roldan, V., Fermas, S., Brewer, P.B., Puech-Pagès, V., Dun, E.A., Pillot, J.P., Rochange, S.F., SL inhibition of shoot branching. *Nature*, *455*, 7210, 189–194, 2008.
39. Stes, E., Depuydt, S., De Keyser, A., Matthys, C., Audenaert, K., Yoneyama, K., Vereecke, D., SLs as an auxiliary hormonal defence mechanism against leafy gall syndrome in Arabidopsis thaliana. *J. Exp. Bot.*, *66*, 16, 5123–5134, 2015.
40. Lv, S., Zhang, Y., Li, C., Liu, Z., Yang, N., Pan, L., Wang, G., SL-triggered stomatal closure requires hydrogen peroxide synthesis and nitric oxide production in an abscisic acid-independent manner. *New Phytol.*, *217*, 1, 290–304, 2018.
41. Saeed, W., Naseem, S., Ali, Z., SLs biosynthesis and their role in abiotic stress resilience in plants: a critical review. *Front. Plant Sci.*, *8*, 279971, 2017.

42. Visentin, I., Vitali, M., Ferrero, M., Zhang, Y., Ruyter-Spira, C., Novák, O., Cardinale, F., Low levels of SLs in roots as a component of the systemic signal of drought stress in tomato. *New Phytol.*, *212*, 4, 954–963, 2016.
43. Rehman, N., Ali, M., Ahmad, M.Z., Liang, G., Zhao, J., SLs promote rhizobia interaction and increase nodulation in soybean (*Glycine max*). *Micro. Patho.*, *114*, 420–430, 2018.
44. Xu, X., Fang, P., Zhang, H., Chi, C., Song, L., Xia, X., Yu, J., SLs positively regulate defense against root-knot nematodes in tomato. *J. Exp. Bot.*, *70*, 4, 1325-1337, 2019.
45. Ren, C.G., Kong, C.C., Xie, Z.H., Role of abscisic acid in SL-induced salt stress tolerance in arbuscular mycorrhizal Sesbania cannabina seedlings. *BMC. Plant Bio.*, *18*, 1–10, 2018.
46. Tian, M.Q., Jiang, K., Takahashi, I., Li, G.D., SL-induced senescence of a bamboo leaf in the dark is alleviated by exogenous sugar. *J. Pest. Sci.*, *43*, 3, 173–179, 2018.
47. Jamil, M., Kountche, B.A., Haider, I., Guo, X., Ntui, V.O., Jia, K.P., Al-Babili, S., Methyl phenlactonoates are efficient SL analogs with simple structure. *J. Exp. Bot.*, *69*, 9, 2319–2331, 2018.
48. Peláez-Vico, M.A., Bernabéu-Roda, L., Kohlen, W., Soto, M.J., López-Ráez, J.A., SLs in the Rhizobium-legume symbiosis: Stimulatory effect on bacterial surface motility and down-regulation of their levels in nodulated plants. *Plant Sci.*, *245,* 119–127, 2016.
49. Cooper, J.W., Hu, Y., Beyyoudh, L., Yildiz Dasgan, H., Kunert, K., Beveridge, C.A., Foyer, C.H., SLs positively regulate chilling tolerance in pea and in Arabidopsis. *Plant signal. Behave.*, *41*, 6, 1298–1310, 2018.
50. Liu, J., He, H., Vitali, M., Visentin, I., Charnikhova, T., Haider, I., Cardinale, F., Osmotic stress represses SL biosynthesis in Lotus japonicus roots: exploring interaction between SLs ABA under abiotic stress. *Planta*, 241, 1435–1451, 2015.
51. Koltai, H. and Kapulnik, Y., SLs as mediators of plant growth responses to environmental conditions. *Plant Signal. Behave.*, *6*, 1, 37–41 2011.
52. Mishra, S., Upadhyay, S., Shukla, R.K., The role of SLs and their potential cross-talk under hostile ecological conditions in plants. *Front Physiol.*, *7*, 234251, 2017.
53. Ha, C.V., Leyva-González, M.A., Osakabe, Y., Tran, U.T., Nishiyama, R., Watanabe, Y., Tran, L.S.P., Positive regulatory role of SL in plant responses to drought and salt stress. *PNAS*, *111*, 2, 851–856, 2014.
54. Kapulnik, Y., Delaux, P.M., Resnick, N., Mayzlish-Gati, E., Wininger, S., Bhattacharya, C., Koltai, H., SLs affect lateral root formation and root-hair elongation in Arabidopsis. *Planta*, *233*, 209–216, 2011.
55. Tsuchiya, Y., Vidaurre, D., Toh, S., Hanada, A., Nambara, E., Kamiya, Y., McCourt, P., A small-molecule screen identifies new functions for the plant hormone SL. *Nat. Chem. Bio.*, *6*, 10, 741–749, 2010.

13

# Strigolactones and Plant Defense: Protection Against Pests and Pathogens

**Bonia Francis[1], C. T. Aravindakumar[1,2] and Sibu Simon[1*]**

*[1]School of Environmental Sciences, Mahatma Gandhi University, Kottayam, Kerala, India*
*[2]Inter University Instrumentation Centre, Mahatma Gandhi University, Kottayam, Kerala, India*

## *Abstract*

Biotic stress is a condition in which detrimental biological factors interact with plants and disrupt their normal growth. These biological factors include organisms like bacteria, fungi, viruses, insects, nematodes, and weeds. They drain nutrients from their host plant, affecting optimal growth and, in severe situations, causing the host plant to die. Biotic stress is a major threat that causes terrific losses in agriculture by affecting crop yield. The primary morphological symptoms of biotic stress are rotting, chlorosis, stunting, and the development of local lesions. In order to combat biotic stresses, plants have developed a wide range of complex defense mechanisms. It includes morphological, biochemical, and molecular adaptive mechanisms triggered by various plant hormonal signaling pathways. Strigolactones (SLs) are plant hormones that influence growth, development, and stress responses in plants. It has been found that SLs play a role in mediating plant defense responses to various biotic stresses. It is found to be involved in defending the host plants against attacks from pathogenic bacteria, fungi, nematodes, aphids, and pests, thus endowing stress tolerance in plants. SLs protect plants from pathogens and pests, primarily by preventing their growth, encouraging stomatal closure to reduce invader access, and increasing the production of hydrogen peroxide. SLs can crosstalk with other defense hormones to activate defense responses in plants. The main focus of this chapter is on SLs' function in activating plant defense against pests and pathogens.

***Keywords*:** Strigolactone, biotic stress, pests, pathogens, defense, signaling

**Corresponding author*: sibusimon@mgu.ac.in

Tariq Aftab and Kaiser Iqbal Wani (eds.) Strigolactones: Emerging Plant Hormones, (281–300) © 2025 Scrivener Publishing LLC

## 13.1 Introduction

Plants can interact with both beneficial and pathogenic organisms above and below ground. Plants modulate their defense mechanisms to thwart harmful interactions while fostering the development of beneficial ones. Plants have strong defense mechanisms to prevent pathogen and pest attacks [1]. These include morphological, physiological, and metabolic adaptations. Plant immune system defends the plants against invading organisms by developing pathogen- and effector-triggered immunity [2].

Plant hormones have a significant role in modulating biotic stress responses in plants. The phytohormones jasmonic acid (JA), salicylic acid (SA), and ethylene (ET) are active in mediating biotic defense responses [3–5]. These hormones often work synergically for the efficient regulation of stress responses in plants [6]. The recently identified plant hormones, strigolactones (SLs), have also been reported to regulate biotic stress responses in plants [7, 8].

SLs are terpenoid lactones that regulate various developmental events and stress responses in plants [9, 10] (Figure 13.1). SLs were discovered 60 years ago in the cotton root exudates that could germinate witch-weed seeds, *Striga lutea Lour* [11]. SLs can inhibit plant shoot branching [12, 13]. However, SLs are involved in stimulating auxin-mediated secondary growth in Arabidopsis [14]. Shoot elongation in rice is promoted by SL-mediated regulation of giberellic acid (GA) metabolism and signaling [15]. Additionally, SLs can regulate leaf senescence in plants [16]. The most important below-ground developmental role of SLs is to regulate root architecture in plants. It promotes primary root and adventitious root elongation and inhibits lateral root formation [17].

SLs promote plant symbiosis with rhizobia and arbuscular mycorrhizal fungi [18]. SLs are also crucial in mediating various abiotic stress responses in plants. SLs regulate plant responses to different abiotic stresses such as nutrient, heat, salinity, drought, cold, and light stress [19–24].

SLs are known to play an important role in regulating various aspects of plant growth and development and act as rhizosphere signaling and response molecules to stress. This chapter focuses on the specific role of SLs in mediating defense against biotic stress in plants, including their ability to defend against pathogens and pests. Gaining greater knowledge about this defense mechanism can help in the development of biotic stress-resistant crop types.

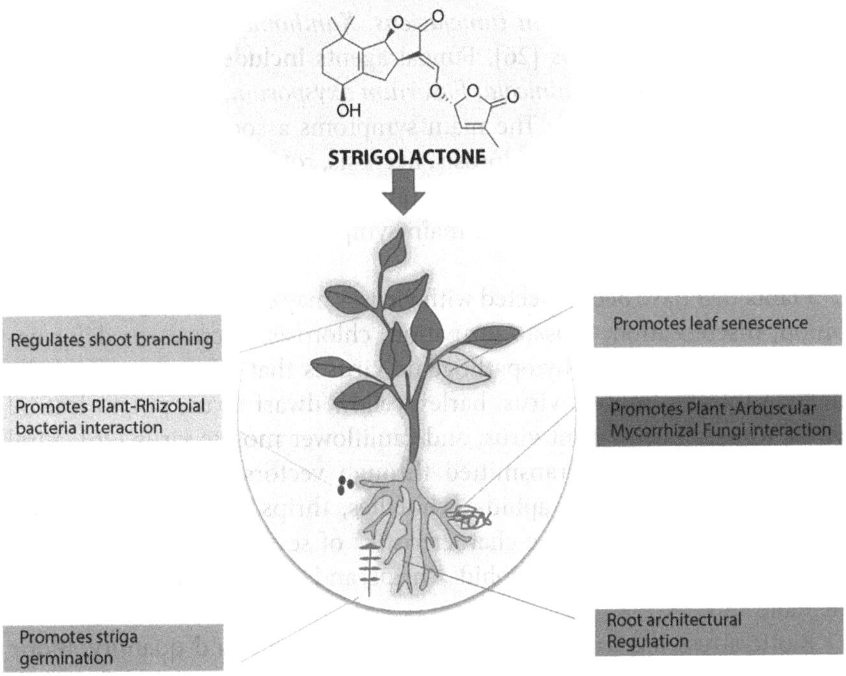


**Figure 13.1** Role of SLs in plant growth and development. SLs regulate shoot branching, leaf senescence, root architecture regulation: primary root growth, lateral root growth, root hair, adventitious root growth, and more. It promotes a symbiotic association with nitrogen-fixing rhizobia and also involved in plant AM fungi interaction.

## 13.2 Biotic Stresses in Plants

Biotic stress is one of the significant environmental factors affecting plant growth and development. A wide range of biotic stress agents can affect plants. Living organisms such as bacteria, fungi, viruses, nematodes, and insects are the major biotic stress agents in plants [8]. Biotic stress agents can attack plants at any stage of their life cycle. They will absorb nutrients from the plant, hindering its growth and sometimes even leading to the death of the host plant [25]. Biotic stress adversely affects plant growth and yield, ultimately impacting food production worldwide. It is one of the significant factors that causes yield loss in crops.

Bacterial stress agents include *Pseudomonas syringae*, *Pectobacterium carotovorum*, *Agrobacterium tumefaciens*, *Xanthomonas axonopodis*, and *Clavibacter michiganensis* [26]. Fungal agents include *Magnaporthe oryzae*, *Mycosphaerella graminicola*, *Fusarium oxysporum*, *Rhizoctonia solani*, and *Ustilago maydis* [27]. The main symptoms associated with bacterial infection in plants include chlorosis, necrosis, rot, galls, wilting, and scab, whereas spotting, rot, wilting, mold, pustules, overgrowth, deformations, mummification in plants are the main symptoms of fungal infection [28] (see Table 13.1).

Plants that have been infected with viruses may experience growth inhibition, discoloration, mosaic formation, chlorosis, necrosis or deformations [62]. The major phytopathogenic viruses that affect crops globally include tobacco mosaic virus, barley yellow dwarf virus, potato leafroll virus, tomato bushy stunt virus, and cauliflower mosaic virus [26]. Viral diseases in plants are transmitted through vectors. These vectors that transmit viruses include aphids, whiteflies, thrips, insects, mites, nematodes, fungi, etc. [63]. The characteristics of several vector-borne plant viral diseases transmitted by aphids, thrips, and white flies are summarized in Table 13.2.

Biotic stress has a major impact on yield quality and quantity. Biotic stress causes both preharvest and postharvest losses in crops. To exemplify, the Australian barley fields are affected mainly by foliar infections, which reduce yield quantity and quality. These foliar infections in barley are caused by net-type net blotch, powdery mildew, scald, spot-type net blotch, and barley leaf rust [80]. The most common diseases damaging wheat production include fusarium head blight, loose smut, leaf rust, stem rust, wheat blast, powdery mildew, and stripe rust [81]. The common bacterial diseases that affect rice include leaf blight, sheath blight, blast, false smut, brown spot, and yellow stem borer [82].

## 13.3 Biotic Stress-Mediated Defense Mechanism in Plants

Regulation of morphological structures serves as the first line of defense against biotic stress. These structures include spines, waxy cuticles, trichomes, and setae. Spines and thorns can protect plants from pests, likewise, trichomes also shield plants from attacking arthropods [83]. Additionally, the secondary metabolites secreted by the glandular trichomes are harmful to these organisms [84].

**Table 13.1** Major classes of plant-infecting bacterial and fungal pathogens and specific diseases they cause with symptoms.

| Causative organism | Disease | Symptoms | Host | Reference |
|---|---|---|---|---|
| Fungi | | | | |
| *Fusarium oxysporum* | Fusarium wilt | Stunted growth, leaf yellowing, leaf wilting | Spinach, lettuce, banana, tomato | [29–33] |
| *Phytophthora infestans* | Late blight | Leaf necrosis, stem necrosis, dark brown lesions, fruit rot | Potato<br>Tomato | [34–36] |
| *Puccinia graminis* | Stem rust | Chlorotic fleck, pustule, black teliospores | Wheat<br>Barley<br>Oats | [37–39] |
| *Ustilago maydis*<br>*Tilletia horrida*<br>*Ustilaginoidea virens*<br>*Ustilago nuda* | Smut<br>Kernel smut<br>False smut<br>Loose smut | Smut galls on tassels, ears, leaf midrib, pustule<br>Black teliospore<br>Velvety yellowish orange to greenish black spore<br>Powdery black spores | Corn<br>Rice<br>Rice<br>Barley | [40–44] |
| Bacteria | | | | |
| *Pseudomonas syringae* | Leaf spot | Brown necrotic spots, leaf yellowing, necrosis, water-soaked lesions | Cucurbits<br>Tomato<br>Coriander<br>Swiss chard | [45–49] |
| *Ralstonia* | Bacterial wilt | Wilting of leaves, vascular browning, browning of pith and cortex | Tomato<br>Potato<br>Pepper | [50–53] |
| *Erwinia carotovora*<br>*Pectobacterium* | Soft rot | Tissue decay, soft and wet rotted tissues | Cabbage<br>Onion<br>Carrot<br>Potato | [54–58] |
| *Xanthomonas Oryzae*<br>*Xanthomonas axonopodis pv. Malvacearum*<br>*Xanthomonas axonopodis pv. phaseoli* | Bacterial blight | Green spots on leaves | Rice<br>Cotton<br>Beans | [59–61] |

**Table 13.2** Major vector-borne viral diseases affecting crops with their symptoms.

| Pest vectors | Causative organism | Symptom | Vector | Host | Reference |
|---|---|---|---|---|---|
| Aphids | Potato leaf roll virus (luteovirus) | Leaf rolling<br>Stunted plants | *Myzus persicae* | Potato | [64] |
| | Turnip mosaic virus (poty virus) | Black spots on leaves, premature leaf drop, leaf vein yellowing, early leaf senescence | *Myzus persicae, Brevicoryne brassicae* | Turnip, cabbage, radish, lettuce | [65, 66] |
| | Bean common mosaic viruses (poty virus) | Leaf mosaic, deformation, necrosis, stunted growth | *Acyrthosiphon pisum, Aphis craccivoracan, Myzus persicae, Aphisfabae* | Beans, pigeonpea, jack bean, chickpea | [67] |
| Thrips | Iris yellow spotted virus (tospovirus) | Diamond-shaped yellow-colored leaf lesions, lesions on flowering scapes, reduced bulb size | *Thrips tabaci* | Bulbs, onion, garlic, chive | [68, 69] |
| | Tomato spotted wilt virus (tospovirus) | Ring spots, chlorosis, stunted growth | *Frankliniella occidentalis* | Tomato, potato, groundnut, bean, pepper, lettuce, tobacco | [70–72] |
| | Capsicum chlorosis virus (tospovirus) | Ring spots, chlorosis, leaf deformations, stunted growth | *Ceratothripoides claratris, Thrips palmi, Frankliniella schultzei* | Tomato, capsicum, chilly | [73, 74] |
| White fly | Tomato torrado virus | Necrotic spots on leaves, chlorosis, stunted growth | *Trialeurodes vaporariorum, Bemisia tabaci* | Tomato, pepper, eggplant | [75–77] |
| | Beet pseudoyellows virus | Yellowing of leaf veins, stunted growth | *Trialeurodes vaporariorum* | Cucurbit, ornamentals | [78] |
| | Tomato yellow leaf curl virus | Curling of leaf margin, reduced leaf size, yellowing of young leaves, stunted growth | *Bemisia tabaci* | Tomato, benthi common bean, lentil | [79] |

Secondary metabolites such as terpenoids, alkaloids, chlorogenic acid, lignin, phenylpropanoids, and quinones effectively participate in defending biotic stress in plants [84]. The terpenoid phytoalexins produced by rice plants are active in fending off *Magnaporthe oryzae infections*; similarly, the phytoalexins produced by maize plants were efficient in combating *Colletotricum graminicola* and *Rhizopus microsporus* attacks in maize plants [85, 86]. Plants are also protected from infection by modifying the cell wall via lignification, callose deposition, and stomata closure [87, 88].

The generation of reactive oxygen species (ROS) is an early response to pathogen attack and a widespread defense mechanism in plant cells [89]. Mitochondria, chloroplasts, and peroxisomes are the primary sources of ROS production. However, ROS are also generated in the cell wall, cell membrane, endoplasmic reticulum, and apoplast in response to stress [90]. Plant mitochondria may be involved in the responses to biotic stress. Interactions between plant pathogens and mitochondrial components have the potential to target and disturb mitochondrial homeostasis, which increases the production of mitochondrial ROS [91].

Plants' immune system is also active in defense against invading pathogens. Plant pattern recognition receptor recognizes the pathogen-associated molecular patterns (PAMPs) and induces PAMP-triggered immunity [1]. A secondary immune response called effector-triggered immunity is activated in plants when resistance proteins recognize the effector molecules secreted by pathogens [2].

Plant hormones are the key players that regulate stress responses in plants. The major phytohormones in plants include auxin, cytokinin (CK), GA, abscisic acid (ABA), JA, SA, ET, brassinosteroids (BR), and SLs. The hormone levels in plants are altered in response to infection with pathogenic organisms [92]. ET, JA, and SA are the phytohormones that are majorly involved in plant biotic stress responses [93]. SA tends to be involved in defending biotrophic pathogens and hemi-biotrophic pathogens, while ET and JA are involved in initiating biotic responses against necrotrophic pathogens and herbivorous insects [94, 95]. To combat necrotrophic pathogen infection, JA crosstalk with ET to activate the expression of defense proteins, PDF1.2 (plant defensin 1.2) [96]. SA activates early defense-related genes, while JA induces late defense-related genes in plants during necrotrophic or hemibiotrophic pathogenic infection [93]. ET reduces the accumulation of ROS, enhancing crop resistance to diseases [97]. The plant hormone SLs can also regulate biotic stress responses in plants (Figure 13.2).

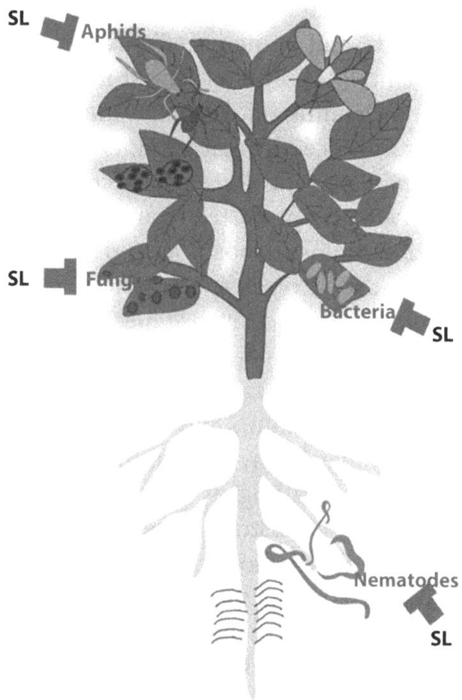


**Figure 13.2** SL-mediated biotic stress response in plants. SL mediates biotic stress response in plants by reducing host sensitivity toward invading pathogens and pests. SLs inhibit pathogenic bacterial and fungal infections. It is also involved in inhibiting attacks from nematodes and aphids.

## 13.4 Strigolactone-Mediated Defense Against Pathogens

SLs can greatly impede the growth of phytopathogenic fungi. The growth of root, shoot, and foliar pathogens experienced growth inhibition when transferred to synthetic SL-containing media. Shoot and foliar pathogens *Botrytis cinerea*, *Alternaria alternata*, *Colletotrichum acutatum*, and the root parasites *Fusarium solani* f. sp. mango, *Macrophomina phaseolina*, *Sclerotinia sclerotiorum*, and *Fusarium oxysporum* f. sp. melonis growth were found to be inhibited when supplemented with synthetic SL GR24 [98]. SL deficient-*ccd8* (carotenoid cleavage dioxygenase) tomato plant mutants were more susceptible to the foliar phytopathogenic fungi *Botrytis cinerea* and *Alternaria alternata*. A drop in JA, SA, and ABA was also found in the *ccd8* mutant, suggesting the role of SLs in providing plant defense via interacting with other hormones [99]. Furthermore, the *ccd7* and *ccd8* mutant moss plants showed increased susceptibility to the phytopathogenic fungus *Sclerotinia sclerotiorum*. The mutants developed more severe symptoms than the wild-type plants, and also the *ccd7* and *ccd8* mutants restored resistance to these fungal pathogens when supplemented with GR24. GR24 application also improved disease resistance in wild-type plants, signifying the role played by SL in plant defense mechanisms [100]. The GR24 can increase resistance to *Botrytis cinerea* infection in grapevine and tobacco plants by increasing ROS levels in *B. cinerea* [101]. SLs also play an important role in conferring resistance to the rice blast disease caused by the phytopathogenic fungus *Magnaporthe oryzae*. The SL biosynthetic mutant *d17* (dwarf17) and the signaling mutant *d14* (dwarf14) showed increased susceptibility to the *Magnaporthe oryzae* infection by altering defense-associated mechanisms in plants [102]. ET biosynthesis and signaling genes, genes involved in hydrogen peroxide ($H_2O_2$) biosynthesis, cell wall biosynthetic, and cell wall modification-related genes were downregulated in *d14* mutants, which might make them more sensitive to the infection [102]. In contrast, SLs negatively regulate resistance towards *Pyricularia oryzae* (*syn. Magnaporthe oryzae*), which causes leaf blasts in rice. Inhibition of SL biosynthesis and signaling significantly decreased rice sensitivity towards *P. oryzae* by increasing the accumulation of soluble sugars, jasmonates, and flavonoid phytoalexins [103]. SLs can resist *Verticillium dahlia*, which causes Verticillium wilt in cotton plants. The *ccd7* and *ccd8* mutants were more susceptible to the infection, whereas the overexpressed mutants increased resistance against *Verticillium dahlia* [104].

The interaction between SLs and phytopathogenic bacteria is less explored. SLs are effective in preventing leafy gall syndrome caused by the *R. fascians*. GR24 supplementation inhibited disease development in *Arabidopsis* upon *R. fascians* infection [105]. SL biosynthesis and signaling mutants of *Arabidopsis* such as *max1* (more axillary growth 1), *max3*, *max4*, and *max2* showed increased sensitivity to the phytopathogenic bacteria *Rhodococcus fascians* compared to wild-type *Arabidopsis* [105]. *max2 Arabidopsis* mutant plants showed increased sensitivity towards *Pectobacterium carotovorum* and *Pseudomonas syringae* [106]. The *max1*, *max2*, *max3*, *max4* and *d14* (dwarf14) were found to be more susceptible to *Pseudomonas syringae* DC3000 [107].

## 13.5 Strigolactone-Mediated Defense Against Pests

The root-knot nematode *Meloidogyne incognita* infection increased SL biosynthesis in tomato plants. SL biosynthesis inhibition increased disease susceptibility, whereas racGR24 application improved resistance against *M. incognita* in tomato plants [108]. The SL biosynthesis *max4-1* mutant increased attraction and invasion of *Heterodera schachtii* beet cyst eel worm to the *Arabidopsis* roots [109]. The SL mutants showed increased sensitivity to aphid infestation than the wild-type pea plants [110]. On the contrary, *Meloidogyne graminicola*, a root-knot nematode, infection promoted SL biosynthesis and signaling and reduced JA accumulation in rice plants. The foliar application of GR24 increased *M. graminicola* infection in rice [103]. Similar to this, reduced SL synthesis in *Vicia faba L* contributes to resistance against the root holoparasites *Orobanche foetida* and *O. crenata* [111].

## 13.6 Strigolactone-Mediated Plant Defense Mechanism

Stomatal closure is a plant's innate immune response that prevents the entry of phytopathogenic microorganisms into plants through stomata [112]. In *Arabidopsis*, stomata are closed in response to bacterial and PAMP exposure [112]. Similarly, chitin oligosaccharide from the fungal cell wall can also induce stomatal closure in *Arabidopsis* [113]. Calcium-dependent protein kinase33 (CPK33) is involved in mediating SL-dependent stomatal closure in *Arabidopsis* [114]. SL-biosynthesis mutants *max1-1*, *max3-9*, and *max4-1* and SL signaling mutants *d14-5* and *max2-1* have wider stomata than the wild-type *Arabidopsis*, whereas exogenous GR24 supplementation promoted stomatal closure in SL-deficient mutants [115]. SL-mediated stomatal closure is linked to $H_2O_2$ and nitric oxide production (NO), which may then trigger the activation of SLAC1 (Slow Anion Associated Channel 1) to induce the closing of stomata [115].

The *max2 Arabidopsis* mutants exhibited greater stomatal conductance than the wild-type plants, and the pathogen-triggered stomatal closing was hindered. The increased susceptibility of *max2* mutants to the pathogens *Pectobacterium carotovorum* and *Pseudomonas syringae* may be due to the impaired stomatal closure in those mutants [106] (Figure 13.3).

SLs can trigger ROS production and are involved in regulating oxidative stress response. The SL biosynthesis *d17* mutant and SL signaling *d14*

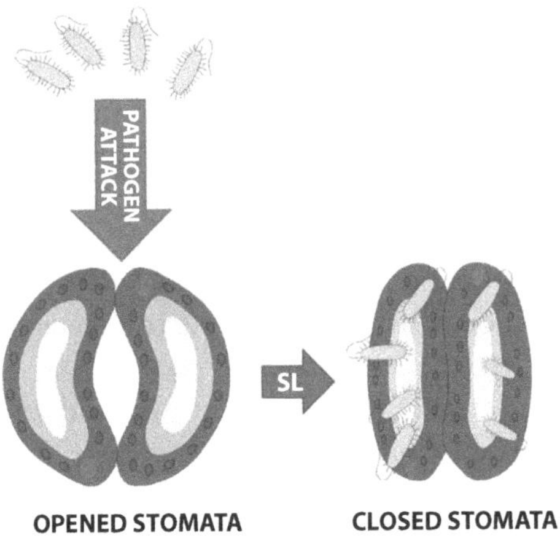


**Figure 13.3** SL-mediated stomatal closure in response to biotic stress in plants. Plants can defend themselves from pathogens by closing their stomata. Stomata can sense pathogens through PAMPs and subsequently close to prevent pathogen entry into the plant. This response can be induced by the plant hormone SLs.

mutants are more vulnerable to *Magnaporthe oryzae* infection compared to wild-type rice plants. They also showed a reduction in $H_2O_2$ accumulation than wild-type plants [102]. The *max2 Arabidopsis* mutants exhibited reduced tolerance to ROS produced in response to *Pseudomonas syringae*. Additionally, the expression of *GST1* (*Glutathione S Transferase 1*), an oxidative stress marker gene, is upregulated in *max2* mutants. These findings highlight the role played by MAX2 in regulating the biotic stress response in *Arabidopsis* [106]. The Mucor species showed increased $H_2O_2$ production, an increased respiration rate, and an inhibition in their mycelium growth when treated with the GR24 [116].

SLs can crosstalk with other defense hormones to elicit biotic stress responses. SLs mediate resistance to *Verticillium dahliae* causing Verticillium wilt in cotton plants via crosstalk with the JA and ABA-signaling pathways [104]. *Slccd8* mutants showed increased sensitivity to fungal pathogens *Botrytis cinerea* and *Alternaria alternata*. The drop in JA, SA, and ABA in the *ccd8* mutant points to the interaction between SLs and other plant defense hormones [99]. SL, SA, ABA, and JA crosstalk was activated in *Solanum lycopersicum* against the root parasitic weed *Phelipanche ramose* [117]. SLs modulate SA-mediated stress response in *Arabidopsis* to *Pseudomonas syringae pv. tomato DC3000* (Pst) infection [118].

## 13.7 Conclusion and Future Perspective

The plant hormone SLs regulates plant growth and development. It mediates shoot branching, leaf senescence, and root growth. It also aids plants in establishing association with rhizobia and arbuscular mycorrhizal fungi. SLs facilitate strong defense responses to abiotic stresses such as nutrient, heat, salinity, drought, cold, and light stress [119, 120]. The SLs cross-talk with other hormones to improve plant resilience against fluctuating environments.

Since biotic stress severely affects plant growth and productivity, it eventually affects food production worldwide. Therefore, finding an appropriate method for improving the survival of plants with greater resilience to biotic stress is essential. The role of SLs in plant development and abiotic stress is well established, although the biotic stress response is still in its infancy. It is imperative to have a deeper comprehension of the role played by SLs in mediating biotic stress response in plants. Gaining a thorough grasp of this fundamental mechanism will pave the way for developing biotic stress-resistant crops in the future.

## Acknowledgment

This work was supported by a grant from KSCSTE-Kerala Biotechnology Commission-Kerala Biotechnology Re-entry Fellowship (KSCSTE/764/2022).

## References

1. Jones, J.D. and Dangl, J.L., The plant immune system. *Nature*, 444, 323, 2006.
2. Muthamilarasan, M. and Prasad, M., Plant innate immunity: an updated insight into defense mechanism. *J. Biosci.*, 38, 433, 2013.
3. Yang, J., Duan, G., Li, C., Liu, L., Han, G., Zhang, Y., Wang, C., The cross-talks between jasmonic acid and other plant hormone signaling highlight the involvement of jasmonic acid as a core component in plant response to biotic and abiotic stresses. *Front. Plant Sci.*, 10, 1349, 2019.
4. Kumar, D., Chapagai, D., Dean, P., Davenport, M., *Elucidation of Abiotic Stress Signaling in Plants: Functional Genomics Perspectives*, pp. 329–346, Springer, New York, 2015.
5. Natalini, A. and Palma, D., Ethylene and biotic stress in crops, in: *The Plant Hormone Ethylene*, pp. 221–232, Academic Press, Cambridge, 2023.

6. Takatsuji, H. and Jiang, C.J., Plant hormone crosstalks under biotic stresses, in: *Phytohormones: a window to metabolism, signaling and biotechnological applications*, pp. 323–350, Springer, New York, 2014.
7. Brewer, P.B., Emerging roles of Strigolactones in plant responses toward biotic stress, in: *Emerging Plant Growth Regulators in Agriculture*, pp. 205–214, Academic Press, Cambridge, 2022.
8. Francis, B., Aravindakumar, C.T., Simon, S., Regulatory role of strigolactones in biotic stress tolerance, in: *Strigolactones*, pp. 189–200, Academic Press, Cambridge, 2024.
9. Pandey, A., Sharma, M., Pandey, G.K., Emerging roles of strigolactones in plant responses to stress and development. *Front. Plant Sci.*, 7, 165491, 2016.
10. Al-Babili, S. and Bouwmeester, H.J., Strigolactones, a novel carotenoid-derived plant hormone. *Annu. Rev. Plant Biol.*, 66, 161, 2015.
11. Cook, C.E., Whichard, L.P., Turner, B., Wall, M.E., Egley, G.H., Germination of witchweed (*Striga lutea Lour.*): isolation and properties of a potent stimulant. *Science*, 154, 1189, 1966.
12. Gomez-Roldan, V., Fermas, S., Brewer, P.B., Puech-Pagès, V., Dun, E.A., Pillot, J.P., Letisse, F., Matusova, R., Danoun, S., Portais, J.C., Bouwmeester, H., Strigolactone inhibition of shoot branching. *Nature*, 455, 189, 2008.
13. Umehara, M., Hanada, A., Yoshida, S., Akiyama, K., Arite, T., Takeda-Kamiya, N., Magome, H., Kamiya, Y., Shirasu, K., Yoneyama, K., Kyozuka, J., Inhibition of shoot branching by new terpenoid plant hormones. *Nature*, 455, 195, 2008.
14. Agusti, J., Herold, S., Schwarz, M., Sanchez, P., Ljung, K., Dun, E.A., Brewer, P.B., Beveridge, C.A., Sieberer, T., Sehr, E.M., Greb, T., Strigolactone signaling is required for auxin-dependent stimulation of secondary growth in plants. *Proc. Natl. Acad. Sci.*, 108, 20242, 2011.
15. Zou, X., Wang, Q., Chen, P., Yin, C., Lin, Y., Strigolactones regulate shoot elongation by mediating gibberellin metabolism and signaling in rice (*Oryza sativa L.*). *J. Plant Physiol.*, 237, 72, 2019.
16. Yamada, Y., Furusawa, S., Nagasaka, S., Shimomura, K., Yamaguchi, S., Umehara, M., Strigolactone signaling regulates rice leaf senescence in response to a phosphate deficiency. *Planta*, 240, 399, 2014.
17. Sun, H., Tao, J., Gu, P., Xu, G., Zhang, Y., The role of strigolactones in root development. *Plant Signal. Behav.*, 11, 1110662, 2016.
18. Marzec, M., Muszynska, A., Gruszka, D., The role of strigolactones in nutrient-stress responses in plants. *Int. J. Mol. Sci.*, 14, 9286, 2013.
19. Foo, E., Yoneyama, K., Hugill, C.J., Quittenden, L.J., Reid, J.B., Strigolactones and the regulation of pea symbioses in response to nitrate and phosphate deficiency. *Mol. Plant*, 6, 76, 2013.
20. Zhang, Y., Lv, S., Wang, G., Strigolactones are common regulators in induction of stomatal closure in planta. *Plant Signal. Behav.*, 13, 1444322, 2018.

21. Ha, C.V., Leyva-González, M.A., Osakabe, Y., Tran, U.T., Nishiyama, R., Watanabe, Y., Tanaka, M., Seki, M., Yamaguchi, S., Dong, N.V., Yamaguchi-Shinozaki, K., Positive regulatory role of strigolactone in plant responses to drought and salt stress. *Proc. Natl. Acad. Sci.*, 111, 851, 2014.
22. Omoarelojie, L.O., Kulkarni, M.G., Finnie, J.F., Pospíšil, T., Strnad, M., Van Staden, J., Synthetic strigolactone (rac-GR24) alleviates the adverse effects of heat stress on seed germination and photosystem II function in lupine seedlings. *Plant Physiol. Biochem.*, 155, 965, 2020.
23. Cooper, J.W., Hu, Y., Beyyoudh, L., Yildiz Dasgan, H., Kunert, K., Beveridge, C.A., Foyer, C.H., Strigolactones positively regulate chilling tolerance in pea and in Arabidopsis. *Plant Cell Environ.*, 41, 1298, 2018.
24. Thula, S., Moturu, T.R., Salava, H., Balakhonova, V., Berka, M., Kerchev, P., Mishra, K.B., Nodzynski, T., Simon, S., Strigolactones stimulate high light stress adaptation by modulating photosynthesis rate in Arabidopsis. *J. Plant Growth Regul.*, 42, 4818, 2023.
25. Agrios, G.N., *Plant Pathology*, Elsevier, Netherlands, 2005.
26. Mansfield, J., Genin, S., Magori, S., Citovsky, V., Sriariyanum, M., Ronald, P., Dow, M.A.X., Verdier, V., Beer, S.V., Machado, M.A., Toth, I.A.N., Top 10 plant pathogenic bacteria in molecular plant pathology. *Mol. Plant Pathol*, 13, 614, 2012.
27. Dean, R., Van Kan, J.A., Pretorius, Z.A., Hammond-Kosack, K.E., Di Pietro, A., Spanu, P.D., Rudd, J.J., Dickman, M., Kahmann, R., Ellis, J., Foster, G.D., The Top 10 fungal pathogens in molecular plant pathology. *Mol. Plant Pathol*, 13, 414, 2012.
28. Nazarov, P.A., Baleev, D.N., Ivanova, M.I., Sokolova, L.M., Karakozova, M.V., Infectious plant diseases: Etiology, current status, problems and prospects in plant protection. *Acta Naturae*, 12, 46, 2020.
29. Katsube, K., Studies on Fusarium Wilt of Spinach [Spinacia oleracea], caused by *Fusarium oxysporum f. sp. spinaciae. Bull. Iwate Agric. Res. Center Jpn.*, 2, 1–60, 2001.
30. G,.TR. and K,.ST., Management of Fusarium wilt of lettuce, in: *Crop protection*, vol. 73, pp. 45–49, Elsevier, Netherlands, 2015.
31. Ploetz, R.C., Fusarium wilt of banana. *Phytopathol.*, 105, 1512, 2015.
32. Maurya, S., Dubey, S., Kumari, R., Verma, R., Management tactics for fusarium wilt of tomato caused by *Fusarium oxysporum f. sp. lycopersici (Sacc.)*: A review. Management. *Int. J. Res. Pharm. Sci.*, 4, 1, 2019.
33. Okungbowa, F.I. and Shittu, H.O., Fusarium wilts: An overview. *Environ. Res. J.*, 6, 83–102, 2012.
34. Majeed, A., Muhammad, Z., Ullah, Z., Ullah, R., Ahmad, H., Late blight of potato (*Phytophthora infestans*) I: Fungicides application and associated challenges. *Turk. Tarim Gida Bilim Teknol. Derg.*, 5, 261, 2017.
35. Nowicki, M., Foolad, M.R., Nowakowska, M., Kozik, E.U., Potato and tomato late blight caused by *Phytophthora infestans*: an overview of pathology and resistance breeding. *Plant Dis.*, 96, 4, 2012.

36. Nowicki, M., Kozik, E.U., Foolad, M.R., Late blight of tomato. Translational Genomics for crop breeding, in: *Biotic stress*, pp. 241–265, Wiley Online Library, New Jersey, 2013.
37. Nazari, K., Mafi, M., Yahyaoui, A., Singh, R.P., Park, R.F., Detection of wheat stem rust (*Puccinia graminis f. sp. tritici*) race TTKSK (Ug99) in Iran. *Plant Dis.*, 93, 317, 2009.
38. Fetch, T., Mitchell Fetch, J., Zegeye, T., Xue, A., Races of *Puccinia graminis* on barley, oat, and wheat in Canada in 2013 and 2014. *Can. J. Plant. Pathol.*, 43, 101, 2021.
39. Leonard, K.J. and Szabo, L.J., Stem rust of small grains and grasses caused by Puccinia graminis. *Mol. Plant Pathol*, 6, 99, 2005.
40. Matei, A. and Doehlemann, G., Cell biology of corn smut disease—*Ustilago maydis* as a model for biotrophic interactions. *Curr. Opin. Microbiol.*, 34, 60, 2016.
41. Draper, M.A., *Common Corn Smut, Fact Sheets, Paper*, vol. 97, 2004.
42. Khanal, S., Gaire, S.P., Zhou, X.G., Kernel Smut and False smut: The old-emerging diseases of rice- A review. *Phytopathol.*, 113, 931, 2023.
43. Woldemichael, M.D., Importance, biology, epidemiology, and management of Loose Smut (*Ustilago nuda*) of barley (*Hordeum vulgare*): A Review. *East Afr. J. Sci.*, 13, 89, 2019.
44. Johnson, S.B., *Loose smut of barley and oats*, Cooperative Extension Publications, The University of Maine Bulletin, Maine, 2014.
45. Newberry, E.A., Jardini, T.M., Rubio, I., Roberts, P.D., Babu, B., Koike, S.T., Bouzar, H., Goss, E.M., Jones, J.B., Bull, C.T., Paret, M.L., Angular leaf spot of cucurbits is associated with genetically diverse *Pseudomonas syringae* strains. *Plant Dis.*, 100, 1397, 2016.
46. Garibaldi, A., Minuto, A., Scortichini, M., Gullino, M.L., First report of syringae leaf spot caused by *Pseudomonas syringae pv. syringae* on tomato in Italy. *Plant Dis.*, 91, 1518, 2007.
47. Cazorla, F.M., Vázquez, M.A., Rosales, J., Arrebola, E., Navarro, J., Pérez-García, A., De Vicente, A., First report of bacterial leaf spot (*Pseudomonas syringae pv. coriandricola*) of coriander in Spain. *J. Phytopathol.*, 153, 181, 2005.
48. Koike, S.T., Henderson, D.M., Bull, C.T., Goldman, P.H., Lewellen, R.T., First report of bacterial leaf spot of Swiss chard caused by *Pseudomonas syringae pv. aptata* in California. *Plant Dis.*, 87, 1397, 2003.
49. Zhao, Y., Damicone, J.P., Demezas, D.H., Rangaswamy, V., and Bender, C.L., Bacterial leaf spot of leafy crucifers in Oklahoma caused by *Pseudomonas syringae pv. maculicola. Plant Dis.*, 84, 1015, 2000.
50. Loreti, S., Fiori, M., Simone, D.D., Falchi, G., Gallelli, A., Schiaffino, A., and Ena, S., Bacterial wilt, caused by *Ralstonia solanacearum*, on tomato in Italy. *Plant Pathol.*, 57, 368, 2008.

51. Patil, V.U., Gopal, J., Singh, B.P., Improvement for bacterial wilt resistance in potato by conventional and biotechnological approaches. *Agric. Res.*, 1, 299, 2012.
52. Kunwar, S., Bamazi, B., Banito, A., Carter, M., Weinstein, S., Steidl, O., Hayes, M.M., Allen, C., Paret, M., First Report of Bacterial Wilt Disease of Tomato, Pepper, and Gboma Caused by *Ralstonia solanacearum* Species Complex in Togo. *Plant Dis.*, 105, 484, 2021.
53. Sarkar, S. and Chaudhuri, S., Bacterial wilt and its management. *Curr. Sci.*, 1439, 2016.
54. Bhat, K.A., Masoodi, S.D., Bhat, N.A., Ahmad, M., Zargar, M.Y., Mir, S.A., Bhat, M.A., Development of soft rot of cabbage (*Brassica oleracea var. capitata*) caused by *Erwinia carotovora sub sp. carotovora*. *J. Phytol.*, 2, 64, 2010.
55. Gore, P.M., Ingle, R.W., Rakhonde, P.N., Management of bulb rot of onion caused by *Erwinia carotovora pv. carotovora*. *J. Pharmacogn. Phytochem.*, 9, 3294, 2020.
56. Parthiban, V.K., Prakasam, V., Prabakar, K., Changes in the biochemical constituents of carrot roots due to bacterial soft rot. *Int. J. Appl. Biol. Pharm. Technol.*, 3, 231, 2012.
57. Czajkowski, R., Pérombelon, M.C.M., Jafra, S., Lojkowska, E., Potrykus, M., Van Der Wolf, J.M., Sledz, W., Detection, identification and differentiation of Pectobacterium and Dickeya species causing potato blackleg and tuber soft rot: a review. *Ann. Appl. Biol.*, 166, 18, 2015.
58. Charkowski, A.O., The changing face of bacterial soft-rot diseases. *Annu. Rev. Phytopathol.*, 56, 269, 2018.
59. Nayak, D., Shanti, M.L., Bose, L.K., Singh, U.D., Nayak, P., Pathogenicity association in *Xanthomonas oryzae pv. oryzae* the causal organism of rice bacterial blight disease. *ARPN J. Agric. Biol. Sci.*, 3, 12, 2008.
60. Jagtap, G.P., Jangam, A.M., Deya, U., Management of bacterial blight of cotton caused by *Xanthomonas axonopodis pv. malvacearum*. *Sci. J. Microbiol.*, 1, 10, 2012.
61. Belete, T. and Bastas, K.K., Common bacterial blight (*Xanthomonas axonopodis pv. phaseoli*) of beans with special focus on Ethiopian condition. *J. Plant Pathol. Microbiol.*, 8, 2–10, 2017.
62. Shkalikov, V.A., Beloshapkina, O.O., Bukreev, D.D., Gorbachev, I.V., Dzhalilov, F.S.U., Korsak, I.V., Minaev, V.Y., Stroykov Yu, M., Plant protection from disease. *M.: Kolos*, 1, 404, 2010.
63. Singh, S., Awasthi, L.P., Jangre, A., Transmission of plant viruses in fields through various vectors, in: *Applied Plant Virology*, pp. 313–334, Academic Press, Cambridge, 2020.
64. Taliansky, M., Mayo, M.A., Barker, H., Potato leafroll virus: a classic pathogen shows some new tricks. *Mol. Plant Pathol*, 4, 81, 2003.
65. Walsh, J.A. and Jenner, C.E., Turnip mosaic virus and the quest for durable resistance. *Mol. Plant Pathol*, 3, 289, 2002.

66. Chung, B.N., San Choi, K., Ahn, J.J., Joa, J.H., Do, K.S., Park, K.S., Effects of temperature on systemic infection and symptom expression of Turnip mosaic virus in Chinese cabbage (*Brassica campestris*). *Plant Pathol. J.*, 31, 363, 2015.
67. Tang, M. and Feng, X., Bean common mosaic disease: etiology, resistance resource, and future prospects. *Agronomy*, 13, 58, 2022.
68. Birithia, R., Subramanian, S., Pappu, H.R., Muthomi, J., Narla, R.D., Analysis of Iris yellow spot virus replication in vector and non-vector thrips species. *Plant Pathol.*, 62, 1407, 2013.
69. Bag, S., Schwartz, H.F., Cramer, C.S., Havey, M.J., Pappu, H.R., Iris yellow spot virus (Tospovirus: *B unyaviridae*): from obscurity to research priority. *Mol. Plant Pathol*, 16, 224, 2015.
70. Parrella, G., Gognalons, P., Gebre-Selassie, K., Vovlas, C., Marchoux, G., An update of the host range of Tomato spotted wilt virus. *J. Plant Pathol.*, 85, 227–264,2003.
71. Gupta, R., Kwon, S.Y., Kim, S.T., An insight into the tomato spotted wilt virus (TSWV), tomato and thrips interaction. *Plant Biotech. Rep.*, 12, 157, 2018.
72. Culbreath, A.K. and Srinivasan, R., Epidemiology of spotted wilt disease of peanut caused by Tomato spotted wilt virus in the southeastern US. *Virus Res.*, 159, 101, 2011.
73. Premachandra, W.T.S.D., Borgemeister, C., Maiss, E., Knierim, D., Poehling, H.M., *Ceratothripoides claratris*, a new vector of a Capsicum chlorosis virus isolate infecting tomato in Thailand. *Phytopathol.*, 95, 659, 2005.
74. Tsai, W.A., Shafiei-Peters, J.R., Mitter, N., Dietzgen, R.G., Effects of elevated temperature on the susceptibility of capsicum plants to Capsicum chlorosis virus infection. *Pathogens*, 11, 200, 2022.
75. Verbeek, M. and Dullemans, A.M., First Report of Tomato torrado virus infecting tomato in Colombia. *Plant Dis.*, 96, 592, 2012.
76. Gambley, C.F., Thomas, J.E., Persley, D.M., Hall, B.H., First report of Tomato torrado virus on tomato from Australia. *Plant Dis.*, 94, 486, 2010.
77. Amari, K., Gonzalez-Ibeas, D., Gómez, P., Sempere, R.N., Sanchez-Pina, M.A., Pendon, J., Moriones, E., Tomato torrado virus is transmitted by Bemisia tabaci and infects pepper and eggplant in addition to tomato. *Arch. Virol.*, 9, 1139, 2017.
78. Boubourakas, I.N., Avgelis, A.D., Kyriakopoulou, P.E., Katis, N.I., Occurrence of yellowing viruses (Beet pseudo-yellows virus, Cucurbit yellow stunting disorder virus and Cucurbit aphid-borne yellows virus) affecting cucurbits in Greece. *Plant Pathol.*, 55, 276, 2006.
79. Moriones, E. and Navas-Castillo, J., Tomato yellow leaf curl virus, an emerging virus complex causing epidemics worldwide. *Virus Res.*, 71, 123, 2000.
80. Angessa, T.T. and Li, C., Exploration and utilization of genetic diversity exotic germplasm for barley improvement, in: *Exploration, identification and utilization of barley germplasm*, pp. 223–240, Academic Press, Cambridge, 2016.

81. Bakala, H.S., Mandahal, K.S., Sarao, L.K., Srivastava, P., Breeding wheat for biotic stress resistance: Achievements, challenges and prospects. *Curr. Trends Wheat Res.*, 12, 11–34, 2021.
82. Singh, P., Verma, R.L., Singh, R.S., Singh, R.P., Singh, H.B., Arsode, P., Kumar, M., Singh, P.K., Biotic stress management in rice (Oryza sativa L.) through conventional and molecular approaches, in: *New frontiers in stress management for durable agriculture*, pp. 609–644, Springer, 2020.
83. Belete, T., Defense mechanisms of plants to insect pests: from morphological to biochemical approach. *Trends. Tech. Sci. Res.*, 2, 30, 2018.
84. Saddique, M., Kamran, M., Shahbaz, M., Differential responses of plants to biotic stress and the role of metabolites, in: *Plant metabolites and regulation under environmental stress*, pp. 69–87, Academic Press, Cambridge, 2018.
85. Schmelz, E.A., Kaplan, F., Huffaker, A., Dafoe, N.J., Vaughan, M.M., Ni, X., Rocca, J.R., Alborn, H.T., Teal, P.E., Identity, regulation, and activity of inducible diterpenoid phytoalexins in maize. *Proc. Natl. Acad. Sci.*, 108, 5455, 2011.
86. Kumar, S., Korra, T., Thakur, R., Arutselvan, R., Kashyap, A.S., Nehela, Y., Chaplygin, V., Minkina, T., Keswani, C., Role of plant secondary metabolites in defence and transcriptional regulation in response to biotic stress. *Plant Stress*, 8, 100154, 2023.
87. Luna, E., Pastor, V., Robert, J., Flors, V., Mauch-Mani, B., Ton, J., Callose deposition: a multifaceted plant defense response. *Mol. Plant-Microbe Interact.*, 24, 183, 2011.
88. Melotto, M., Underwood, W., Sheng, Y.H., Role of stomata in plant innate immunity and foliar bacterial diseases. *Annu. Rev. Phytopathol.*, 46, 101, 2008.
89. Bolwell, G.P., Bindschedler, L.V., Blee, K.A., Butt, V.S., Davies, D.R., Gardner, S.L., Gerrish, C., Minibayeva, F., The apoplastic oxidative burst in response to biotic stress in plants: a three-component system. *J. Exp. Bot.*, 53, 1367, 2002.
90. Suman, S., Bagal, D., Jain, D., Singh, R., Singh, I.K., Singh, A., Biotic stresses on plants: Reactive oxygen species generation and antioxidant mechanism, in: *Frontiers in plant-soil interaction*, pp. 381–411, Academic Press, 2021.
91. Amirsadeghi, S., Robson, C.A., Vanlerberghe, G.C., The role of the mitochondrion in plant responses to biotic stress. *Physiol. Plant.*, 129, 253, 2007.
92. Robert-Seilaniantz, A., Navarro, L., Bari, R., Jones, J.D., Pathological hormone imbalances. *Curr. Opin. Plant Biol.*, 10, 372, 2007.
93. Bari, R. and Jones, J.D., Role of plant hormones in plant defence responses. *Plant Mol. Biol.*, 69, 473, 2009.
94. Loake, G. and Grant, M., Salicylic acid in plant defence—the players and protagonists. *Curr. Opin. Plant Biol.*, 10, 466, 2007.
95. Wasternack, C. and Hause, B., Jasmonates: biosynthesis, perception, signal transduction and action in plant stress response, growth and development. An update to the 2007 review in Annals of Botany. *Ann. Bot.*, 111, 1021–1058, 2013.

96. Pieterse, C.M., Van der Does, D., Zamioudis, C., Leon-Reyes, A., Van Wees, S.C., Hormonal modulation of plant immunity. *Annu. Rev. Cell Dev. Biol.*, 28, 489, 2012.
97. Jasrotia, S. and Jasrotia, R., Role of ethylene in combating biotic stress. *Ethylene Plant Biol.*, 388–397, 2022.
98. Dor, E., Joel, D.M., Kapulnik, Y., Koltai, H., Hershenhorn, J., The synthetic strigolactone GR24 influences the growth pattern of phytopathogenic fungi. *Planta*, 234, 419, 2011.
99. Torres-Vera, R., García, J.M., Pozo, M.J., López-Ráez, J.A., Do strigolactones contribute to plant defence? *Mol. Plant Pathol*, 15, 211, 2014.
100. Decker, E.L., Alder, A., Hunn, S., Ferguson, J., Lehtonen, M.T., Scheler, B., Kerres, K.L., Wiedemann, G., Safavi-Rizi, V., Nordzieke, S., Balakrishna, A., Strigolactone biosynthesis is evolutionarily conserved, regulated by phosphate starvation and contributes to resistance against phytopathogenic fungi in a moss, *Physcomitrella patens. New Phytol.*, 216, 455, 2017.
101. Vogel, D., Hills, P., Moore, J.P., Strigolactones GR-24 and Nijmegen applications result in reduced susceptibility of tobacco and grapevine plantlets to Botrytis cinerea infection. *Plants*, 12, 3202, 2023.
102. Nasir, F., Tian, L., Shi, S., Chang, C., Ma, L., Gao, Y., Tian, C., Strigolactones positively regulate defense against *Magnaporthe oryzae in* rice (Oryza sativa). *Plant Physiol. Biochem.*, 142, 106, 2019.
103. Lahari, Z., Ullah, C., Kyndt, T., Gershenzon, J., Gheysen, G., Strigolactones enhance root-knot nematode (*Meloidogyne graminicola*) infection in rice by antagonizing the jasmonate pathway. *New Phytol.*, 224, 454, 2019.
104. Yi, F., Song, A., Cheng, K., Liu, J., Wang, C., Shao, L., Wu, S., Wang, P., Zhu, J., Liang, Z., Chang, Y., Strigolactones positively regulate Verticillium wilt resistance in cotton via crosstalk with other hormones. *Plant Physiol.*, 192, 945, 2023.
105. Stes, E., Depuydt, S., De Keyser, A., Matthys, C., Audenaert, K., Yoneyama, K., Werbrouck, S., Goormachtig, S., Vereecke, D., Strigolactones as an auxiliary hormonal defence mechanism against leafy gall syndrome in *Arabidopsis thaliana. J. Exp. Bot.*, 66, 5123, 2015.
106. Piisilä, M., Keceli, M.A., Brader, G., Jakobson, L., Jõesaar, I., Sipari, N., Kollist, H., Palva, E.T., Kariola, T., The F-box protein MAX2 contributes to resistance to bacterial phytopathogens in *Arabidopsis thaliana. BMC Plant Biol.*, 15, 1, 2015.
107. Kalliola, M., Jakobson, L., Davidsson, P., Pennanen, V., Waszczak, C., Yarmolinsky, D., Zamora, O., Palva, E.T., Kariola, T., Kollist, H., Brosché, M., Differential role of MAX2 and strigolactones in pathogen, ozone, and stomatal responses. *Plant Direct*, 4, e00206, 2020.
108. Xu, X., Fang, P., Zhang, H., Chi, C., Song, L., Xia, X., Shi, K., Zhou, Y., Zhou, J., Yu, J., Strigolactones positively regulate defense against root-knot nematodes in tomato. *J. Exp. Bot.*, 70, 1325, 2019.

109. Sikder, M.M., Vestergård, M., Kyndt, T., Kudjordjie, E.N., Nicolaisen, M., Phytohormones selectively affect plant parasitic nematodes associated with Arabidopsis roots. *New Phytol.*, 232, 1272, 2021.
110. Swiegers, H.W., Karpinska, B., Hu, Y., Dodd, I.C., Botha, A.M., Foyer, C.H., The Effects of High CO2 and Strigolactones on Shoot Branching and Aphid–Plant Compatibility Control in Pea. *Int. J. Mol. Sci.*, 23, 12160, 2022.
111. Trabelsi, I., Kaori, Y., Zouhaier, A., Moez, A., Xie, X., Kisugi, T., Kim, H., II, Mohamed, K., Characterization of strigolactones produced by *Orobanche foetida* and *Orobanche crenata* resistant faba bean (Vicia faba L.) genotypes and effects of phosphorous, nitrogen, and potassium deficiencies on strigolactone production. *S. Afr. J. Bot.*, 108, 15, 2017.
112. Melotto, M., Underwood, W., Koczan, J., Nomura, K., He, S.Y., Plant stomata function in innate immunity against bacterial invasion. *Cell*, 126, 969, 2006.
113. Ye, W., Munemasa, S., Shinya, T., Wu, W., Ma, T., Lu, J., Kinoshita, T., Kaku, H., Shibuya, N., Murata, Y., Stomatal immunity against fungal invasion comprises not only chitin-induced stomatal closure but also chitosan-induced guard cell death. *Proc. Natl. Acad. Sci.*, 117, 20932, 2020.
114. Wang, X., Lv, S., Han, X., Guan, X., Shi, X., Kang, J., Zhang, L., Cao, B., Li, C., Zhang, W., Wang, G., The calcium-dependent protein kinase CPK33 mediates strigolactone-induced stomatal closure in Arabidopsis thaliana. *Front. Plant Sci.*, 10, 493936, 2019.
115. Lv, S., Zhang, Y., Li, C., Liu, Z., Yang, N., Pan, L., Wu, J., Wang, J., Yang, J., Lv, Y., Zhang, Y., Strigolactone-triggered stomatal closure requires hydrogen peroxide synthesis and nitric oxide production in an abscisic acid-independent manner. *New Phytol.*, 217, 290, 2018.
116. Rozpądek, P., Domka, A.M., Nosek, M., Ważny, R., Jędrzejczyk, R.J., Wiciarz, M., Turnau, K., The role of strigolactone in the cross-talk between *Arabidopsis thaliana* and the endophytic fungus Mucor sp. *Front. Microbiol.*, 9, 307553, 2018.
117. Torres-Vera, R., García, J.M., Pozo, M.J., López-Ráez, J.A., Expression of molecular markers associated to defense signaling pathways and strigolactone biosynthesis during the early interaction tomato-*Phelipanche ramosa. Physiol. Mol. Plant Pathol.*, 94, 100, 2016.
118. Kusajima, M., Fujita, M., Soudthedlath, K., Nakamura, H., Yoneyama, K., Nomura, T., Akiyama, K., Maruyama-Nakashita, A., Asami, T., and Nakashita, H., Strigolactones modulate salicylic acid-mediated disease resistance in *Arabidopsis thaliana. Int. J. Mol. Sci.*, 23, 5246, 2022.
119. Wani, K., II, Abdullah, M.N., Aftab, T., Understanding the Interplay Between Strigolactone and Nitric Oxide in Alleviating Cadmium-Induced Toxicity in *Artemisia annua. Environ. Exp. Bot.*, 225, 105856, 2024.
120. Wani, K., II, Naeem, M., Khan, M.M.A., Aftab, T., Insights into strigolactone (GR24) mediated regulation of cadmium-induced changes and ROS metabolism in *Artemisia annua. J. Hazard. Mater.*, 448, 130899, 2023.

# Index

Printed in the USA/Agawam, MA
September 16, 2026

910661.076